Integrated Analytical Systems

Series Editor
Radislav A. Potyrailo
Niskayuna, USA

This comprehensive and interdisciplinary series offers the most recent advances in all key aspects of development and applications of modern instrumentation for chemical and biological analysis on the microscale.

These key aspects will include (1) innovations in sample introduction through micro- and nano-fluidic designs, (2) new types and methods of fabrication of physical transducers and ion detectors, (3) materials for sensors that became available due to the breakthroughs in combinatorial materials science and nanotechnology, and (4) innovative data processing and mining methodologies that provide dramatically reduced rates of false alarms.

Clearly, a true multidisciplinary effort is required to meet objectives for a system with previously unavailable capabilities. This cross-discipline fertilization is driven by the expanding need for chemical and biological detection and monitoring and leads to the creation of instruments with new capabilities for new demanding applications. Indeed, instruments with more sensitivity are required today to analyze ultra-trace levels of environmental pollutants, pathogens in water, and low vapor pressure energetic materials in air. Sensor devices with faster response times are desired to monitor transient in-vivo events and bedside patients. More selective instruments are wanted to analyze specific proteins in vitro and analyze ambient urban or battlefield air. For these and many other applications, new features of modern microanalytical instrumentation are urgently needed. This book series is a primary source of both fundamental and practical information on both the current state of the art and future directions for microanalytical instrumentation technologies. This book series is addressed to the rapidly growing number of active practitioners and developers and those who are interested in starting research in this direction, directors of industrial and government research centers, laboratory supervisors and managers, students and lecturers.

More information about this series at http://www.springer.com/series/7427

Per Jonsson • Göran Olofsson • Torbjörn Tjärnhage
Editors

Bioaerosol Detection Technologies

Springer

Editors
Per Jonsson
Division of Sensor and EW Systems
FOI—Swedish Defence Research Agency
Linköping
Sweden

Torbjörn Tjärnhage
Division of CBRN Defence and Security
FOI—Swedish Defence Research Agency
Umeå
Sweden

Göran Olofsson
Division of CBRN Defence and Security
FOI—Swedish Defence Research Agency
Umeå
Sweden

ISSN 2196-4475 ISSN 2196-4483 (electronic)
ISBN 978-1-4419-5581-4 ISBN 978-1-4419-5582-1 (eBook)
DOI 10.1007/978-1-4419-5582-1
Springer New York Heidelberg Dordrecht London

Library of Congress Control Number: 2014942060

Printed on acid-free paper

Springer is part of Springer Science+Business Media (www.springer.com)

Preface

Bioaerosols, in the context of this book, are defined as airborne dispersions of particles containing whole or parts of biological entities and debris, such as bacteria, viruses, dust mites, biological toxins, or fungal components. Bioaerosols travel in the atmosphere and may have an important impact on both environmental processes and human health. In some cases, there is a desire to follow the concentration of a certain biological substance in the air. One noticeable example is from the military and anti-terrorist arena, where different ways to warn people about exposure to deliberately disseminated dangerous biological substances is critical. Apart from military or terrorist threats, monitoring of other bioaerosols that may have an impact on our health and environment is of great interest. During the last decades we have seen examples of disease spreading events in the society, e.g., the foot and mouth disease outbreak in the UK in 2001, the SARS outbreak in 2002–2003, and H1N1 (swine flu) in 2009. Preventive actions, such as exposure reduction and medical treatment are generally more effective the quicker they are initiated. Therefore, the impact of these incidences could be reduced with better and more rapid methods of bioaerosol monitoring. Finally, a direct way of monitoring the real-time biological fluctuations in the atmosphere can assist in understanding fundamental aspects in the indoor or outdoor environment.

Over the years there has therefore been an increasing interest in developing specific technologies and methods that can detect and characterize the population of the bioaerosols. An ideal bioaerosol detection system should be able to monitor, in real-time, the actual concentration, down to single organism level, and recognize it with very high specificity. Presently, no single sensor or detection technology reaches these requirements.

Traditional molecular biotechnological methods, such as genetic and/or immunological technologies, can provide the high specificity, but the response times reach from many minutes to hours and even days since they may need sample preparation and wet chemistry or culturing. To reach high sensitivity these methods may also need purification and amplification steps, e.g., polymer chain reaction (PCR), which also takes time.

Detectors based on physical phenomena such as elastic and inelastic scattering, laser-induced fluorescence, emission and mass spectroscopy can give information about size, shape and molecular or elemental composition of the aerosol particles. This information can be used, to a certain degree, to determine biological properties of an aerosol particle. These direct detection technologies can both be sensitive and, more importantly, provide very fast response times.

Biotechnological methods are well covered in the literature as well as traditional methods for sampling and analyzing bioaerosols. However, very few books give an overview of the different rapid detection technologies allowing near real-time detection and monitoring of bioaerosols, especially for wide area surveillance.

This book is intended to give technological background and practical examples, but also to give general insight into the on-going technology development in the area of biodetection. The content is therefore suitable for different stakeholders (decision makers, purchasing officers, etc.) and end-users of biodetection equipment within the areas of health, environment, safety and security, and military preparation. In the end, we hope that the reader of this book will gain more knowledge about the different biodetection technologies and thus better judge their capabilities in relation to desired applications.

The first section of the book is an introduction to the whole area of bioaerosol detection and monitoring. It includes the fundamental physical and biological properties of bioaerosols, how bioaerosols are dispersed and transported in the atmosphere and how the aerosols can be sampled and transferred into detection systems. Also covered is the terminology used in this research area, as well as a historical survey of the early attempts to construct bioaerosol detectors which led to the first fluorescence-based detector.

The second section of the book contains more detailed information about the most commonly used detection principles for continuously monitoring bioaerosol content. The section thoroughly describes the basic principles of elastic and inelastic light scattering, absorption, laser-induced fluorescence, atomic emission, and mass spectroscopy. The basic principles are described, and practical examples are given, as well as discussions about their potential use as biodetectors.

The third section is especially devoted to technologies that have been used in standoff applications. After a short introduction, the section is focused on fluorescence-based techniques and a technique based on multiple light scattering in an aerosol cloud.

The last section of the book gives an overview of trends in bioaerosol detection both with respect to challenges, technologies and applications.

The chapter authors in the book have provided their material and it has been processed by the book editors. The views stated do not necessarily reflect the official position of the organizations to which the authors belong.

FOI – Swedish Defence Research Agency
Umeå February 2014

Per Jonsson
Göran Olofsson
Torbjörn Tjärnhage

Acknowledgments

As editors of this book, we want to express our sincere gratitude to all the chapter contributors for their efforts to compile the chapters but also for their patience as the project stretched out in time. The same gratitude is valid for the Springer and particularly the series editors, Kenneth Howell and Radislav Potyralio. Deadlines have been passed many times and you have been waiting long for the completion of the book. We also want to thank all the reviewers for their valuable assistance. Finally, we wish to thank the Swedish Defence Research Agency, FOI for providing us resources to work on this book over the years.

Per Jonsson
Göran Olofsson
Torbjörn Tjärnhage

Contents

Part I Introduction to Bioaerosol Detection

1 **Introduction and Bioaerosol Detection Terminology** 3
Torbjörn Tjärnhage, Per Jonsson and Yannick Morel

2 **History of the Early Biodetection Development** 9
Jim Ho

3 **Physical and Biological Properties of Bioaerosols** 33
Jakob Löndahl

4 **Dispersion in the Atmosphere** 49
Lennart Thaning and Leif Persson

5 **Aerosol Sampling and Transport** 63
Jorma Keskinen and Marko Marjamäki

Part II Principles and Technologies for Bioaerosol Detection

6 **Light Scattering and Particle Charge Techniques for the Detection of Biological Warfare Agents** 87
James M. Clark

7 **Bioaerosol Detection with Fluorescence Spectroscopy** 111
Per Jonsson and Fredrik Kullander

8 **Bioaerosol Detection with Atomic Emission Spectroscopy** 143
Nicolas Leone, Damien Descroix and Salam Mohammed

9 **Mass Spectrometry Techniques in the Analysis of Bioaerosols: Development and Advancement** 169
Rabih E. Jabbour, Samir V. Deshpande, A. Peter Snyder and Mary M. Wade

10 Detection of Bioaerosols Using Raman Spectroscopy 203
Hilsamar Félix-Rivera and Samuel P. Hernández-Rivera

11 Biological Detection with Terahertz Spectroscopy 241
Tatiana Globus and Boris Gelmont

Part III Standoff Sensor Systems for Bioaerosol Detection

12 Introduction to Stand-Off Detection of Biological Warfare Agents 267
Per Jonsson and Göran Olofsson

13 Spectrally Resolved Laser-Induced Fluorescence Lidar Based Standoff Biodetection System 269
Jean-Robert Simard, Sylvie Buteau and Pierre Lahaie

14 Standoff Aerosol Size Determination based on Multiple-Field-Of-View of Elastic Scattering 301
Gilles Roy and Nathalie Roy

Part IV Outlook and Challenges

15 Trends in Biological Detection 317
Per Jonsson and Torbjörn Tjärnhage

Index 323

Contributors

Sylvie Buteau Defence Research and Development Canada: Valcartier Research Centre, Quebec, QC, Canada

James M. Clark Salisbury, UK

Damien Descroix Physical Detection Department, DGA CBRN Defence, Vert le Petit, France

Samir V. Deshpande Science and Technology Corporation, Edgewood, MD, USA

Tatiana Globus University of Virginia, Charlottesville, VA, USA

Boris Gelmont University of Virginia, Charlottesville, VA, USA

Jim Ho Defence R&D Canada–Suffield, Medicine Hat, Alberta, Canada

Rabih E. Jabbour U.S. Army Edgewood Chemical Biological Center Aberdeen Proving Ground, MD, USA

Per Jonsson Division of Sensor and EW Systems, FOI—Swedish Defence Research Agency, Linköping, Sweden

Jorma Keskinen Department of Physics, Tampere University of Technology, Tampere, Finland

Fredrik Kullander Division of Sensor and EW Systems, FOI—Swedish Defence Research Agency, Linköping, Sweden

Jakob Löndahl Division of Ergonomics and Aerosol Technology, Department of Design Sciences, Lund University, Lund, Sweden

Nicolas Leone Physical Detection Department, DGA CBRN Defence, Vert le Petit, France

Pierre Lahaie Defence Research and Development Canada: Valcartier Research Centre, Quebec, QC, Canada

Marko Marjamäki Department of Physics, Tampere University of Technology, Tampere, Finland

Salam Mohammed Division of CBRN Defence and Security, FOI—Swedish Defence Research Agency, Umeå, Sweden

Yannick Morel DGA Maîtrise NRBC—CBRN Defence, Direction Générale de l'Armement, Vert le Petit, France

Göran Olofsson Division of CBRN Defence and Security, FOI—Swedish Defence Research Agency, Umeå, Sweden

Leif Persson Division of CBRN Defence and Security, FOI—Swedish Defence Research Agency, Umeå, Sweden

Gilles Roy Defence Research and Development Canada—Valcartier, Quebec, QC, Canada

Nathalie Roy Defence Research and Development Canada—Valcartier, Quebec, QC, Canada

Hilsamar Félix-Rivera Center for Chemical Sensors Development/Chemical Imaging Center (CCSD/CIC), ALERT DHS—Center of Excellence for Explosives Research, Department of Chemistry, University Puerto Rico-Mayagüez, Mayagüez, USA

Samuel P. Hernández-Rivera Center for Chemical Sensors Development/ Chemical Imaging Center (CCSD/CIC), ALERT DHS—Center of Excellence for Explosives Research, Department of Chemistry, University Puerto Rico-Mayagüez, Mayagüez, USA

Jean-Robert Simard Defence Research and Development Canada: Valcartier Research Centre, Quebec, QC, Canada

A. Peter Snyder U.S. Army Edgewood Chemical Biological Center, Aberdeen Proving Ground, MD, USA

Torbjörn Tjärnhage Division of CBRN Defence and Security, FOI—Swedish Defence Research Agency, Umeå, Sweden

Lennart Thaning Division of CBRN Defence and Security, FOI—Swedish Defence Research Agency, Umeå, Sweden

Mary M. Wade U.S. Army Edgewood Chemical Biological Center, Aberdeen Proving Ground, MD, USA

List of Abbreviation

ACPLA	Agent containing particle per liter of air
ADC	Analogue to digital converers
AES	Atomic emission spectroscopy
AI	Artificial intelligence
ANN	Artificial neural network
APD	Avalanche photo-diodes
APS	Aerodynamic particle sizer
ARL	Army Research Laboratory
ASD	Aerosol sodium detector
ATOF-MS	Aerosol time-of-flight mass spectrometry system
BA	*Bacillus anthracis*
BAMS	Bioaerosol Mass Spectrometer
BARTS	Biological Agent Real Time Sensor
BAWS	Biological Agent Warning Sensor
BC	*Bacillus cereus*
BG	*Bacillus atrophaeus* (previously known as *B. globigii* and *B. subtilis var Niger*)
BioACE	Biological aerosol-capture-enrichment
BPAT	Buffered antigen plate agglutination test
BT	*Bacillus thuringiensis*
BTWC	Biological and Toxin Weapons Convention
BW	Biological warfare
BWA	Biological warefare agents
CAM	Cell adhesion molecules
CARS	Coherent anti-stokes Raman spectroscopy
CBRN	Chemical biological radiological and nuclear
CCD	Charge-coupled device
CEA	Carcinoembryonic antigen
CELISA	Competitive enzyme immunoassay
CERS	Cavity enhanced Raman spectroscopy
CFT	Complement fixation test
CFU	Colony forming unit

CPC	Condensation particle counting
CSL	Crocus sativus lectin
cw	Continuous wave
DA	Discriminant analysis
DARPA	Defense Advanced Research Projects Agency
DFA	Discriminant factor analysis
DGA	Direction Générale de l'Armement
DHB	2, 5-dihydroxybenzoic acid
DISC	Differential scattering
DMA	Differential mobility analyzer
DNA	Deoxyribonucleic acid
DOP	Diocyl phtalate
DPA	Dipicolinic acid
DPG	Dugway Proving Grounds
DRDC	Defence Research and Development Canada
Dstl	Defence Science and Technology Laboratory
DTRA	Defense Threat Reduction Agency
EC	*Escherichia coli*
ECBC	Edgewood Chemical Biological Center
EDA	European Defence Agency
EEM	Excitation-emission matrix
EH	*Erwinia herbicola* (presently known as *Pantoea agglomerans*)
ELPI	Electrical low pressure impactor
ESA	European Space Agency
ESI	Electrospray ionization
FAB	Fast atom bombardment
FABIOLA	Fluorescence applied to biological agents detection
FAD	Flavin adenine dinucleotide
FAK	Focal adhesion kinase
FES	Flame emission spectroscopy
FLAPS	Fluorescence aerodynamic particle sizer
FLASAS	Fluorescence aerosol shape analysis system
FOI	Swedish Defence Research Agency
FOV	Field-of-view
FPD	Flame photometric detection/detector
FRET	Fluorescence resonance energy transfer
GC	Gas chromatography
GL	Glycyrrhizin
HA	Hemagglutinin
HCA	Hierarchical cluster analysis
HIV	Human immunodeficiency virus
HPD	Hybrid photodiode detector
HSV-1	Herpes simplex virus type 1
ICCD	Intensified charge-coupled device
ICP	Induction-coupled plasma
IELISA	Indirect enzyme immunoassay

IMD	Instantaneous microbial detection
IR	Infrared spectroscopy
JBSDS	Joint biological standoff detection system
kNN	K-nearest neighbor classifier
KTH	Royal Institute of Technology
LC	Liquid chromatography
LDA	Linear discriminant analysis
LDH	Lactate dehydrogenase
LED	Light-emitting diode
LIBS	Laser-induced breakdown spectroscopy
Lidar	Light detection and ranging
LIF	Laser-induced fluorescence
LIPS	Laser-induce plasma spectroscopy
LPS	Lipopolysaccharides
LRR	Leucine-rich repeat
LTRS	Laser trapping Raman spectroscopy
MAB	Moniteur d'Alarme Biologique (Biological Alarm Monitor)
MALDI	Matrix-assisted laser-desorption-ionization
MAMP	Microorganism-associated molecular patterns
MASU	Mobile aerosol sampling unit
MCMV	Murine cytomegalovirus
MFOV	Multiple-field-of-view
MIP	Microwave induced plasma
MRT	Milk ring test
MS	Mass spectrometry
MS2	Male-specific bacteriophage 2
MSL	Mars Science Laboratory
Mur	Muramic acid
MVDA	Multi-variate data analysis
NADH	Nicotineamide adenine dinucleotide
NASA	National Aeronautics and Space Administration
NBA	3-nitrobenzyl alcohol, 3
NRL	Naval Research Laboratory
ONPG	O-nitrophenyl ß-galactopyranoside
OPC	Optical partical counter
OPO	Optical parametric oscillator
OV, OA	Ovalbumin
PA	*Pantoea agglomerans* (previously known as *Erwinia herbicola*)
PA	Picolinic acid
PAMP	Pathogen associated molecular patterns
PAMS	Particle analysis by mass spectrometry
PCA	Principal component analysis
PCR	Polymerase chain reaction
PFU	Plaque-forming unit
Phe	Penylalanine
PLS	Partial least squares

PLS-DA	Partial least squares-discriminant analysis
PMT	Photomultiplier tube
poly I:C	Polyinosine-deoxycytidylic acid
ppl	Particles per liter
PRF	Pulse repetition frequency
PRR	Pattern recognition receptors
Py	Pyrolysis
QDA	Quadratic discriminant analysis
QE	Quantum efficiency
RBAD	Real-time biochemical aerosol detector
RGP	Resorufin-ß-D-galactopyranoside
RNA	Ribonucleic acid
ROC	Receiver operating characteristic
RR	Resonance Raman
RSV	Respiratory syncytial virus
SA	Shape analyser
SA	Sinapinic acid
SARS	Severe acute respiratory syndrome
SBS	Sick building symdrome
SERS	Surface-enhanced raman spectroscopy
SINBAHD	Standoff integrated bioaerosol active hyperspectral detection
SNR	Signal-to-noise ratio
SORS	Spatially offset Raman spectroscopy
SPAMS	Single particle aerosol mass spectrometry
SPFA	Single particle fluorescence analyzer
STA	Slit-to agar impactor
STAG	Spinning top aerosol generator
SUVOS	Semiconductor ultraviolet optical source
SVM	Support vector machines
SWIR	Short-wavelength infrared
TAOS	Two-dimensional angular scattering
THz	Teraherz
TLR	Toll-like receptor
TOF-MS	Time-of-flight mass spectrometry
TRELIBS	Time resolved laser-induced breakdown spectroscopy
TRL	Technological readiness level
Trp	Tryptophan
TUCBE	Targeted ultraviolet chemical, biological, and explosives
TUT	Tampere University of Technology
Tyr	Tyrosine
UVRR	Ultraviolet resonance Raman
VI	Virtual impactor
WIBS	Wide issue bioaerosol sensor
VOAG	Vibrating orifice aerosol generator
YP	*Yersinia pestis*

About the Editors

Per Jonsson received his M.Sc. in Electrical Engineering from KTH—Royal Institute of Technology, Stockholm, Sweden in 1996 and his Ph.D. in Photonics from KTH in 2002. The title of his thesis is "Generation, Detection and Application of Single Photons". Dr. Jonsson joined FOI—Swedish Defence Research Agency in 2002 and is presently a senior scientist within the competence group Laser Systems in the Department of Sensors and EW systems in Linköping. His main research area at FOI has been detection of biological and chemical warfare agents. He has mainly worked with systems based on ultraviolet laser-induced fluorescence, but has also investigated other technologies such as elastic scattering, Raman spectroscopy and laser-induced breakdown spectroscopy. The work includes systems both for point (local) and standoff (distance) detection.

Göran Olofsson received an M.Sc. and engineering in Physical engineering at the Linköping Institute of Technology in 1976. Mr. Olofsson joined the Division for CBRN Defence and Security at the Swedish Defence Research Agency, FOI (then FOA) in 1979, where he since then has been active in the area of CWA and BWA detection, often as project manager and since 2002 holding the position as Research Director. Mr. Olofsson was in 1991 an inspector for UNSCOM in Iraq. He also served at the Assistance and Protection Branch of OPCW, for a shorter period in 2002 and was during 2005/2006 for 10 months stationed at CEB (now DGA Maîtrise NRBC) in *VERT-LE-PETIT*, France. Mr. Olofsson is presently mainly working with UV-Raman spectroscopy for chemical and biological detection.

Torbjörn Tjärnhage received his B.Sc. in chemistry from Umeå University, Sweden in 1989 and his Ph.D. in Analytical Chemistry from Umeå University in 1996. Dr. Tjärnhage joined FOI—Swedish Defence Research Agency in 1995. Since 2000 his research areas has been focused towards real-time spectroscopic characterization of biological aerosols in combination with multivariate data extraction techniques. This also includes more applied work directed to support to the Swedish Armed Forces development of biodetection capabilities. In 2006 Dr. Tjärnhage was promoted senior scientist and has been project manager for biological detection projects.

Part I
Introduction to Bioaerosol Detection

Chapter 1
Introduction and Bioaerosol Detection Terminology

Torbjörn Tjärnhage, Per Jonsson and Yannick Morel

The Biological Warfare Background

As described in the preface of this book, there has been an increase in the interest in detecting and characterizing biological aerosols in the past decades. Now the use of biodetectors is entering environmental, medical, and purely scientific bioaerosol studies, and the development has been largely driven by requirements in the biodefense and security arena. This is reflected in this book as many of the technologies have their background and development in defense applications.

Biological warfare (BW) could, in short, be described as the deliberate use of disease-causing agents, such as bacteria, viruses, fungi, or biological toxins, to kill or incapacitate humans, animals, or plants as an act of war [1]. BW agents (BWAs) can be isolates of naturally occurring endemic foci or the result of man-made manipulations. The success in using microorganisms and toxins to deliberately cause disease, death or incapacity among humans depends on a variety of factors, the most important being the agent itself and its formulation as well as the technology of dispersion used to generate bioaerosols. The Biological and Toxin Weapons Convention (BTWC), which entered into force in 1975,

> Bans the development, production, stockpiling, acquisition and retention of microbial or other biological agents or toxins, in types and in quantities that have no justification for prophylactic, protective or other peaceful purposes. It also bans weapons, equipment or means of delivery designed to use such agents or toxins for hostile purposes or in armed conflict. [2]

T. Tjärnhage (✉)
Division of CBRN Defence and Security, FOI—Swedish Defence Research Agency, Cementvägen 20, 901 82 Umeå, Sweden
e-mail: torbjorn.tjarnhage@foi.se

P. Jonsson
Division of Sensor and EW Systems, FOI—Swedish Defence Research Agency, Olaus Magnus väg 42, 581 11 Linköping, Sweden

Y. Morel
DGA Maîtrise NRBC—CBRN Defence, Direction Générale de l'Armement, 5 rue Lavoisier, BP 3, 91710, Vert le Petit, France

P. Jonsson et al. (eds.), *Bioaerosol Detection Technologies*, Integrated Analytical Systems, DOI 10.1007/978-1-4419-5582-1_1, © Springer-Verlag New York 2014

In the past, several states have had extensive biological weapon programs, which are today either closed or converted into institutes for homeland security and public health. It is not within the scope of this book to address the current threat level for the potential use of BWAs.

As most nations today have adopted the BTWC, the risk that a single state will use BWAs in a massive aggressive way is dramatically reduced. However, a few nations have chosen not to follow the convention and together with the demonstrated possibility that terrorist organizations and other groups can gain knowledge of how to produce and handle BWAs, the threat is not completely removed. As shown in the Anthrax letter event in the USA in 2001, even a relatively small amount released will generate consequences that have devastating outcome for individuals, society, and infrastructure.

Aerosolized BWA is the major threat since inhalation is the most effective route for infection. Furthermore, even a small amount of BWA dispersed as aerosols can infect people in a large area. Aerosols are therefore the main target when biodetectors are developed for defense applications. Aerosols in the size range of approximately 1–10 μm are also often mentioned as typical in BW threat situation [3]. These facts are essential inputs in the design of biodefense detector system.

Detection vs Identification

From the military perspective, *detection* is, according to the European Defence Agency (EDA), *the capability to quickly discover of the presence of health threatening levels of Chemical Biological Radiological and Nuclear (CBRN) substances before they have negative impact on mission effectiveness and provision of timely information.* The detection response will permit forces to adopt an appropriate level of physical protection. This is often denoted as the "detect-to-warn" concept. The definition within the North Atlantic Treaty Organization (NATO) is similar.

The main purpose is to provide a timely warning and raise awareness that countermeasures including protective actions must be taken or that a specific area must be avoided. This detect-to-warn concept is derived from chemical agents detection for which deadly effects are immediate and medical therapeutics may be too late. For BWAs, infection only develops after several hours (symptoms appear at least 24 h postexposure). If therapeutic treatment is administered before symptom onset, recovery can be expected (detect-to-treat concept). However, efficient medical countermeasures are not available against all the possible BWAs and, therefore, detect-to-warn systems are desirable.

It is important to mention that the detection of a certain bioaerosol, whether a pathogenic bacteria in biodefense application or a specific target microorganism in an environmental application, needs to be performed in the presence of a great number of other biological particles. This natural "background" is highly variable and its composition may vary both in quantity (total concentration of microorganisms per liter of air) and in quality (occurrence of different microbial species). These

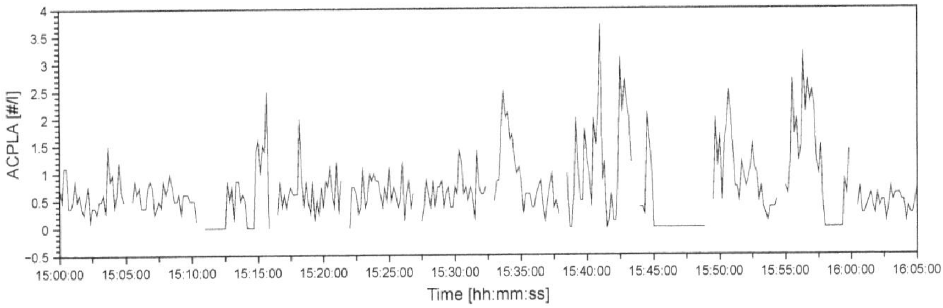

Fig. 1.1 Typical fluctuations of the aerosol biological background during a 1 h time period. The graph shows the content of agent containing particles per liter of air (ACPLA), i.e., cultivable particles as a function of time. Sampling was performed with a slit-sampler in a peri-urban environment [4]

fluctuations may be very rapid, even within seconds as can be seen in Fig. 1.1. In addition, the background composition and concentration may be very different according to the location (urban area, countryside, etc.) and season. Random concentration changes may occur in an otherwise rather flat sequence due to various reasons such as wind, human, or animal activity. The content of the biobackground can also be significantly altered by climatic events, e.g., heavy rain showers or lightning storms [5].

Owing to the great variability and random fluctuations, it is impossible to accurately model and predict the background of bioaerosol. This lack of knowledge may become a critical problem if the biodetector is too "generic" and only measures biological particles with poor selectivity. A biodefense detector may interpret a natural change in concentration as a change in a threat bioaerosol and thereby provide a false positive alarm. Historical knowledge of the natural fluctuations of the background in the monitored area in combination with sufficient selectivity is therefore required to perform biodetection with confidence. If the detector response is under manual evaluation by an operator, additional observation of a nearby activity known to generate aerosols can be taken into account before an alarm is raised. It is a major concern that detector response from the background level can be more than the detector response of infectious concentrations of the BWAs. Hence, triggering a biodetector at too high a level will result in the nondetection of deadly situations. Conversely triggering at low threshold levels will generate many false alarms that may induce operators not to rely on the system. The main problem thus lies in the trade-off between sensitivity, false alarms, and time to result.

To be certain about the true nature of the biological aerosol, a more detailed (usually laboratory based) examination of the material is required. This analysis can include traditional molecular biotechnological methods, such as genetic and immunological assays. They provide high specificity and possibly a higher degree of *identification,* but the response time ranges from a number of minutes to hours or even days since they may need sample preparation, purification, and amplification (e.g., polymerase chain reaction [PCR]) or other biochemical reaction steps. To reach even higher confidence and sensitivity, these methods may also need additional complementary analysis or culturing that can further increase the time to result.

In biodefense applications, it is common that biodetectors are used in combination with identification systems to provide both quick awareness of a potential threat and, a bit later, a better confirmed answer provided by the identification step.

However it is well to acknowledge that there is a technological and functional distinction between systems that are used for biological detection and systems that will perform biological identification. Ideally, one would desire a system (or technology) that can provide identification within just a few seconds, but the currently available technologies do not meet such requirements.

This book is devoted to describing the technologies that are exploited in detectors that can directly monitor the presence of bioaerosols and can tentatively be defined as "*A biological detector (or biodetector or B-detector) can be defined as a system that continuously and in near real-time analyzes particles in the air to assess and possibly classify their biological properties and/or origin.*" Since the real-time requirement is rather important, this has to be typically done by using physical, mainly spectroscopic, techniques. The strategy is to exploit some characteristic property, direct or indirect, of the bioaerosol that is known to be related to biological activity or origin. That could, for instance, be finding the presence of certain target molecules, functional structures, or elementary composition. However, more basic parameters such as size, shape, and electrostatic charge of the aerosol particles have also been used to classify their origin.

The biological difference between a pathogenic microorganism and its completely benign close neighbor can be very small. There might be just some small alteration in the genetic code that may or may not cause the virulence. The question is then how much information can be extracted from an aerosol particle while passing through the measuring cell of a detector that is based on physical principles?

Levels of Detection

In some fora different categories of detection capability are listed. For example, at the time of finalizing this book, the EDA test and evaluation biological detection, identification and monitoring project (EDA T&E BioDIM), is proposing a biodetection capability-level definition based on several sources [1]. The final version may differ, but essentially Table 1.1 just lists an increasing degree of specificity in the outcome of the detector or analyzer response and other lists or definitions may exist. However, such lists can be illustrative and may be used to clarify that the word *detector* by itself is not sufficient to describe the capability or function.

The technologies described in this book are mainly belonging to capability levels 1–3, possible touching level 4 for some technologies (in ideal conditions).

The reason biodetectors have an important purpose is their inherent capability to operate continuously and provide a time-resolved response to the biological content of air. This monitoring capability is particularly useful in biodefense applications, but, as mentioned previously, is also useful for other applications. It is important to understand the degree of capability a certain biodetector is actually providing. This

Table 1.1 Different levels of detection confidence and corresponding common nomenclature

Capability level	Level of confidence or resolution	Common names used for the capacity
1	Capability to react to changes in airborne particle concentration and composition	Detect to react, trigger detection
2	Capability to react to changes in airborne particle concentration of biological origin	Detect to warn, trigger detection, generic detection
3	Capability to react to changes in airborne particle concentration of biological particles belonging to a group (class) of microorganisms that may contain both pathogens and nonpathogens	Detect to warn, trigger detection, generic detection, classifier (detect to classify)
4	Capability to perform provisional identification at species level	Detect to treat, identification, specific detection, provisional identification
5	Capability to perform confirmed identification at strain level	Detect to treat, confirmed identification

is most prevalent in biodefense applications where the outcome of an "alarm" may be very dramatic and thus the output from the detector has to be of high confidence. As will be seen in the following chapters, biodetector technologies based on physical principles will have certain restrictions in their selectivity and specificity. That is, most of the techniques presented in this book do not qualify for real *identification* of the exact nature of the biological content (capability level 4 and 5 in Table 1.1).

Key Parameters and Terminologies of Bioaerosol Detection Systems

An optimal bioaerosol monitoring system is sensitive to infectious doses, ideally a single particle level, and has high specificity, i.e., recognizes specific species of bioaerosols with a low probability of false alarms and a fast response time (close to real-time detection). No single sensor or detection technology today fulfills these requirements for sensitivity, specificity, and speed. It is therefore important to have knowledge about the possibilities that current biodetection technologies offer, and, more importantly, the limitations each technology may possess. A biological detector will have some characteristic properties such as:

- sensitivity (limit of detection, dynamic range, probability of detection, etc.)
- selectivity (specificity, false alarm rates)
- speed/response time
- size and weight

Most detection systems are designed as *point detectors*, i.e., they analyze the aerosol content where the detector is placed. This is done simply by extracting the air directly into a measurement cell in the detector system. However, in some cases, it is also possible to probe a distant aerosol cloud. Such *stand-off detectors* use a passive or active optical technique to gain information from a significant distance. In other cases, point detectors can be placed at a certain distance either alone or in cluster and transfer their response to a central control function. In that case, one speaks of *remote detection.*

A potential user of a biodetection system must ensure that the technology has specifications that meet the requirements of the selected purpose. The technologies described in this book are demonstrated to be able to detect biological particles as expected, but there are still considerations that they may also be sensitive to other nonrelevant particles in the air.

In certain field applications, the (non)-complexity of the detector and its size and weight are of great interest and may, in some cases, be operationally more important than the actual detection parameters.

This book presents the most commonly used detection principles. The basic principles are described in theory, practical examples are given, as well as discussions about their potential use as bioaerosol detectors. The goal of this book is that the readers will acquire knowledge about the different technologies and their potentials and limitations and thus better understand the capability of bioaerosol detectors in different applications.

References

1. Tjärnhage T, Jonsson P, Sagerfors E, Wästerby P, Blatny JM, Skogan G, Humppi T (2011) Detection of airborne biological agents: technological review and outlook. FOI-R–3267–SE. FOI, Umeå
2. Biological and Toxin Weapons Convention (BTWC) website. www.opbw.org. Accessed 28 Januari 2014
3. Walt DR, Franz DR (2000) Peer Reviewed: Biological Warfare Detection. Anal Chem 72 (23):738A–746A. doi:10.1021/ac003002a
4. EU FP7 project “IMPACT”, Airborne biological background measurements. WP400, deliverable D400.5
5. Ho J, Tjärnhage T, Burke J, Stadnyk L (2004) Background aerosol sampling: optical characteristics of live particles associated with thundershowers in Umeå, Sweden in September 2003. Paper presented at the 8th International Symposium on Protection against Chemical and Biological Warfare Agents, Gothenburg, Sweden, 2–6 June 2004

Chapter 2
History of the Early Biodetection Development

Jim Ho

Introduction

The user of a biological detector simply wishes to be notified whether the invisible population of aerosol particles that is about to be inhaled is potentially infectious. On top of that, the military has made it known that there is a requirement for a biological detector that is small enough for handheld applications. The detector has to run on battery power, responds within seconds or minutes to warn of the passage of a potential threat aerosol cloud without false alarms. The information it provides has to be suitable for a medical officer to treat exposed personnel. In addition, the cost and maintenance of each instrument must fall within strict guidelines. At the time of preparation of this chapter (mid-2013), such an instrument did not exist. This review will explore what historically useful contributions have been attempted to solve the problem.

Definition of Terms

There is a need to clarify the term "detection" which is used in the context of biological aerosol threats. The user's simplistic desire is for an instrument that unambiguously notifies him or her of the presence of an infectious or virulent agent. Knowing this will permit him or her to take evasive/protective action and/or seek for medical assistance. For a variety of technical reasons (testing with animals being one) that will become apparent, it is not feasible to measure infectivity given the technical and time limitations [1]. If time is not an issue, clinical methods can be employed to obtain taxonomic results that only imply if an unknown agent was infectious [2]. The accepted "gold standard" to determine virulence and infectivity is the use of the animal model [3]. There are no detection "short cuts" in the tightly regulated medical and public health environments [4]. Nevertheless, as a compromise, the closest

J. Ho (✉)
Defence R&D Canada–Suffield, 578 1st Street, NW, Medicine Hat, Alberta T1A 6H8, Canada
e-mail: jimhoho@shaw.ca

P. Jonsson et al. (eds.), *Bioaerosol Detection Technologies*, Integrated Analytical Systems,
DOI 10.1007/978-1-4419-5582-1_2, © Springer-Verlag New York 2014

to achieving this ideal has been to adopt rapidly responding optical techniques to reveal if the threat agent has "live" characteristics since the ability of an agent to replicate is a prerequisite to infectivity. Following this train of thought, it is clear that current expedient approaches to biological detection cannot be considered as substitute for classical "identification" or "taxonomic" methodologies. This may appear obvious to trained microbiologists but throughout the literature, it is not unusual to encounter papers describing the use of optical approaches that claim to provide capabilities either in identification [5–7] or taxonomy [8]. We respectfully suggest that future workers in biological detection refrain from using the term "identification" if they have not followed the classical medical/clinical methodologies. As an alternative, we suggest using terms such as "segregation" or "sorting" as in segregating or sorting biological particles from background noise material [9].

In the military context, it has been proposed [10] that the primary aim in biological detection is to (1) provide warning to initiate protection and (2) assist the medical officer in prophylaxis and therapy. The latter part of this aim encompasses terminologies in common with the medical profession. Thus, the terms "identification" and "taxonomy" in the clinical setting have very specific meanings as the performance of each task and the results derived thereof can have life and death implications. Simply put, a registered clinical microbiologist has to follow very strict procedures when performing these. At the risk of appearing pedantic, here is the outline of the preliminary identification steps to be executed presuming the availability of a pure bacterial unknown isolate:

1. Determine if the bacterium being identified is Gram-positive or Gram-negative.
2. Is the bacterium round (a coccus) or rod-shaped?
3. Is the bacterium "acid-fast" or not?
4. Does the bacterium produce spores?
5. Does the bacterium grow in the presence of air?

If the answers to these questions were: (1) Gram-positive, (2) rod, (3) no, (4) yes, (5) yes then the bacterium being identified belongs to the genus *Bacillus*. Further biochemical tests will be required to determine the taxonomic species. Even after all these procedures had been performed, there will be uncertainties with respect to the accepted identity and taxonomy of the unknown as has been well illustrated for the pathogen *Burkholderia spp* [11] and a non-pathogen [12]. Using nucleic acid techniques, Danin-Poleg and colleagues [13] demonstrated the difficulties in differentiating between harmless and virulent strains of *Escherichia coli, Listeria monocytogenes and Vibrio cholerae*. In addition, it should not be surprising that all microorganisms being living things, can evolve with varying rates dependent on environmental pressures [14], making the tasks of "identification" and "taxonomy" much more demanding. In light of these caveats, the strategic approach to biological detection has been proposed in a review chapter [10] and will not be reiterated here. In brief, a biological detector, primarily relying on hardware and software is unlikely to credibly perform "identification" and "taxonomy" tasks to the satisfaction of the medical community.

As in most technological advances, the history of biological detection has been punctuated by a variety of discrete events. At times, these events may be the availability of components, reagents, methodologies or sometimes rediscovery of any of the above. There may be some elements of serendipity associated with the developments where new discoveries are put to use to help solve the biological detection problems. In this review, an attempt will be made to link the events that ultimately led to the results of efforts expended in the quest of detecting the illusive infectious aerosol agents.

Early Attempts Using Light Scatter Optical Particle Counting

In the mid-1800s, John Henry Bell, known as a "British anthrax investigator", was taking notes of anthrax disease contracted by Bradford wool sorters in the UK. He wrote: "The workers pointed to the fleeces they sorted as the source of the disease and tried to avoid working with "bad" bales of fleece. They suspected that imported fleeces, the alpaca and "van" mohair from the Middle Eastern Levant region, had some kind of dusty poison in them" [15]. Bell himself [16] observed that anthrax infection was most probable from wool with dry dusty content compared to sticky material as shown in this quote: "Alpaca and mohair are not the only infective materials. Many fatal cases have occurred from the manipulation of other dry and dusty hairs and wools. "Greasy" wools are less dangerous; it is probable that the "yolk" in them fixes the infective particles. All hairs and wools (excepting Algerian) may at times contain uncleansed fleeces from animals which have died from anthrax and may occasionally communicate the disease to those who come in contact with them. I have seen such cases from the sorting of British wools, and have heard from my medical friends of others, but the proof is not yet complete". The reference to "dust" invoked a condition conducive to aerosol generation. The word "aerosol" was coined on page 600, line seven in a paper by Whytlaw-Gray et al. [17] who first introduced the word in print and attributed its origin to Professor F.G. Donnan [18]. So it is not surprising that subsequent workers in biological detection naturally directed their attention and efforts to measuring particles in air, aerosol as is now the common term.

Gentry in a review [19], attributed John Tyndall (1820–1893) to the development of dark field illumination (Tyndall effect) and hence to demonstrating the presence of microorganisms in air. It was noted that the Tyndall effect was also used to detect airborne pollutants. Knowledge from having done these measurements inspired Tyndall to develop a respirator with a charcoal layer to filter out gas vapors and a cotton wool layer to remove particles. His measurements also led to the disproving "Spontaneous Generation" which paved the way to the concept of an explanation for the origin of microorganisms. Tolman and Vliet built an optical instrument called a "Tyndallmeter" primarily to examine smoke particles [20]. It was also shown that light scatter intensity was related to particle size [21].

In an introduction to the biological detector paper by Gucker et al. [22], it was noted that Arthur Guyton [23] at Camp (present day Fort) Detrick between June and October 1945, developed a particle counter. The device caused a stream of particles to impact on a metal plate and apparently particle size was gauged from charge characteristics. It was not sensitive to particles less than about 2.5 μm in diameter. Due to this and other deficiencies, instrument development was abandoned.

In collaboration with the US Army, Gucker's team working in the Chemistry Department of Northwestern University (Evanston, IL), built the first optical particle counter (OPC) for the sole purpose of detecting bacterial spores [22]. It has been known for a long time that anthrax in spore form was a significant military, civilian and agricultural threat [24, 25]. The instrument employed what is now commonly known as sheath flow to control particle transport. The air stream containing the particulate material was confined to the center of a larger sheath stream that passed through the focal point of a dark-field microscope. Particles passing through the system scattered light into a collection lens, eventually producing electrical signals from a RCA 931A photomultiplier tube (PMT). The instrument could just barely resolve the optical signal from a moving bacterial spore in the size range of 0.6 μm in diameter. Of historical significance, the use of *Bacillus globigii* (BG) aerosol as a bacterial spore simulant was first mentioned by Gucker et al. [22]. Being an electronics engineer, he had to rely on scientists at Harvard University to develop the BG aerosol dissemination procedure [26]. For aerosol generation, a medical nebulizer was used. The original glass device, subsequently characterized at the US Army Medical Research Institute of Infectious Diseases, Frederick, MD, was found to produce droplets of about 1.7 μm mass median diameter, whereas a latter day modern plastic model yielded larger ones at about 2.3 μm [27]. It should be noted that the 1947 Gucker instrument had no sizing capabilities; a calibration method for the OPC had not been invented yet.

The limited success of the Gucker instrument in detecting spores relied on the fortuitous availability of certain key components. Shapiro commented that the light source with sufficient brightness to make the Gucker project possible came from a Ford headlight [28]. The 931A PMT came from the Lancaster, Pennsylvania facility opened by the US Navy in 1942 and operated by RCA for the manufacture of radio and microwave tubes. However, Smyth examined a number of 931A PMTs made in the late 1940s and found that their sensitivities were highly variable [29]. Edels and Gambling found that these early production tubes suffered from spatial variations causing signal instability from unfixed sources like inadequately focused aerosol particles [30]. For all these difficulties, it was truly remarkable that Gucker's team could manage to obtain data from their machine. By incorporating 90° light scatter measurement, Gucker and Okonski made improvements to the instrument with the hope of resolving BG and sulfur particles [31]. It was noted that uneven particle dryness caused variability in the signal consistency. The interesting item from this work was that by using a polarizing filter for the light source, BG counts were enhanced.

The next advancement in OPC was registered by Okonski and Doyle [32] which started to measure aerosolized uniform spherical particles of polystyrene and polyvinyltoluene (latex beads) in the size range from 0.1 to 1 μm which had just been produced by Dow Chemical Co. (Midland, MI). It was reported that the particles, when dispersed individually as an aerosol, were ideal for calibration and evaluation of counting instruments because of their high degree of uniformity and low vapor pressures.

By the early 1950s, Gucker moved from Northwestern University to Indiana University at Bloomington where he calculated the Mie scattering properties of spherical particles over a wide range of angles and refractive index [33]. A few years later, he reported on angular scattering characteristics of particles measured with a new instrument. Gucker and Egan measured large portions of the scattering diagram of single aerosol particles (Dioctyl phthalate, DOP) suspended electrostatically in a measuring chamber and viewed by a photocell that moved around the particle (40° to 140°) in 15 or 1.5 min [34]. In later implementation, angular scatter measurement was made more rapidly by using a moving mirror [35]. It is interesting to note that the Harvard University group did not adopt the Gucker detector for their subsequent biological characterization work [36]. However, Gucker's angular scattering measurement instrument inspired Phillip Wyatt who, years later, applied this technique in characterizing bacterial cells in liquid [5].

Wyatt was a physicist who competently exploited historical theoretical particle scattering work of Mie [37], Lorenz [38] and Rayleigh [39] in characterizing batches of bacterial cells in liquid [40]. Many years later, Bronk's group also carried out similar measurements with bacterial suspensions [41]. Bottiger who worked in the same basement laboratory at Edgewood as Bronk, commissioned Wyatt to build a multi-sensor globe for capturing multiple angles of light scatter from a moving stream of aerosol particles [42]. It was discovered that a series of rod shaped bacterial cells produced inconsistent signals when measured at six angles. The authors noted that although 58 channels of light scatter could be potentially monitored, the lack of data handling electronics to match the requirement caused the project to fall short of expectations. The instrument was used to characterize non-biological submicron sized aerosol particles by Peter McMurry's group from the Department of Mechanical Engineering, University of Minnesota in the early 1990s [43]. They concluded that for the small particles examined, the counting efficiency was about 50 %. In some instances, it was suspected that aspheric particles compromised data gathering and electronic noise hindered reflective index estimates with the less than optimal seven channels available. Given these experiences, it can be concluded that the power of multiple angle light scatter as a way for bacterial aerosol characterization had not been fairly tested. The biggest challenge that future workers in this field will encounter is that biological particles are inherently aspheric as discussed in the next section. Thus, attempting to fit light scatter data from randomly shaped material can become difficult to manage.

Bacterial Shape Analysis

Microbiologists have known for a long time that bacteria display a variety of cell shapes, including round, rod, spiral and amorphous. In a recent review, Young described over 24 morphological structures that bacteria may assume [44]. Maintenance of cell shape is vital for cell growth and cell division in most bacteria. The distinct shape of most bacteria is retained by a peptidoglycan layer, enveloping bacterial cells as a single structural macromolecule [45]. Analogous to actins in eukaryotes, the bacterial tubulin, FtsZ (standard gene designation has no full name equivalent), has an important function in cell division [46] and contributes to cell morphogenesis. FtsZ protein forms a ring-like structure at mid-cell, which serves as a scaffold for other cell division proteins. Another bacterial cytoskeletal protein MreB determines the length of the short axis of the cell, whereas Mbl determines the length of the long axis of the cell. Both MreB and Mbl form a helix or filament just beneath the cytoplasmic membrane along the long axis of the cell in *B. subtilis* [47]. It can be safely assumed that the bacterial shape is fundamental to the distinctiveness of the native or non-mutated version of the species. It is natural for those workers with microbial training to exploit this property in biological detection.

In the mid-1980s Cox who later wrote a book on biological aerosols [48] asked Bexon who worked on aerosol holography [49] how to design an instrument for measuring the shape of a single bacterial aerosol particle. Bexon, a physicist, probably had Mie in mind, he theorized that by monitoring three axes of light scatter, particle shape resolution may be possible. Cox commissioned the Laser Systems Research Group at the Hatfield Polytech, Hatfield, Herts, UK to build a small battery powered "shape analyser" (SA). Eventually, a much larger version than the one initially envisioned was built. In characterizing spheres and fibers the term "asymmetry factor" was introduced [50]. The instrument was demonstrated to produce spherical scatter data close to Mie predictions [51]. However, it was discovered that the SA instrument was mostly suited to characterizing elongated fibers rather than bacterial shapes [52–54]. In the late 1990s the Hatfield group combined a 266 nm excitation light source to the original SA concept with the hope of supplementing shape analysis with intrinsic protein fluorescence to provide more robust biological particle resolution [55]. However, by mid-2000s, the group abandoned the SA concept altogether to concentrate their efforts on using 260–280 nm excitation to produce fluorescence signals for characterizing proteins. Much later, 370 nm excitation, pioneered by Suffield in the 1990s, was adopted for probing biological metabolic molecules [56]. More recently, the SA concept was used to characterize atmospheric ice crystals [57]. As a footnote, in a recent paper describing multiple angle light scatter by bacterial spores, it was concluded that the measured signals did not correspond to modeling predictions using T-matrix to simulate a spore with three different particle shapes [58]. Gogoi et al. reported that for complex particles like fresh water diatoms, multiple angle light scatter data did not correlate to Mie predictions [59]. Interestingly, in 1978, Latimer setting out to define "asphericity" of single particles, concluded that data reliability could be compromised by random particle orientation and irregularity of shapes [60]. After over a quarter century of putting Mie to the test, the consensus is that relying on light scatter measurement of single biological particles as a distinguishing characteristic for detection purposes will remain illusive. Nevertheless, the technology has been actively used to detect sudden changes in the aerosol composition and to trigger subsequent sampling and analysis steps. An artificially made aerosol will introduce a fraction of aerosols in the natural background that can be detected by the SA technology. Classification algorithms can continuously analyze the scatter profile and detect deviation from normal conditions.

Origins of Luminol

Luminol (5-amino-2, 3-dihydro-1-4-phthalazinedione) was first synthesized in Germany in 1902, but the compound was not named luminol until the late 1920s. Metcalfe and Quickenden [61] attributed much of the early studies to H.O. Albrecht [62] who synthesized and worked with the compound in 1928 but it was difficult to obtain the original paper for verification. Subsequent workers found that the compound produced a strong glow (chemiluminescence) in the presence of an iron contain material like hemoglobin [63]. Neufeld et al. from the US Army Biological Laboratories, Fort Detrick, Frederick, MD, discovered that the compound was effective for measuring other iron bound proteins [64]. Soon after the appearance of this paper, the Russians also published the paper on this subject [65]. Later, Miller and Vogelhut reported that the assay had a limit of sensitivity of about 10^5 to 10^6 viable bacteria per ml [66]. At about this time, in the early to mid-1970s, probably based on these and other studies [67], the US Army contracted Bendix, Environmental and Process Instrument Division in Baltimore, MD to manufacture a biological detector (XM19) based on the luminol chemistry (Fig. 2.1).

As a historical footnote, there was a similar Russian instrument that was probably produced slightly ahead of the XM19 although verification information is difficult to obtain.

The XM19 was made possible by a number of converging technologies. To detect the presence of hazardous biological particles in air, it was necessary to concentrate the material sufficiently for downstream measurements. Marple from the University of Minnesota reviewed the historical fundamentals of virtual impactor (VI) technologies that were used for such a purpose [68]. Carl Peterson (SCP Dynamics, Inc., Minneapolis, MN) manufactured the VI used for the XM19 that had a flow rate of 1000–1350 l/min at the intake end and 1 l/min at the output or delivery end. A companion sample collector called XM2 (Fig. 2.2) used for aerosol sampling, was described in the open literature [69]. A modern version called the "SCP 1021" was characterized by Kesavan and Doherty [70] and more recently by Bergman et al. [71].

In the XM19, the delivered aerosol particles were impacted on a moving plastic tape, part of a cassette transport system [72] that was calibrated to shift from

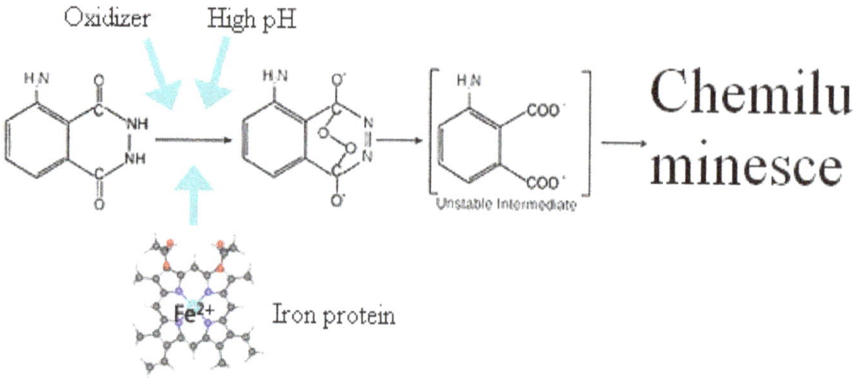

Fig. 2.1 XM19 biological detector showing virtual impactor, tape cassette, liquid chemistry, electronic sensor and analogue electronic panel

the collection site to the liquid chemistry site, timed precisely to ensure sufficient particles were trapped on the tape surface. At the liquid chemistry site, luminol reagents were applied to the particles as a flowing stream. The swirling liquid stream dislodged some or most of these particles off the tape surface. Clever plumbing was employed to extract a fraction of the biological material; delivery rate was designed to ensure an appropriate reaction time, eventually presenting the reacted material to

Fig. 2.2 XM2 was a liquid based sampler that sequentially impinged aerosol particulates into two liquid containing plastic containers. The system could take a series of two samples after being triggered by the XM19

a PMT for chemiluminescence detection. The optical signal strength could be translated to indicate if there were sufficient particles measured to warrant an alarm. On alarming, a trigger was transmitted to a companion VI-based liquid impingement collector (XM2) as shown in Fig. 2.2. Under chamber test conditions, the instrument couple performed as expected when presented with BG aerosol. However, when tested outdoors, random false alarms became an issue [73]. In hindsight, it became clear why the machine responded to the non-specific background aerosol materials. Subsequent to the inception of the instrument design, it has been known that air contains a variety of iron containing particles from marine and terrestrial sources [74]. Most particles in ambient air may have a soil origin and it has been shown that soil humic acid forms a complex with iron providing substrate for the luminol reaction [75]. After a few years of environmental testing in a variety of locations, the US Army rejected the XM19 for the task of biological detection [76]. Currently, luminol has found acceptance in forensic applications for detection of blood splatter in crime scenes where the occasional false positives can be mitigated by other technologies [77]. By present day standards, the reagents are relatively inexpensive and in forensic applications, its high sensitivity compensates for its other deficiencies.

Fig. 2.3 XM19 (*left*), the detector and XM2 (*right*) the sample collector were designed to work electrically linked together. When the detector triggered an alarm from analysis of a biological aerosol presence, a signal was sent to the XM2 to initiate the collection of liquid samples. Note the provision of handles for mobility. In practice, hand carrying each unit was a two man effort

Lessons Learnt

Some valuable lessons could be gleaned from this financially significant failure. First, measuring one parameter, biological iron, as the sole determinant for the presence of a biological threat was insufficient for minimizing false alarms, in view of the ubiquity of the interfering material in ambient air. Second, the XM19 relied on continuous operation to be effective, thus required 24/7 consumption of liquid reagents that over time, may become expensive and have limited shelf live. Support and maintenance of a network of systems can translate into enormous logistical and cost requirements. Third, the combined weight of the XM19 and XM2 exceeded 150 kg, probably more than an average man can comfortably transport (Fig. 2.3). As the two VI concentrators operated a high flow rates, power requirements became significant. Fourth, for developing more sophisticated alarming technologies, there was an urgent need to understand background air, for example, detailed knowledge of aerosol particle characteristics.

Revisiting Light Scatter Measurements and Optical Particle Counting

For studying ambient aerosol characteristics, the aerodynamic particle sizer (APS, TSI St Paul, MN) that was commercialized in 1982 [78] with real time software driven data display, turned out to be a good choice. In a review, Baron has provided some background on the development of the APS in addition to describing the instrument characteristics with respect to solid and liquid particles [79]. Volckens and Peters exploring the measurement efficiencies reported that for solid particles, counting efficiencies ranged between 85% and 99% [80]. Ananth and Wilson calculated correction factor as a function of particle density and APS derived

Fig. 2.4 The Mobile Aerosol Sampling Unit (MASU) that was deployed to the first Gulf war. It housed an APS, dichotomous sampler, large volume aerosol liquid sampler with ruggedized computer controller system. Positive pressure air supplied to the interior was filtered by a bank of military canisters that removed gases as well as particulates. Temperature control was provided by a heat pump. Four modules could fit into a Hercules C130 aircraft

aerodynamic diameter [81]. This was an important consideration because biological aerosol particles like BG spores have significantly different density values, for wet (1.2 g ml^{-1}) versus dry (1.45 g ml^{-1}) materials [82]. Failure to correct for particle density can potentially incur >40% error in sizing.

Preliminary APS measurements of ambient air suggested that the major background number contribution was derived from those particles <1 μm in size [83]. This observation was later confirmed by other workers [84]. It was also demonstrated that military threat aerosols, best represented by artificially generated biological aerosols using agricultural sprayers, were predominantly larger than 2.5 μm in diameter [85]. Coincidentally, while the APS was being used as a tool to explore point detection technologies, a light detection and ranging (LIDAR)[1] system was simultaneously used for remote detection from 1985 to 1990 [86]. Data from remote measurements strongly suggested that biological particles tend to form aggregates that were mostly larger than 2.5 μm in diameter. Combining these two bits of information, it was postulated that by using software technologies, it might be possible to distinguish hazardous biological aerosols from background material [87]. The detection concept using the APS as a trigger controlled by proprietary alarming algorithms was deployed during the first Gulf war from the beginning of January to the end of February 1991 (Fig. 2.4).

[1] Light detection and ranging.

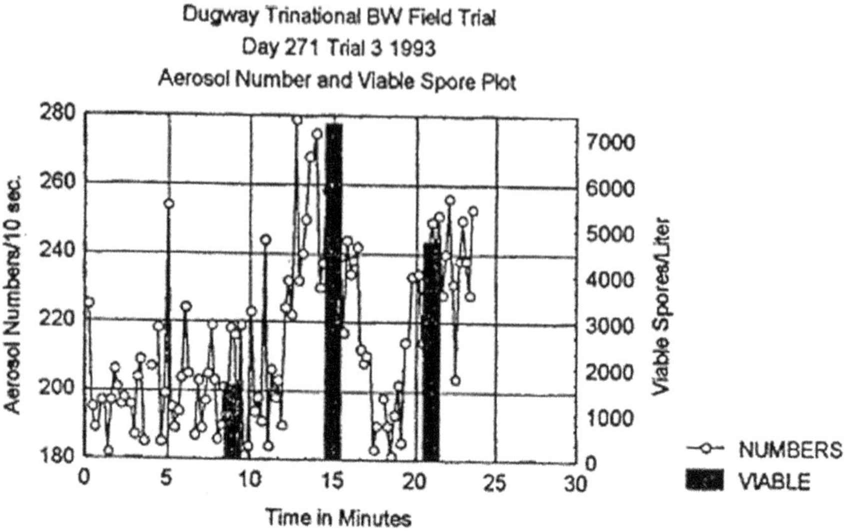

Fig. 2.5 Aerodynamic particle sizer (APS) measurement of biological aerosol concentrations. Results from an early attempt at capturing artificially generated biological aerosol containing spore particles using a slit to agar sampler and correlating that with a real time APS. The bar graph shows live particles measured in discrete time segments whereas the aerosol numbers are presented as 10 s resolution counts [83]

The APS as a sensing technology was used to participate in the first international biological field trial [83] held at Dugway Proving Grounds (DPG), Utah, in 1993. This was where artificially generated BG aerosol, sprayed with an agricultural device was used to challenge detectors as an exercise in distinguishing between background particulates and real hits (Fig. 2.5). In this figure, it can be seen that the presence of culturable BG aerosol material could be correlated with light scatter signals. Under ideal conditions, 24/7 particle sizing using OPCs requiring no expensive reagents, served adequately as surrogate biological sensors. But in reality, interfering background particulate material presented unacceptable false alarm episodes. Moreover, Landrin et al. reported that conventional microbiological growth assays and optical particle counting did not correlate [88]. This observation may have been due to interference by inanimate particles in the air. Clearly, a more robust optical technology with enhanced signal to noise characteristics was needed. To achieve that, it became apparent that fundamental science must be carried out. If optical technologies were to be effective, the process would have to provide more information from each and every particle interrogated as a continuous stream.

Flow Cytometry

Biologists have been using commercial flow cytometers (FC) for characterizing liquid suspended mammalian cells since the mid 1970s [89]. These instruments measured light scatter and dye based fluorescence to reveal cellular physiology, studies primarily driven by the cancer research community. In the late 1970s Steen and Lindmo described a, microscope-based, FC designed specifically to resolve particles in the size range typical of bacteria (0.5–2 μm) [90]. This instrument could measure light scatter of particles that were focused with a high pressure liquid stream projected on a glass cover slip aligned with an oil immersion lens that also detected fluorescence signals. By coincidence, as will be explained later, the excitation light source was a 100 W mercury arc lamp. In combination with modern electronics, computer interfacing and real time displaying analytical software, the system was able to analyze up to 1×10^4 bacteria per second [91]. The requirement to examine many cells in a short time is essential for characterizing bacterial populations that are inherently heterogeneous [92]. This is a point that differs from the fields in engineering or chemistry where discrete signals observed from a series of events are assumed to have come from a fairly homogeneous sample. Brehm-Stecher and Johnson described optical measurements with single bacterial cells and cautioned difficulties and complexities encountered in interpreting results obtained from batch cellular measurements [93].

It is appropriate to digress to explain why the mercury light source was crucial to Suffield observations made with the Steen designed FC. In the late 1950s, a Dutch group [94] was studying the physiology of photosynthetic microorganisms and they wanted to measure the reduced form of intracellular nicotinamide adenine dinucleotide (NADH; previously known as diphosphopyridine nucleotide or pyridine nucleotide) a critical electron transport agent linked to the light harvesting centers. Up until this time, this redox molecule was measured in yeast by absorption spectrophotometry [95]. Mayevsky and Rogatsky have reviewed the importance of NADH as the electron donor during respiration in all cellular life forms [96]. It was realized in those days that the ability to measure intracellular NADH changes could reveal information on life regulating biochemistry.

Duysens and Amesz built a spectrofluorometer hoping to excite the NADH molecules within the actively photosynthesizing bacterial cells [94]. A Philips 125 W mercury arc lamp was selected primarily for its strong 366 nm line. The sample was excited with the appropriately filtered wavelength of light and the fluorescence signal was sorted by a monochromator. For the first time, they reported success in resolving the signal output from 400–600 nm, producing a fairly broad NADH emission spectrum. Unlike fluorescing organic molecules, biological molecules of metabolizing live systems exhibit broad emission spectral. The authors also cautioned that 313 nm exciting light could potentially induce an interfering fluorescence signal that masked the one for NADH.

Intrinsic Fluorescence Measurement in a Single Spore

By the early 1990s, it was possible to obtain a commercial of the shelf FC driven by an IBM PC to obtain useful data soon after it was calibrated. This was fortunate as prior to that, FC instruments were big, costly and required specialized personnel

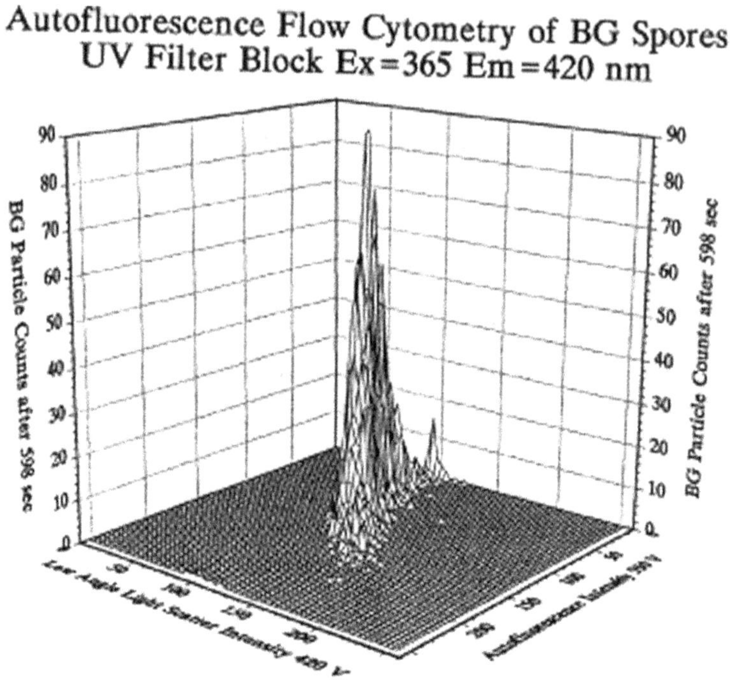

Fig. 2.6 Flow cytometric analysis of BG spores in water showing intrinsic fluorescence. Excitation was with 366 nm line from a Hg arc lamp. Emission signal was collected from 400–600 nm [97]

to maintain the various water cooled lasers and mainframe computing components. Ho and Fisher [97], using the Steen FC, demonstrated that BG spores could be induced to fluoresce when excited with 366 nm light (Fig. 2.6). Since the experimental conditions for this work were similar to those of Duysens and Amesz [94], it was speculated that the fluorescing biological molecule could be NADH. Previous workers have attempted to measure NADH in spores but the presence of significant quantities of this material has not been unequivocally demonstrated [98]. However, spore fluorescence monitoring of NADH in *B. megaterium* was attributed to synthesis of ATP during germination [99]. This suggested the presence of an endogenous amount of NADH within a spore as highly probable. The immediate implication was that intrinsically fluorescing spores were potentially live and capable of germinating. Many years later, this point was shown to be credible in a paper by Laflamme et al. [100]. These authors demonstrated that by using a sorting FC, intrinsically fluorescing spores had higher probability of being culturable. This observation was also confirmed with metabolic dye binding methods that revealed viability characteristics in spores that fluoresced. The same technique was also used to demonstrate spore germination [101] and in the present day, the validity of this method for spore germination measurements was confirmed by Setlow's group [102].

In the year 1993, based on these observations and assumptions, it was decided to construct a particle-in-air instrument to replicate the results obtain with the FC.

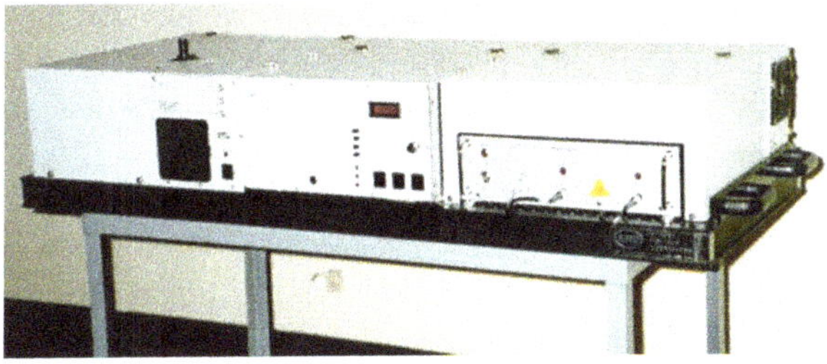

Fig. 2.7 The prototype Fluorescence Aerodynamic Particle Sizer, also called version 1 (FLAPS1). In the *center* is seen the APS electronics with its familiar face plate. On the *right* is the large power supply used to drive a HeCd laser which was located at the back of the unit, essentially occupying the length of the system. Additional optical elements were located on the *left* side to focus and direct the UV beam to the APS aerosol box. The aerosol intake can be seen protruding from the *left top panel*. A Newport optical bench provided stability to the optical and electronic elements. This was essential for long distance transport to field trial locations

If this proves to be successful, the anticipated signal to noise enhancement might suggest the presence of potentially "live" particles in an aerosol, analogous to what could be done for particles in liquid via FC. Ironically, this project vaguely recalled what Gucker did in the late 1940s when he attempted biological detection with an OPC. Benefiting from over half a century of technological progress, the prototype instrument was based on the red laser diode optical box of an existing APS with an added HeCd CW laser for exciting (352 nm) the individual particles as they exit the nozzle. For each particle that crossed the optical beams, two items of data were obtained, namely, size in micrometres and fluorescence intensity in arbitrary units. As a technological footnote, data handling was orchestrated by "firmware" or machine code which addressed a dedicated microprocessor. Sensors were monitored by digital electronics using analogue to digital convertors of high enough speed to resolve fluorescence signal pulses at high frequency. More critically, PMT sensors had the sensitivity to resolve the faint fluorescence signal from a single BG spore [103].

A simplistic approach was adopted for naming of this instrument. As the particles were sized by the APS optical frontend and fluorescence was measured after that, it was called the fluorescence aerodynamic particle sizer (FLAPS1) . As this machine was built on an optical bench, the whole system was measured to be $60 \times 150 \times 35$ cm^3 weighing 100 kg (Fig. 2.7). For transporting the FLAPS1 to trial locations, a purpose built trailer was constructed to house it along with supporting equipment.

It was anticipated that live background biological particles may not be in sufficiently high enough concentration to permit sampling at 3 s per sample. A virtual impactor modified from an XM2 was attached to the input end of the APS, effectively providing a potential to concentrate the particle content in 1300 l of air down to 1 l. In practice, this was attenuated to about 400–500 l flow rate to provide optimal signal to noise characteristics for 3 s sampling rate. This instrument was transported to Dugway, Utah from Alberta to participate in an open air biological field trial held in 1995. The initial success of this endeavor was reported at an aerosol conference [104] in which the detection of BG spores as fluorescing particles was described for the first time. The significant achievement for FLAPS1 was that not only the instrument was able to perform in the laboratory, it also worked well in the field. This distinction set it apart from many instruments that were never field tested.

A second generation FLAPS (FLAPS2) was designed to be smaller, power efficient and field portable by TSI in 1995 and 1996. In this version, the HeCd CW laser was replaced with a pulsed tripled Nd:YAG at a wavelength 355 nm. The early development of FLAPS and the initial testing are presented in two articles [10, 105]. This early work influenced the development of many subsequent fluorescence-based bioaerosol detectors which is presented in the chapter on fluorescence spectroscopy in this book.

Ambient Air Interference with Optical Measurements

Jones and Harrison estimated the biological content found in the air over global terrestrial surfaces [106]. They figured that, this amounted to a quarter of the total airborne biological particulate and these consisted of pollen, fungal spores, bacteria, viruses, or fragments of plant and animal matter. Traidl-Hoffmann et al., stated that most of these affect humans as allergens consisting of mostly proteins [107]. Teale and Weber [108], who attributed early fluorescence studies of aromatic amino acids to Ley and von Englehardt [109], measured the fluorescence spectra peaks of phenylalanine (282 nm), tyrosine (303 nm) and tryptophan (348 nm) using short wavelength UV excitation (200–320 nm). These molecular constituents of proteins were found to possess high quantum yields [108]. This optical property inspired Bronk's laboratory to use short UV wavelength excitation to characterize intrinsic fluorescence of particles like BG spores [110] as part of their biological aerosol detection strategy. Interestingly, the Russians also caught on to this approach at about the same time [111].

Earlier in this review, the importance of measuring "live" particles as a proxy for infectivity was discussed. The presence of everything else in the detection milieu was considered as potentially degrading to "signal to noise" characteristics. Illuminating protein-based aerosol particles the majority of which consist of "dead" material may provide an abundance of signals but this can constitute an unacceptable noise burden, considering that only about < 1 % of these may be culturable and of bacterial in origin [112]. For this reason, as hinted by Jones and Harrison [106], 99 % of the atmospheric particles produced fluorescent noise as a result of short UV excitation. Moreover, in quoting Cheng et al. [113], Sivaprakasam et al. [114], commenting on short UV excitation measurements, noted that "it is now reasonably well known that fluorescence spectra of different types of bacteria are very similar, and that actual species-level identification based on these spectra is unlikely". Pan et al. report of another group of interfering particles of soil origin, humic/fulvic acids and humic-like substances that made up 28–43 % of fluorescent aerosol particles [115]. For this reason many original practitioners of short UV excitation measurements have included multiple excitation wavelengths around 266 nm and 355 nm [114, 116, 117] or used

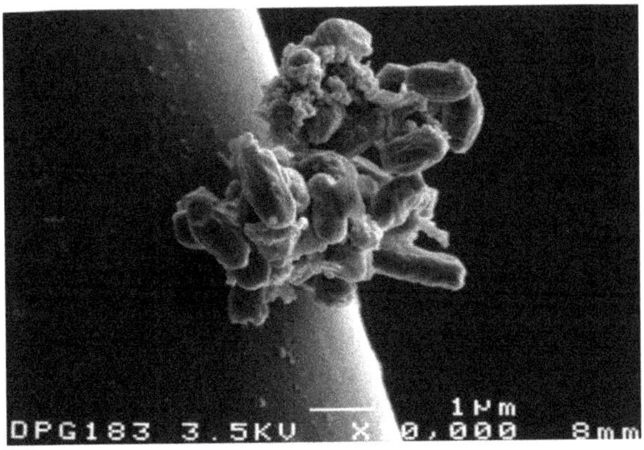

Fig. 2.8 A 6 μm spore particle aggregate trapped on a strand of spider web. Used with permission from the Life Sciences Test Facility, US Army, Dugway Proving Ground, Utah

supplemental strategies [118]. Short UV excitation of aromatic amino acid has also found limited applicability in liquid-based capillary and microchip electrophoresis measurements of separated purified protein material [119].

Conclusions

This historical excursion looked at the early attempts of using optical technologies as a means to reveal useful information from examining single biological aerosol particles. It has been shown that, early attempts at employing purely physical definitions to characterize biological agents proved inadequate. But much was gained from observing light scatter signals from biological particles especially when Lidar signal analysis suggested that the scattering elements strongly depolarized light [86]. In combination with other measurements, it was concluded that the biological aerosol particles of interest were those in the >2.5 μm size range consisting of aggregates. Subsequently, a switch to metabolic centric approaches in revealing "live" characteristics gave satisfyingly meaningful answers. The same optical principle was extended to UV Lidar measurements which confirmed that obtained from point detection [120].

As a bonus, measuring fluorescence signals related to metabolic cellular molecules in single particles also led to the discovery that naturally occurring aerosol particles possess viable or culturable properties. It is believed that, this fortuitous switch marked the turning point for all previous detection efforts which, due to lack of ground rules, were based on intuitively simple minded assumptions as to what detectors needed to measure in a biological aerosol threat. It has finally now been realized that a potential biological threat aerosol may exist only as a small fraction of a larger population of atmospheric particulate material. Furthermore, the real agents may be masked by a complex agglomerate of mixed sized material containing growth media impurities mixed with other unknown background interfering proteins. It is now believed that, the composite of a threat agent will have the appearance as shown in Fig. 2.8 depicting a typical biological particle consisting of a large number of individuals (spores in this example), loosely bound together by polysaccharides associated with unspecified contaminant material [121]. To put it plainly, all the laboratory measurements reported in 99 % of biological detection papers that have experimented with purified single particle aerosol material will have to be verified in the field challenges presented with representative threat material. In this historical review, organized field trials held periodically in Dugway, Utah have been described. For future detection, technologies to gain recognition and acceptance as a viable approach, credible performance at this or similar venues is crucial. This simplistic vignette may form the basis for future workers who will hopefully approach the problem with fresh perspectives. This review of biological detection has revealed that optical approaches provide the most promising way for providing a trigger mechanism that can serve as early or rapid responding system. This response will alert those who may be exposed to potential live agents to don protective respiratory equipment. It is ironic that up to this point, expensive protective equipment has been given to the military but there had been no trigger mechanism to let them know when to use it. This trigger mechanism is embodied in the FLAPS and other spectroscopic technology.

Acknowledgements Many colleagues have helped in making it possible to achieve the work described in this review. Peter Hairston of TSI Inc. was instrumental in the providing engineering expertise required to design and build the fluorescence optical detector. A few who have been involved with the US Army, for example, Jeff Mohr, Douglas Andersen and Robert McGhin made it possible for testing detection system in Dugway, Utah and Florida. Finally, the commercial success of the instrument is due in no small part to the engineering and marketing skills of TSI Inc., St. Paul, MN and Dycor Technologies, Edmonton AB.

References

1. Koch R (1893) Ueber den augenblicklichen Stand der bakteriologischen Choleradiagnose. Z Hyg Infekt 14 (1):319–338. doi:10.1007/BF02284324
2. May ML, Robson J (2007) Microbiological diagnostic procedures in respiratory infections: suppurative lung disease. Paediatr Respir Rev 8 (3):185–194. doi:10.1016/j.prrv.2007.08.009
3. Evans AS (1976) Causation and disease: the Henle-Koch postulates revisited. Yale J Biol Med 49 (2):175–195
4. van Schaik E, Tom M, DeVinney R, Woods DE (2008) Development of novel animal infection models for the study of acute and chronic *Burkholderia pseudomallei* pulmonary infections. Microbes Infect 10 (12–13):1291–1299. doi:10.1016/j.micinf.2008.07.028
5. Wyatt PJ (1968) Differential Light Scattering: a Physical Method for Identifying Living Bacterial Cells. Appl Opt 7 (10):1879–1896. doi:10.1364/AO.7.001879
6. Ammor MS (2007) Recent advances in the use of intrinsic fluorescence for bacterial identification and characterization. J Fluoresc 17 (5):455–459. doi:10.1007/s10895-007-0180-6
7. Courvoisier F, Bonacina L, Boutou V, Guyon L, Bonnet C, Thuillier B, Extermann J, Roth M, Rabitz H, Wolf J-P (2008) Identification of biological microparticles using ultrafast depletion spectroscopy. Faraday Discuss 137:37–49. doi:10.1039/B615221J
8. Tourkya B, Boubellouta T, Dufour E, Leriche F (2009) Fluorescence Spectroscopy as a Promising Tool for a Polyphasic Approach to Pseudomonad Taxonomy. Curr Microbiol 58 (1):39–46. doi:10.1007/s00284-008-9263-0

9. Janda JM, Abbott SL (2002) Bacterial Identification for Publication: When Is Enough Enough?. J Clin Microbiol 40 (6):1887–1891. doi:10.1128/JCM.40.6.1887-1891.2002
10. Ho J (2002) Future of biological aerosol detection. Anal Chim Acta 457 (1):125–148. doi:10.1016/S0003-2670(01)01592-6
11. van Pelt C, Verduin CM, Goessens WHF, Vos MC, Tümmler B, Segonds C, Reubsaet F, Verbrugh H, van Belkum A (1999) Identification of Burkholderia spp. in the Clinical Microbiology Laboratory: Comparison of Conventional and Molecular Methods. J Clin Microbiol 37 (7):2158–2164
12. Cullimore DR, McCann AE (1977) The identification, cultivation and control of iron bacteria in ground water. In: Skinner FA, J.M. S (eds) Aquatic Microbiology. Academic Press, London, pp 219–261
13. Danin-Poleg Y, Somer L, Cohen LA, Diamant E, Palti Y, Kashi Y (2006) Towards the definition of pathogenic microbe. Int J Food Microbiol 112 (3):236–243. doi:10.1016/j.ijfoodmicro.2006.04.010
14. Maughan H, Birky CW, Nicholson WL (2009) Transcriptome Divergence and the Loss of Plasticity in *Bacillus subtilis* after 6,000 Generations of Evolution under Relaxed Selection for Sporulation. J Bacteriol 191 (1):428–433. doi:10.1128/jb.01234-08
15. Jones SD, Teigen PM (2008) Anthrax in transit: practical experience and intellectual exchange. Isis 99 (3):455–485
16. Bell JH (1881) On Woolsorters' Disease. Br Med J 1 (1067):915–916
17. Whytlaw-Gray R, Speakman J, Campbell J (1923) Smokes: Part I. A Study of their Behaviour and a Method of Determining the Number of Particles they Contain. Proc R Soc Lond A 102 (718):600–615. doi:10.1098/rspa.1923.0018
18. Taylor H (1961) Frederick George Donnan. J Am Chem Soc 83 (14):2979–2981. doi:10.1021/ja01475a001
19. Gentry JW (1997) The legacy of John Tyndall in aerosol science. J Aerosol Sci 28 (8):1365–1372. doi:10.1016/S0021-8502(97)00008-6
20. Tolman RC, Vliet EB (1919) A Tyndallmeter for the examination of disperse systems. J Am Chem Soc 41 (3):297–300. doi:10.1021/ja01460a001
21. Tolman RC, Gerke RH, Brooks AP, Herman AG, Mulliken RS, Smyth HD (1919) Relation between intensity of Tyndall beam and size of particles. J Am Chem Soc 41 (4):575–587. doi:10.1021/ja01461a008
22. Gucker FT, O'Konski CT, Pickard HB, Pitts JN (1947) A Photoelectronic Counter for Colloidal Particles. J Am Chem Soc 69 (10):2422–2431. doi:10.1021/ja01202a053
23. Guyton A (1946) Electronic counting and size determination of particles in aerosols. J Ind Hyg Toxicol 28 (4):133–141
24. Kenefic LJ, Pearson T, Okinaka RT, Schupp JM, Wagner DM, Ravel J, Hoffmaster AR, Trim CP, Chung W-K, Beaudry JA, Foster JT, Mead JI, Keim P (2009) Pre-Columbian Origins for North American Anthrax. PLoS ONE 4 (3):e4813.1–21. doi:10.1371/journal.pone.0004813
25. Wallin A, Luksiene Z, Zagminas K, Surkiene G (2007) Public health and bioterrorism: renewed threat of anthrax and smallpox. Medicina (Kaunas) 43 (4):278–84
26. Ferry RM, Farr LE, Hartman MG (1949) The Preparation and Measurement of the Concentration of Dilute Bacterial Aerosols. Chem Rev 44 (2):389–417. doi:10.1021/cr60138a010
27. Larson EW, Young HW, Walker JS (1976) Aerosol evaluations of the De Vilbiss No. 40 and Vaponefrin nebulizers. Appl Environ Microbiol 31 (1):150–151
28. Shapiro HM (2003) Practical flow cytometry. 4th edn. John Wiley & Sons, Hoboken, New Jersey
29. Smyth M (1952) Notes on the 931-A photomultiplier. Mon Not R Astron Soc 112:88–93
30. Edels H, Gambling WA (1954) Spatial variations of the spectral response of photomultiplier cathodes. J Sci Instrum 31 (4):121. doi:10.1088/0950-7671/31/4/303
31. Gucker Jr FT, O'Konski CT (1949) An improved photoelectronic counter for colloidal particles, suitable for size-distribution studies. J Colloid Sci 4 (6):541–560. doi:10.1016/0095-8522(49)90052-5

32. O'Konski CT, Doyle GJ (1955) Light-Scattering Studies in Aerosols with New Counter-Photometer. Anal Chem 27 (5):694–701. doi:10.1021/ac60101a002
33. Gucker Jr FT, Cohn SH (1953) Numerical evaluation of the Mie scattering functions; table of the angular functions πn and τn of orders 1 to 32, at 2.5° intervals. J Colloid Sci 8 (6):555–574. doi:10.1016/0095-8522(53)90045-2
34. Gucker FT, Egan JJ (1961) Measurement of the angular variation of light scattered from single aerosol droplets. J Colloid Sci 16 (1):68–84. doi:10.1016/0095-8522(61)90064-2
35. Gucker FT, Tůma J (1968) Influence of the collecting lens aperture on the light-scattering diagrams from single aerosol particles. J Colloid Interface Sci 27 (3):402–411. doi:10.1016/0021-9797(68)90177-X
36. Ferry RM, Maple TG (1954) Studies of the Loss of Viability of Stored Bacterial Aerosols I. *Micrococcus Candidus*. J Infect Dis 95 (2):142–159. doi:10.1093/infdis/95.2.142
37. Mie G (1908) Articles on the optical characteristics of turbid tubes, especially colloidal metal solutions. Ann Phys 25 (3):377–445
38. Logan N (1965) Survey of some early studies of the scattering of plane waves by a sphere. Proc IEEE 53 (8):773–785. doi:10.1109/PROC.1965.4055
39. Twersky V (1964) Rayleigh Scattering. Appl Opt 3 (10):1150–1150. doi:10.1364/AO.3.001150
40. Wyatt PJ (1969) Identification of Bacteria by Differential Light Scattering. Nature 221 (5187):1257–1258. doi:10.1038/2211257a0
41. Van De Merwe WP, Huffman DR, Bronk BV (1989) Reproducibility and sensitivity of polarized light scattering for identifying bacterial suspensions. Appl Opt 28 (23):5052–5057. doi:10.1364/AO.28.005052
42. Wyatt PJ, Schehrer KL, Phillips SD, Jackson C, Chang Y-J, Parker RG, Phillips DT, Bottiger JR (1988) Aerosol particle analyzer. Appl Opt 27 (2):217–221. doi:10.1364/AO.27.000217
43. Dick WD, McMurry PH, Bottiger JR (1994) Size- and Composition-Dependent Response of the DAWN-A Multiangle Single-Particle Optical Detector. Aerosol Sci Technol 20 (4):345–362. doi:10.1080/02786829408959690
44. Young KD (2006) The Selective Value of Bacterial Shape. Microbiol Mol Biol Rev 70 (3):660–703. doi:10.1128/mmbr.00001-06
45. Vollmer W, Bertsche U (2008) Murein (peptidoglycan) structure, architecture and biosynthesis in *Escherichia coli*. Biochim Biophys Acta 1778 (9):1714–1734. doi:10.1016/j.bbamem.2007.06.007
46. Margolin W (2005) FtsZ and the division of prokaryotic cells and organelles. Nat Rev Mol Cell Biol 6 (11):862–871. doi:10.1038/nrm1745
47. Jones LJF, Carballido-López R, Errington J (2001) Control of Cell Shape in Bacteria: Helical, Actin-like Filaments in *Bacillus subtilis*. Cell 104 (6):913–922. doi:10.1016/S0092-8674(01)00287-2
48. Cox CS, Wathes CM (eds) (1995) Bioaerosols handbook. Lewis, London
49. Bexon R, Gibbs J, Bishop GD (1976) Automatic assessment of aerosol holograms. J Aerosol Sci 7 (5):397–407. doi:10.1016/0021-8502(76)90026-4
50. Kaye PH, Eyles NA, Ludlow IK, Clark JM (1991) An instrument for the classification of airborne particles on the basis of size, shape, and count frequency. Atmos Environ A 25 (3–4):645–654. doi:10.1016/0960-1686(91)90062-C
51. Kaye PH, Hirst E, Clark JM, Micheli F (1992) Airborne particle shape and size classification from spatial light scattering profiles. J Aerosol Sci 23 (6):597–611. doi:10.1016/0021-8502(92)90027-S
52. Hirst E, Kaye PH, Guppy JR (1994) Light scattering from nonspherical airborne particles: experimental and theoretical comparisons. Appl Opt 33 (30):7180–7186. doi:10.1364/AO.33.007180
53. Hirst E, Kaye PH (1996) Experimental and theoretical light scattering profiles from spherical and nonspherical particles. J Geophys Res 101 (D14):19231–19235. doi:10.1029/95JD02343
54. Kaye PH, Alexander-Buckley K, Hirst E, Saunders S, Clark JM (1996) A real-time monitoring system for airborne particle shape and size analysis. J Geophys Res 101 (D14):19215–19221. doi:10.1029/96JD00228

55. Kaye PH, Barton JE, Hirst E, Clark JM (2000) Simultaneous light scattering and intrinsic fluorescence measurement for the classification of airborne particles. Appl Opt 39 (21):3738–3745. doi:10.1364/AO.39.003738
56. Kaye PH, Stanley WR, Hirst E, Foot EV, Baxter KL, Barrington SJ (2005) Single particle multichannel bio-aerosol fluorescence sensor. Opt Express 13 (10):3583–3593. doi:10.1364/OPEX.13.003583
57. Kaye PH, Hirst E, Greenaway RS, Ulanowski Z, Hesse E, DeMott PJ, Saunders C, Connolly P (2008) Classifying atmospheric ice crystals by spatial light scattering. Opt Lett 33 (13):1545–1547. doi:10.1364/OL.33.001545
58. Auger J-C, Aptowicz KB, Pinnick RG, Pan Y-L, Chang RK (2007) Angularly resolved light scattering from aerosolized spores: observations and calculations. Opt Lett 32 (22):3358–3360. doi:10.1364/OL.32.003358
59. Gogoi A, Buragohain AK, Choudhury A, Ahmed GA (2009) Laboratory measurements of light scattering by tropical fresh water diatoms. J Quant Spectrosc Radiat Transfer 110 (14–16):1566–1578. doi:10.1016/j.jqsrt.2009.03.008
60. Latimer P, Brunsting A, Pyle BE, Moore C (1978) Effects of asphericity on single particle scattering. Appl Opt 17 (19):3152–3158. doi:10.1364/AO.17.003152
61. Metcalf WS, Quickenden TI (1965) Di-imide and the Chemiluminescence of Luminol. Nature 206 (4983):506–507. doi:10.1038/206506b0
62. Albrecht HO (1928) Uber die Chemiluminescenz des Aminophthalsaurehydrazids. Z Phys Chem 136:321–330
63. Drew HDK, Garwood RF (1938) Chemiluminescent organic compounds. Part VI. The isolation of peroxide derivatives of phthalaz-1: 4-diones. J Chem Soc:791–793. doi:10.1039/JR9380000791
64. Neufeld HA, Conklin CJ, Towner RD (1965) Chemiluminescence of luminol in the presence of hematin compounds. Anal Biochem 12 (2):303–309. doi:10.1016/0003-2697(65)90095-3
65. Sotnikov GG (1970) Detection of iron-porphyrin proteins with a biochemiluminescent method in search of extraterrestrial life. Life Sci Space Res 8:90–8
66. Miller CA, Vogelhut PO (1978) Chemiluminescent detection of bacteria: experimental and theoretical limits. Appl Environ Microbiol 35 (4):813–816
67. Ewetz L, Thore A (1976) Factors affecting the specificity of the luminol reaction with hematin compounds. Anal Biochem 71 (2):564–570. doi:10.1016/S0003–2697(76)80025-5
68. Marple VA (2004) History of Impactors—The First 110 Years. Aerosol Sci Technol 38 (3):247–292. doi:10.1080/02786820490424347
69. Brenner KP, Scarpino PV, Clark CS (1988) Animal viruses, coliphages, and bacteria in aerosols and wastewater at a spray irrigation site. Appl Environ Microbiol 54 (2):409–415
70. Kesavan J, Doherty RW (2001) Characterization of the SCP 1021 Aerosol Sampler. ECBC-TR-211. Edgewood Chemical Biological Center Aberdeen Proving Ground MD. Available from: http://handle.dtic.mil/100.2/ADA397460
71. Bergman W, R LJS, Sawyer S, Milanovich F, Mariella Jr R (2005) High air flow, low pressure drop, bio-aerosol collector using a multi-slit virtual impactor. J Aerosol Sci 36 (5–6):619–638. doi:10.1016/j.jaerosci.2004.12.010
72. Barrett WJ, Miller HC (1975) Investigation of Luminol and Collection Tape Components and the Effects of Airborne Interferents on the XM19 Detector. AD-AO07 274. Southern Research Institute, Birmingham, Alabama. Available from: http://oai.dtic.mil/oai/oai?verb=getRecord&metadataPrefix=html&identifier=ADA007274
73. Putscher RE, McCrone WC (1975) Characterization of air particles giving false responses with biological detectors. AD-A015 519. Walter C Mccrone Associates Inc., Chicago Il. Available from: http://www.dtic.mil/cgi-bin/GetTRDoc?Location=U2&doc=GetTRDoc.pdf&AD=ADA015519
74. Vong L, Laës A, Blain S (2007) Determination of iron–porphyrin-like complexes at nanomolar levels in seawater. Anal Chim Acta 588 (2):237–244. doi:10.1016/j.aca.2007.02.007
75. Laglera LM, van den Berg CM (2009) Evidence for geochemical control of iron by humic substances in seawater. Limnol Oceanogr 54 (2):610–619

76. Hinte J (2004) A Short History of Detectors. CB Quartely. Edgewood Chemical Biological Center, Aberdeen Proving Ground, MD
77. Creamer JI, Quickenden TI, Apanah MV, Kerr KA, Robertson P (2003) A comprehensive experimental study of industrial, domestic and environmental interferences with the forensic luminol test for blood. Lumin 18 (4):193–198. doi:10.1002/bio.723
78. Agarwal JK, Remiarz RJ, Quant FR, Sem GJ (1982) Real time aerodynamic particle size analyzer. J Aerosol Sci 13:222–223
79. Baron PA (1986) Calibration and Use of the Aerodynamic Particle Sizer (APS 3300). Aerosol Sci Technol 5 (1):55–67. doi:10.1080/02786828608959076
80. Volckens J, Peters TM (2005) Counting and particle transmission efficiency of the aerodynamic particle sizer. J Aerosol Sci 36 (12):1400–1408. doi:10.1016/j.jaerosci.2005.03.009
81. Ananth G, Wilson JC (1988) Theoretical Analysis of the Performance of the TSI Aerodynamic Particle Sizer—The Effect of Density on Response. Aerosol Sci Technol 9 (3):189–199. doi:10.1080/02786828808959207
82. Carrera M, Zandomeni RO, Sagripanti JL (2008) Wet and dry density of *Bacillus anthracis* and other *Bacillus species*. J Appl Microbiol 105 (1):68–77. doi:10.1111/j.1365–2672.2008.03758.x
83. Ho J, Spence M, Fisher GR (1995) Detection of BW Agents: Dugway Trinational BW Field Trial 20–30 October 1993, Dugway, Utah, Mobile Aerosol Sampling Unit Data Analysis. DRES-619. Defence Research Establishment Suffield, Ralston, Alberta, Canada. Available from: http://cradpdf.drdc.gc.ca/PDFS/zba7/p150526.pdf
84. Held A, Zerrath A, McKeon U, Fehrenbach T, Niessner R, Plass-Dülmer C, Kaminski U, Berresheim H, Pöschl U (2008) Aerosol size distributions measured in urban, rural and high-alpine air with an electrical low pressure impactor (ELPI). Atmos Environ 42 (36):8502–8512. doi:10.1016/j.atmosenv.2008.06.015
85. Ho J (1991) Characteristics of simulant aerosols for study of the BCD inlet nozzle. DRES-543. Defence Research Establishment Suffield, Ralston, Alberta, Canada. Available from: http://cradpdf.drdc.gc.ca/PDFS/zbc81/p70649.pdf
86. Evans BTN, Yee E, Roy G, Ho J (1994) Remote detection and mapping of bioaerosols. J Aerosol Sci 25 (8):1549–1566. doi:10.1016/0021-8502(94)90224-0
87. Yee E, Ho J (1990) Neural network recognition and classification of aerosol particle distributions measured with a two-spot laser velocimeter. Appl Opt 29 (19):2929–2938. doi:10.1364/AO.29.002929
88. Landrin A, Bissery A, Kac G (2005) Monitoring air sampling in operating theatres: can particle counting replace microbiological sampling? J Hosp Infect 61 (1):27–29
89. Shapiro HM (2004) The evolution of cytometers. Cytometry Part A 58A (1):13–20. doi:10.1002/cyto.a.10111
90. Steen H, Lindmo T (1979) Flow cytometry: a high-resolution instrument for everyone. Science 204 (4391):403–404. doi:10.1126/science.441727
91. Steen HB, Boye E, Skarstad K, Bloom B, Godal T, Mustafa S (1982) Applications of flow cytometry on bacteria: Cell cycle kinetics, drug effects, and quantitation of antibody binding. Cytometry 2 (4):249–257. doi:10.1002/cyto.990020409
92. Shapiro HM (2000) Microbial analysis at the single-cell level: tasks and techniques. J Microbiol Methods 42 (1):3–16. doi:10.1016/S0167-7012(00)00167-6
93. Brehm-Stecher BF, Johnson EA (2004) Single-Cell Microbiology: Tools, Technologies, and Applications. Microbiol Mol Biol Rev 68 (3):538–559. doi:10.1128/mmbr.68.3.538-559.2004
94. Duysens LNM, Amesz J (1957) Fluorescence spectrophotometry of reduced phosphopyridine nucleotide in intact cells in the near-ultraviolet and visible region. Biochim Biophys Acta 24:19–26. doi:10.1016/0006-3002(57)90141-5
95. Theorell H, Chance B (1951) Studies on liver alcohol dehydrogenase. II. The kinetics of the compound of horse liver alcohol dehydrogenase and reduced diphosphopyridine nucleotide. Acta Chem Scand 5:1127–1144. doi:10.3891/acta.chem.scand.05-1127
96. Mayevsky A, Rogatsky GG (2007) Mitochondrial function in vivo evaluated by NADH fluorescence: from animal models to human studies. Am J Physiol Cell Physiol 292 (2):C615–C640. doi:10.1152/ajpcell.00249.2006

97. Ho J, Fisher G (1993) Detection of BW agents: flow cytometry measurement of *Bacillus subtillis* (BG) spore fluorescence. DRES-M-1421. Defence Research Establishment Suffield, Ralston, Alberta, Canada. Available from: http://cradpdf.drdc.gc.ca/PDFS/zbc83/p135828.pdf
98. Setlow B, Setlow P (1977) Levels of oxidized and reduced pyridine nucleotides in dormant spores and during growth, sporulation, and spore germination of *Bacillus megaterium*. J Bacteriol 129 (2):857–865
99. Sano K, Otani M, Uehara R, Kimura M, Umezawa C (1988) Primary role of NADH formed by glucose dehydrogenase in ATP provision at the early stage of spore germination in *Bacillus megaterium* QM B1551. Microbiol Immunol 32 (9):877–885
100. Laflamme C, Verreault D, Lavigne S, Trudel L, Ho J, Duchaine C (2005) Autofluorescence as a viability marker for detection of bacterial spores. Front Biosci 10 (2):1647–1653
101. Laflamme C, Ho J, Veillette M, Latrémoille M-C, Verreault D, Mériaux A, Duchaine C (2005) Flow cytometry analysis of germinating Bacillus spores, using membrane potential dye. Arch Microbiol 183 (2):107–112. doi:10.1007/s00203-004-0750-9
102. Magge A, Setlow B, Cowan AE, Setlow P (2009) Analysis of dye binding by and membrane potential in spores of *Bacillus species*. J Appl Microbiol 106 (3):814–824. doi:10.1111/j.1365-2672.2008.04048.x
103. Hairston PP, Ho J, Quant FR (1997) Design of an instrument for real-time detection of bioaerosols using simultaneous measurement of particle aerodynamic size and intrinsic fluorescence. J Aerosol Sci 28 (3):471–482. doi:10.1016/S0021-8502(96)00448-X
104. Ho J (1996) Real time detection of biological aerosols with fluorescence aerodynamic particle sizer (FLAPS). J Aerosol Sci 27 (1001):581–582. doi:10.1016/0021-8502(96)00363-1
105. Ho J, Spence M, Hairston P (1999) Measurement of biological aerosol with a fluorescent aerodynamic particle sizer (FLAPS): correlation of optical data with biological data. Aerobiol 15 (4):281–291. doi:10.1023/A:1007647522397
106. Jones AM, Harrison RM (2004) The effects of meteorological factors on atmospheric bioaerosol concentrations—a review. Sci Total Environ 326 (1–3):151–180. doi:10.1016/j.scitotenv.2003.11.021
107. Traidl-Hoffmann C, Jakob T, Behrendt H (2009) Determinants of allergenicity. J Allergy and Clin Immunol 123 (3):558–566. doi:10.1016/j.jaci.2008.12.003
108. Teale F, Weber G (1957) Ultraviolet fluorescence of the aromatic amino acids. Biochem J 65 (3):476–482
109. Ley H, von Engelhardt K (1910) Ultraviolet flourescence and chemical constitution in cyclic compounds. Z Phys Chem 74 (1):1–64
110. Faris GW, Copeland RA, Mortelmans K, Bronk BV (1997) Spectrally resolved absolute fluorescence cross sections for bacillus spores. Appl Opt 36 (4):958–967. doi:10.1364/AO.36.000958
111. Kalinin I, Vorob'ev SA, Khramov EN, Vorob'eva EA, Kuznetsov AP, Kiselev OS (2000) (Use of laser flow-type fluorescence aerosol particle counter to evaluate the concentration of microbes in the surface air under high dust content). Vestn Ross Akad Med Nauk (10):16–19
112. Amann RI, Ludwig W, Schleifer KH (1995) Phylogenetic identification and in situ detection of individual microbial cells without cultivation. Microbiological Reviews 59 (1):143–169
113. Cheng YS, Barr EB, Fan BJ, Hargis J, P. J., Rader DJ, O'Hern TJ, Torczynski JR, Tisone GC, Preppernau BL, Young SA, Radloff RJ (1999) Detection of Bioaerosols Using Multiwavelength UV Fluorescence Spectroscopy. Aerosol Sci Technol 30 (2):186–201. doi:10.1080/027868299304778
114. Sivaprakasam V, Huston AL, Scotto C, Eversole JD (2004) Multiple UV wavelength excitation and fluorescence of bioaerosols. Opt Express 12 (19):4457–4466. doi:10.1364/OPEX.12.004457
115. Pan Y-L, Pinnick RG, Hill SC, Rosen JM, Chang RK (2007) Single-particle laser-induced-fluorescence spectra of biological and other organic-carbon aerosols in the atmosphere:

Measurements at New Haven, Connecticut, and Las Cruces, New Mexico. J Geophys Res 112 (D24):D24S19.1–15. doi:10.1029/2007jd008741
116. Huang HC, Pan Y-L, Hill SC, Pinnick RG, Chang RK (2008) Real-time measurement of dual-wavelength laser-induced fluorescence spectra of individual aerosol particles. Opt Express 16 (21):16523–16528. doi:10.1364/OE.16.016523
117. Pan Y-L, Hill SC, Pinnick RG, Huang H, Bottiger JR, Chang RK (2010) Fluorescence spectra of atmospheric aerosol particles measured using one or two excitation wavelengths: Comparison of classification schemes employing different emission and scattering results. Opt Express 18 (12):12436–12457. doi:10.1364/OE.18.012436
118. Jeys TH, Herzog WD, Hybl JD, Czerwinski RN, Sanchez A (2007) Advanced Trigger Development. Linc Lab J 17 (1):29–62
119. Schulze P, Belder D (2009) Label-free fluorescence detection in capillary and microchip electrophoresis. Anal Bioanal Chem 393 (2):515–525. doi:10.1007/s00216-008-2452-7
120. Buteau S, Simard J-R, Lahaie P, Roy G, Mathieu P, Déry B, Ho J, McFee J (2008) Bioaerosol Standoff Monitoring Using Intensified Range-Gated Laser-Induced Fluorescence Spectroscopy. In: Kim YJ, Platt U (eds) Advanced Environmental Monitoring. Springer Netherlands, Dordrecht, pp 203–216. doi:10.1007/978-1-4020-6364-0_16
121. Han YW (1990) Microbial levan. Adv Appl Microbiol 35:171–194

Chapter 3
Physical and Biological Properties of Bioaerosols

Jakob Löndahl

Introduction

Bioaerosols make up a diverse group of airborne particles with biological origins. They include bacterial cells and spores, viruses, pollen, fungi, algae, detritus, allergens and cell fragments. Sometimes the term is also used to include secondary particles in the atmosphere that are formed from the condensation of gaseous molecules released by biological organisms. In this chapter, though, bioaerosols refer to primary (directly emitted) aerosol particles.

Bioaerosol particles are usually a small fraction of all aerosol particles in our surroundings, but their impact can be critical—a single viable microbial pathogen could, for instance, be sufficient to spread an infection. Bioaerosols are a means for transmission of disease between humans, crops and livestock. They can cause allergic reactions and affect indoor air quality. They have to be controlled and minimized in hospital operating theatres, the food industry, pharmaceutical manufacturing and cleanrooms. They are important in ecology and biodiversity. They can damage our cultural heritage and in the worst case, be used for biological terrorism or warfare.

The physical characteristics of bioaerosols are highly varied. The smallest particles, such as cell fragments, may consist of a few molecules and be 0.001 μm in diameter while pollen and algae can have sizes of 100 μm or more (Figs. 3.1 and 3.2). This range, which spans over five orders of magnitude, is comparable to the difference between a pinhead and the St. Peter's Basilica in Rome.

Aerosol particles, from 0.001–100 μm, have different mechanisms for transport in the air, deposition on surfaces and light scattering—properties that are important both for their effects and for their detection. The size also has implications for the amount of material carried by the particle, since the volume is proportional to the cubed diameter ($\propto d_p^3$) and thus spans over 15 orders of magnitude.

Characteristics other than physical size are significant for aerosol movement. These include density, shape and electrical charge. The density of biological

---

J. Löndahl (✉)
Division of Ergonomics and Aerosol Technology, Department of Design Sciences, Lund University, Box 118, 221 00 Lund, Sweden
e-mail: jakob.londahl@design.lth.se

P. Jonsson et al. (eds.), *Bioaerosol Detection Technologies*, Integrated Analytical Systems, DOI 10.1007/978-1-4419-5582-1_3,

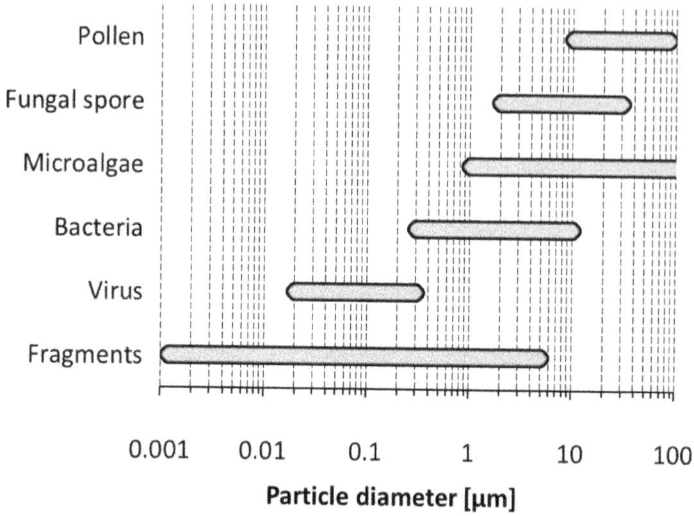

Fig. 3.1 Approximate size range of various individual bioaerosol particles. Fragments include, for instance, parts of cells, plants and biofilms or the crust on sand and rocks

material typically is around 1.0–1.5 g cm^{-3}, but bioaerosol particles are often mixed with other matter with higher density, such as mineral dust or sea salt. The shape of bioaerosols may vary from spherical to elongated structures, from a single cell to chains or complex aggregates. The electrical charge is presumably close to the "neutral" Boltzmann charge distribution. Some studies have suggested that airborne bacteria can carry a high charge, but this is likely due to the methods used for spraying them [1].

A key to understanding the impact of microorganisms in the air is their viability, that is, their capacity to survive and grow. Many microbes have developed systems that protect them against the harsh conditions in the atmosphere where exposure to UV radiation may be high, temperature and oxygen pressure low, and the availability of liquid water and nutrition limited [2]. Formation of resting stages, pigmentation, biofilms and psychrotolerance are among the mechanisms that are involved to preserve viability.

By mass, the major part of microbial cells generally consists of water (~70%), although at atmospheric conditions they may desiccate. The remainder mainly is comprised of proteins, lipids, carbohydrates and other macromolecules.

Ambient concentrations of bioaerosol particles remain uncertain due to a limited amount of data and the lack of measurement standard. Until recently, culture-based techniques have been most common and it is well known that these methods capture only a fraction of the total diversity of the biological particles present in the air. Culture-independent methods are rapidly developing, but data from these on environmental bioaerosols are still scarce.

This chapter provides an overview of the main types of bioaerosol particles, their sources, transport and sinks, and their potential effects on health and atmosphere.

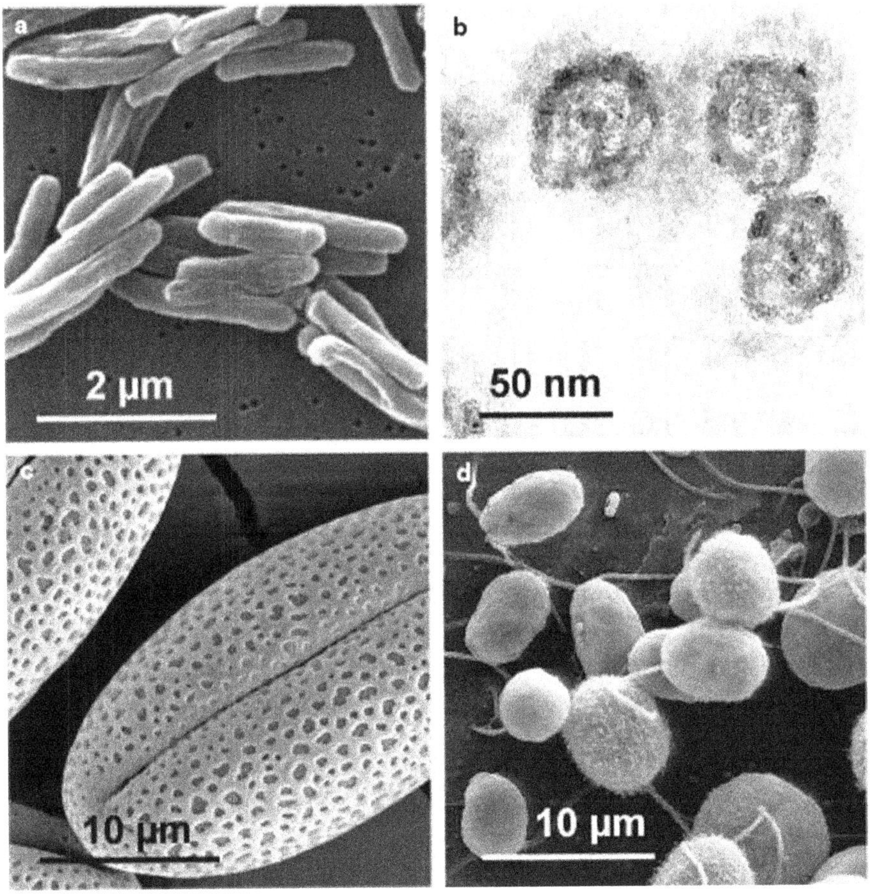

Fig. 3.2 Electron microscope images of (**a**) *Mycobacterium tuberculosis* (image credit: Janice Carr), (**b**) Flavivirus, (**c**) Redbud pollen and (**d**) *Chlamydomonas* sp., green microalgae

Types of Bioaerosols

Bacteria and Archea

Bacteria are unicellular, prokaryotic, microorganisms found in almost all environments on Earth. Humans carry ten times more bacteria by number on their skin and in their gut flora than there are cells in the body. Airborne bacteria may be transported as single cells, but typically they are attached to other particles such as leaf, soil or skin fragments [3]. Spherical bacteria (*cocci*, e.g. *Micrococcus* or *Streptococcus*) often occur in pairs, chains or clusters. Rod-shaped bacteria (*bacilli,* e.g. *Bacillus anthrasis*) may be single or form chains.

Based on the structure of the cell wall, most bacteria can be grouped as either Gram-positive or Gram-negative. The cell walls of Gram-positive bacteria generally have a thick layer of peptidoglycan, while Gram-negative have several thin layers of peptidoglycan surrounded by an outer membrane containing lipopolysaccharides (endotoxin). Some Gram-positive bacteria, such as *Bacillus* and *Clostridium*, can develop endospores. These are dormant forms with high resistance to radiation, heat, freezing and other types of environmental stress. Endospores can be both smaller and larger than the vegetative cell.

Outdoor particles with bacteria are normally found to have an aerodynamic diameter of approximately 4 μm except at coastal sites and over oceans where the diameter is about half of this or less [3]. Continental bacteria are normally larger than marine bacteria, but the difference may be partly due to the formation process, where airborne bacteria produced from the water sometimes carry less residual material [4]. The reported sizes of indoor bacteria are about 1–3 μm [5].

Archaea were previously believed to be specific type of bacteria, but are now recognized as a separate domain of life. Originally, archaea were found mostly in extreme environments, but over time they have been detected in many common habitats. The knowledge about archaea in the air, however, is very limited.

Viruses

Viruses are parasitic infectious agents that only can reproduce in living cells of animals, plants or bacteria. Those specialized on infecting bacteria are called bacteriophages. Since viruses need a host cell to multiply and carry out metabolic processes, they are usually not considered organisms. Viruses are the most common biological entity in ecosystems. A viral particle that is released from a cell and capable of infecting other cells is called a virion.

All viruses carry nucleic acid, either as DNA or RNA (but not both). A protein shell called a capsid surrounds the genetic material, which is the source of infection. The capsid allows recognition of receptors on the host cell and helps the virion in some cases to enter the host cell. In addition, some viruses have an envelope of lipoproteins originating from the cell membrane.

Viruses are normally found in the size range 0.02–0.3 μm, although a few exceptions exist such as mimivirus around 0.4 μm and the exotic *Megavirus chilensis*, which is even larger [6, 7]. Viruses are typically shaped either as rods (helical) or spheres (icosahedral).

The literature on airborne viruses is comparatively scarce. This is most likely due to inefficient measurement methods—especially before the development of molecular techniques such as PCR. It is assumed that viruses, at least those that are viable, normally are transported attached to larger particles. Since they have no repair systems they are usually more easily inactivated by heat, radiation and other types of environmental stress than live microorganisms. The relationship between relative humidity and survival of viruses in the air is complex and varying depending on virus type [8].

Pollen

Pollen is produced by plants to transfer the male gametes (sperm cells) to the pistils or female structures where fertilization takes place. The size range of the pollen grains is approximately 10–100 μm, although many that are found in the atmosphere are between 20 and 60 μm. Most pollen that contains allergens is anemophilous ("wind-loving"). Anemophilous plants (e.g. grass, conifers, birches and ragweed) use wind as their primary dispersal vector. However, not all pollens are transferred through the air.

Pollen grains vary not only in size, but also in surface structure, density and shape. Normally they have a hard shell that protects the seed and prevents deterioration of the genetic material present in the nuclei from environmental stress during transport through the atmosphere. The shell may rupture into smaller fragments between 0.1 and 5 μm at high humidity [9]. The shape and structure of the pollen grains are characteristic for different plant species and their origin can often be identified under microscope.

The dispersal, resuspension and transport of pollen into the atmosphere are largely determined by meteorological factors. For instance, emissions decrease when wind speeds are low and during rainfall. There also appears to be an association between transport distances and temperature, with an increase in distances travelled as the temperature increases [10].

The concentration of pollen in the air follows a seasonal cycle that primarily is a function of climate and has a limited shift due to meteorological factors. Climate change may adjust the onset of these pollination seasons and alter their duration.

Fungi

Fungi are among the most common organisms on Earth and include molds, rusts, smuts, yeasts, mildews and mushrooms. In many ecosystems they play a key role as decomposers. They may be independent or live in symbiotic or parasitic associations with plants and animals. Fungi are a source of severe problems when they cause crop diseases, infectious disease or allergies, but can also be beneficial when used in biological pesticides or for the production of antibiotics.

Some fungi are unicellular, such as yeasts, but usually they consist of multicellular tubular filaments called hyphae. The hyphae branch into a network called mycelium, which may develop reproductive spores. These spores can be released by passive processes, using wind or other external forces, or by active processes where an ejection is produced by, for instance, osmotic pressure or surface tension.

It is commonly assumed that spores constitute the bulk of the bioaerosols produced from fungi. These are found in the size range 1–50 μm, but more typically between 2 and 10 μm [11, 12]. Frequent genera in the atmosphere are *Cladosporium*, *Alternaria* and the actively released ascospores and basidiospores. It has been shown that emission of the spores may be accompanied by significant amounts of fungal fragments [13].

Some fungi produce mycotoxins, which are secondary metabolites associated with both acute and chronic health effects. Fungal spores are coated with ergosterol and sometimes sugar alcohols (e.g. arabitol and mannitol). These compounds can be used as chemical tracers for the total spore concentration in the air.

Microalgae and Cyanobacteria

Microalgae are eukaryotic and predominantly photosynthetic organisms that are present in both aquatic and many non-aquatic habitats. In aquatic environments, microalgae play an important role in the microbial food web.

Cyanobacteria, also called blue-green algae, are bacteria which, similar to most algae, can produce chemical energy through photosynthesis. Algae and cyanobacteria together produce the major part of the atmospheric oxygen.

Algae can be both unicellular and multicellular and range in size from about 1 μm to several meters. Few studies have investigated airborne microalgae and cyanobacteria, but they have nevertheless been found in both indoor and outdoor air around the globe [14, 15].

Allergens, Lichens, Fragments and Other Bioaerosols

A variety of biologically derived particles in the air can be added to the list. Crusts on rocks and sand often carry biological material that may be suspended into the air as a result of erosion from wind and rainfall [16]. Lichens, which are a symbiotic union between fungi and microalgae or cyanobacteria, are widespread on Earth and have also been detected in the atmosphere [17]. Plant fragments are, by mass, among the most common bioaerosol materials in the atmosphere [18]. Other particle types that can be important, especially from a health point of view as they may cause allergic reactions, are skin fragments, fur fibers, insect secretions and dandruffs.

Sources, Transport and Sinks

Dispersal is fundamental to life. After survival and reproduction it can be regarded as the third most important need for an organism. Since the air provides an efficient pathway for fast transport over long distances, it is thus not surprising that many microbes have developed mechanisms to spread into the atmosphere, stay viable for long periods in its inhospitable environment and deposit on a new location to begin life. It has even been suggested that through millions of years of evolution microbes have adapted at the group level to become airborne by, for example, changing the local climate and increasing wind speeds [19, 20]. Transport by the wind may very well be the most common route for microbial dispersal [21]. This section reviews emission, aging and deposition of bioaerosols (Fig. 3.3).

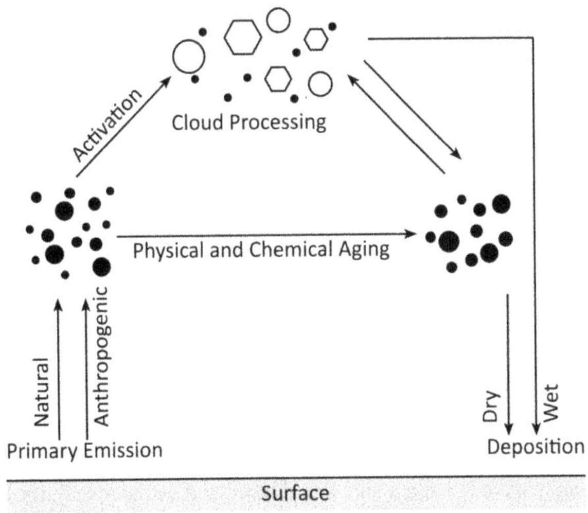

Fig. 3.3 Cycling of primary aerosols in the atmosphere (modified from [22])

Sources—Primary Emission

Bioaerosols are produced by a range of processes including bubble bursting, erosion, active dispersal and mechanical interaction between surfaces. The particles may be airborne as individual cells, but often they are attached to other material such as dust, plant fragments or as a group of cells trapped in biofilm.

Globally, open water is a major source of ambient bioaerosols, since water covers about 70 % of the Earth. The surface of the water has an almost omnipresent microlayer, up to some hundreds of μm thick, which is in contact with the atmosphere. This surface microlayer is a unique habitat and compared to subsurface waters, it is enriched by orders of magnitude with bacteria, viruses and other microorganisms as well as lipids, carbohydrates, polysaccharides, amino acids and proteins. The aerosol particles dispersed from the surface are even further concentrated with biological material [23, 24].

There are three main types of droplets formed from bodies of water: spume drops, film drops and jet drops (Fig. 3.4) [25]. Spume drops, which generally have a minimum diameter of about 40 μm, are torn off from wave crests when wind speed exceeds 7–11 m s^{-1}. Film and jet drops originate from the bursting of bubbles. Film drops, usually with diameters between 1 and 10 μm, are released horizontally when the thin film on top of the bubble ruptures, while jet drops, with diameters of 6–100 μm or about one-tenth of the initial bubble, are ejected in an upward direction when the bubble collapses. Bubbles larger than 8 μm in diameter are required for jet drops [26]. Major sources of bubbles are breaking waves, rain and boat traffic.

Over dry land, plants release bioaerosol particles as a result of decay processes or for reproduction—the latter including pollen and spores. Particles may be emitted from surfaces of vegetation, soil and rock because of erosion, wind or other types of mechanical interaction. Droppings from birds and rodents can transmit viral and microorganism material. In populated areas, the concentration of bioaerosols is often

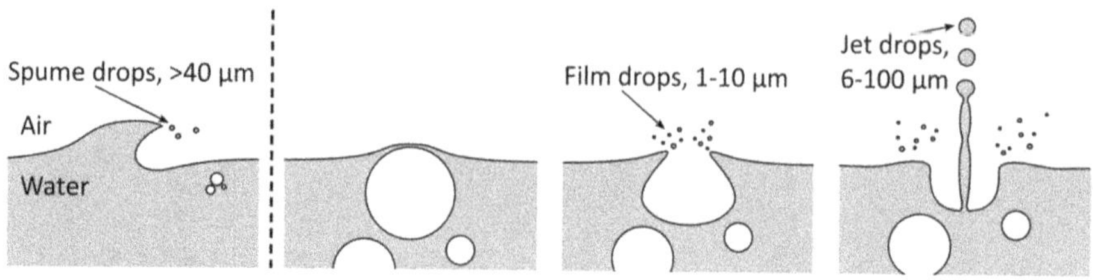

Fig. 3.4 Aerosol production from water and typical droplet diameters

enhanced due to agriculture, horticulture, waste disposal, sewage treatment works, traffic and other human activities.

The outdoor emissions of bioaerosols vary with time and location [18]. Over water, the production of bioaerosols is influenced by ice cover and wind speed, which can vary over the year. Over land, the emissions largely depend on the vegetation cycles and meteorological conditions, which changes daily and weekly as well as seasonally. Temporal variation is most pronounced in rural and remote regions. In densely populated areas, the variation over time is limited as human activity is relatively constant. Geographically, bioaerosol emissions in uninhabited regions may be homogenous over vast areas, while the emissions in cities often vary strongly with spatial location since they often are dominated by a few point sources.

Indoor bioaerosols may originate from the outdoor air, but often other sources are more significant. Especially if water or moisture is available, microbial growth can be excessive in a building structure, or its ventilation and air conditioning systems. Particles directly emitted from humans and animals also constitute a large fraction of the indoor bioaerosols.

In some indoor environments, bioaerosol particles may be detrimental, such as in operating theatres, industrial cleanrooms, isolation rooms and the food industry. In these locations efficient ventilation, clean procedures and airlocks usually minimize most particle sources. However, some particles generated inside the facilities such as skin fragments or emissions from specific processes may be difficult to remove completely and can cause severe problems.

Biological warfare and terrorism have been behind much of the development of new detection techniques for bioaerosols in the past decades. Agents intended for airborne transmission include bacteria such as *Bacillus anthracis*, *Francisella tularensis* and *Yersinia pestis*, viruses such as *Variola major* (smallpox) and Ebola, toxins such as botulinum toxin (produced by the bacteria *Clostridium botulinum*) and many mycotoxins. The consequences can be disastrous if a virus that is not normally spread through the air is modified to allow airborne transmission—a scenario which has become less unlikely with the advancement of biotechnology [27].

Transformation and Aging during Transport

Aerosol particles often stay for weeks in the atmosphere before deposition on the ground. Some particles reach the stratosphere, which further prolongs their atmospheric lifetime. During this time, they can be transported thousands of kilometers, but also transformed and inactivated since particles in the air are exposed to both physical and chemical aging processes.

The size and structure of aerosols are modified mainly by coagulation, restructuring and evaporation or condensation. During coagulation, particles collide and form larger units. The coagulation rate increases proportionally to the square of the number concentration ($\sim N^2$) and is thus most rapid in polluted environments. Particle restructuring may especially occur as a result of water uptake at high humidity because of the increasing surface tension or due to droplet formation. The exchange of semivolatile compounds by evaporation and condensation is continuous and highly related to the chemical reactions in the air. Common chemical reactions in the air include oxidation, nitration, hydrolysis and photolysis. The reactions are often driven by UV radiation as well as oxidants such as OH, ozone, NO_3 or acids such as H_2SO_4 and water.

Despite the environmental stress in the air, it is possible for many microbes to remain metabolically active, grow and maintain reproductive capacity [28, 29]. As mentioned in the introduction, these microbes often have protection mechanisms that help them preserve viability. Many microorganisms survive by formation of endospores, but also non-spore-forming bacteria have the means to maintain viability such as repair mechanisms and pigmentation. Some biological agents without repair mechanism, such as viruses and fungal spores, may be shielded by attachment to dust, pollen, biofilms or other material. For instance, the foot-and-mouth disease virus was found to cross the English Channel and the concentration of the influenza virus increased when dust from China reached Taiwan [30, 31]. It has even been claimed that bacteria can remain viable more than 50 km above the Earth's surface [32]. The atmosphere may very well be a habitat of its own for some microorganisms and not only a means for transportation [2].

Deposition

Particles are removed from the air by either wet or dry deposition on a surface. Wet deposition, when particles are washed out from the atmosphere by precipitation, is the main sink for atmospheric particles (between ~0.1 and 10 μm). The particles may be scavenged in a cloud by formation of or collision with cloud droplets, or below the cloud if they come in contact with rain or snowfall.

In dry deposition, particles collide and stick to the ground as a result of sedimentation, diffusion, impaction or other processes such as interception, turbulent eddies and thermophoresis. Particles larger than about 0.5 μm, as most bioaerosols, transport relative to the air primarily due to gravitational settling. The settling velocity is proportional to the squared diameter ($\propto d_p^2$) and thus increases dramatically with size. Large particles that are not removed by sedimentation may land on a surface by impaction when the air stream changes direction, or be intercepted on it if some part of the particle comes in contact. Particles smaller than 0.5 μm move mainly by diffusion due to Brownian motion. The diffusion speed is approximately proportional to the inverse of the squared diameter ($\propto 1/d_p^2$). In narrow air spaces, such

Fig. 3.5 Respiratory tract deposition fractions of inhaled hydrophobic particles in a healthy adult. Values are calculated for nasal breathing during relaxed conditions (12 breaths per minute with a tidal volume of 0.63 L) by the MPPD model

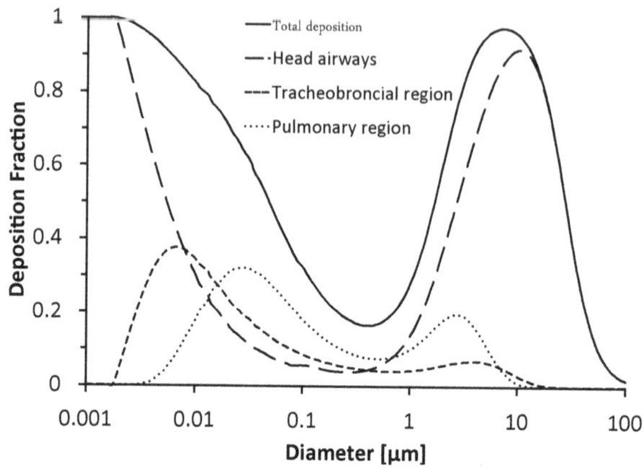

as the respiratory tract, deposition due to diffusion can efficiently remove the main fraction of particles below 0.1 µm within a few seconds.

In the range 0.1–1 µm, neither diffusion nor sedimentation or impaction act strongly on the aerosol particles. Thus, particles in this size span accumulate and stay longest in the air. It can also be noted that certain bioaerosols, such as some pollen and fungal spores, are aerodynamically shaped to be buoyant in the air, which facilitates long-distance transport.

From a health point of view, deposition of bioaerosol particles in the respiratory tract is of special concern. Figure 3.5 illustrates the regional deposition of inhaled particles in a normal nose breathing adult during relaxed conditions as calculated with the Multiple-Path Particle Dosimetry Model (MPPD, version 2.11, Chemical Industry Institute of Toxicology, Research Triangle Park, NC). As expected, the deposition probability has a minimum for particles between 0.1 and 1 µm. Bioaerosol particles larger than 10 µm—such as many pollen, fungal spores and algae—most likely deposit in the nose and mouth, although some may reach further down into the respiratory tract. Bacteria, viruses, cell fragments and other small bioaerosol particles can easily reach the alveoli where gas exchange takes place: around 30–60 % of the particles in the range 0.01–2 µm that deposit in the respiratory tract end up in the pulmonary region. The pulmonary region is the most sensitive part of the lungs.

Health Effects

Historical Background

Observations of adverse health outcomes from bioaerosols have accompanied humanity throughout history. An early notion of harmful contagions, possibly molds, and measures for decontamination is found in Leviticus 14 in the Bible. More devastating was the bubonic plague, caused by the bacterium *Yersinia pestis*, which likely was the reason for the Black Death that killed at least one third of the population of Europe in the fourteenth century [33]. Bubonic plague is a zoonotic disease, transmitted from animals to humans, but can also spread between humans via coughing, for instance. Later, in the mid-nineteenth century, the Irish potato famine gave rise to one million deaths and large-scale emigration. The starvation was caused by a potato blight from the fungus-like microorganism *Phytophthora infestans*, which can be rapidly transmitted through the air.

Over the last century, several major outbreaks of influenza have been noted [34]. The pandemic of 1918–1920, known as the Spanish flu, with a total of 40–50 million deaths was by far the worst. Notorious are also the 1957 Asian flu and the 1968 Hong Kong flu, each with a total mortality of about one million, and the less destructive but more recent Swine flu in 2009. Judging from history, we will have more influenza pandemics in the future—perhaps next time if the highly pathogenic H5N1 virus, known as bird flu, develops an ability to spread in the air from person to person.

Health Effects from Exposure

There are many health effects associated with bioaerosols related to toxic and allergenic materials, and to infectious diseases from pathogenic biological agents. Respiratory infections, which often can be linked to inhaled pathogens, are the most severe diseases worldwide in terms of mortality. They are the source of about 7 % of the total deaths, which in 2002 corresponded to almost 4 million people [35]. Tuberculosis, a lung disease usually caused by airborne *Mycobacterium tuberculosis* (Fig. 3.2a), is the reason for another 3 % of the yearly loss of lives in the world, primarily in low- and middle-income countries [36]. Infectious diseases linked to bioaerosols are essentially caused by viruses (e.g. influenza, measles), bacteria (e.g. tuberculosis, Legionnaire's disease) or fungal spores (e.g. aspergillosis, blastomycosis).

A biological agent must be viable in order to cause an infection, but non-viable bioaerosols can also have implications for health if they carry toxic or allergenic material. Allergens comprise a range of molecular structures from chemicals with low molecular weight to proteins with high molecular weight [37]. Common allergens include: proteases (enzymes important in protein catabolism), some proteins present in dander, saliva and urine from cats and dogs, glycopeptides from fungi, and industrially produced enzymes from fungi and bacteria.

Asthma, which is a chronic inflammation of the airways, is sometimes associated with exposure to allergens from dust mites, dander, cockroaches and mold. Onset can also be due to exposure to non-allergenic biological material such as endotoxins.

Many toxins with well-known health effects are produced from bacteria and fungi. Bacterial endotoxins can be strongly pro-inflammatory and inhalation studies have linked them to symptoms such as fever, neutrophilic airway inflammation, cough and lung function impairment [38]. Knowledge about airborne exposure to fungal toxins—mycotoxins—is still limited. Notable is the carcinogenic aflatoxin from *Aspergillus flavus*, but the primary route for exposure is ingestion. Other compounds that can be used as biomarkers for toxic effects are ergosterol, which is a component in fungal membranes and yeasts, and β-(1→3)-D-glucans, which are cell wall components of most fungi and some bacteria.

Indoor and Workplace Bioaerosols

A building and its ventilation system can facilitate survival of microorganisms since it provides a closed environment where many of the natural inactivation mechanisms are inhibited, such as UV-light, ozone and large temperature variations. This is amplified by efforts to save energy by decreased ventilation and increased insulation. Many indoor environments also accommodate potent sources of biological agents from industrial processes, waste handling, animals or people.

Nonspecific health symptoms related to a building where people reside are often described as sick building syndrome (SBS). The cause of the symptoms is not known, but often linked with poor air quality and nearby biological or chemical contaminants. SBS is probably common, but in absence of clear criteria for diagnosis, the prevalence remains uncertain.

Workers may sometimes be exposed to high concentrations of bioaerosols. Occupations at high risk for infectious diseases are health care workers, farmers and veterinarians. Of increasing concern is, for instance, multidrug-resistant *Staphylococcus aureus* (MRSA), which can affect hospital workers and patients. Around 5–10% of all patients admitted to hospitals in Europe and the United States are affected by health care associated or nosocomial infections, whereof many are spread through the air [39].

Hygiene Hypothesis

It has long been noted that a certain exposure early in life to microbial agents may decrease the risk of developing allergies [40, 41]. This is referred to as the hygiene hypothesis and the basic idea is that contact with a diverse range of biological agents during childhood is essential for the development of the immune system. Although the hypothesis is well supported by epidemiological findings, there are contradictory data that remain to be explained.

Atmospheric and Ecosystem Impact

Biological material is one of the most important groups of aerosols in the atmosphere, although both its concentration and impact on the environment remains largely unclear. It has been suggested that up to 1000 Tg of bioaerosols are emitted

Table 3.1 Estimated global emissions of primary bioaerosol particles and magnitude of concentrations over vegetated regions [18, 44]. Primary emitted aerosols from sea salt, mineral dust and fossil fuels are inserted for comparison

	Global emissions [Tg year^{-1}] median (min–max)	Number concentration [m^{-3}]	Mass concentration [μg m^{-3}]
Total bioaerosol	132 (< 10–1000)		
Bacteria	1.66 (0.7–28)	~ 10^4	~ 0.1
Viral particles		~ 10^4	~ 10^{-3}
Pollen	66 (47–84)	~ 10 (up to 10^3)	~ 1
Fungal spores	31 (8–186)	~ 10^3–10^4	~ 0.1–1
Algae		~ 100 (up to 10^3)	~ 10^{-3}
Sea salt	3340		
Mineral dust	2150		
Fossil fuel	35		

each year (compared to 3300 Tg for sea salt and 2000 Tg for mineral dust) [42, 18] (Table 3.1). Proteins, amino acids and related amino compounds alone may constitute up to 10 % of the fine particulate matter in the atmosphere [43].

The most critical effects of atmospheric aerosols are their influence on the global mean temperature. Airborne particles are essential to the Earth's energy balance—the absorption and reflection of incoming radiation from the sun and outgoing radiation from the Earth's system. Overall, aerosol particles cool the planet, mainly by backscattering of light from the sun. The backscattering may be either direct (i.e., reflection of the light directly by the particles) or indirect, in which case the aerosol particles alter light scattering through interactions with clouds. At atmospheric conditions, cloud droplets are formed by condensation of water vapor on already existing particles, referred to as cloud condensation nuclei (CCN) or ice nuclei (IN). Aerosol particles are, in other words, necessary for cloud formation. The average albedo of Earth aerosols—the fraction of light reflected back into space—is about 20 % and most of this is from clouds.

Bioaerosols represent a relatively small fraction of the total amount of particles in the atmosphere and thus have a limited effect on the directly scattered light on a global scale. However, an increasing number of scientific studies are showing that bioaerosols may have a significant influence on clouds and precipitation [18, 19]. As sources of cloud condensation nuclei biological particles seem to behave similarly to other particles of equal size, but as ice nuclei they may be extremely efficient.

The formation of ice crystals in clouds at temperatures between 0 and −38 °C originates on ice nuclei, which only are a minor fraction of all aerosol particles. When the ice crystals have formed, they may quickly grow and thereby initiate precipitation. The most efficient ice nuclei particles in the atmosphere appear to be bioaerosols, as some bacterial strains and fungi are found to form ice at high subzero temperatures [45, 46]. Notable are *Pseudomonas syringae, Pseudomonas fluorescens* and *Erwinia herbicola.*

Other suggested atmospheric effects of bioaerosols include modification of cloud chemistry. Metabolically active bacteria in cloud droplets can contribute to the transformation of chemical compounds in the air [43]. The extent of this effect is uncertain.

In addition, the transport of biological particles through the atmosphere can sometimes have important consequences for ecosystems by changing biodiversity. The introduction of new species can affect community diversity and modify the population's genetic structure (e.g. [47]).

Conclusion

Bioaerosols play a key role for our health and environment through a variety of processes. Inversely, anthropogenic activity also has an extensive impact on bioaerosols. We do not yet have the full picture of all effects from airborne particles with biological origins, but the developments in molecular microbiology and aerosol technology over the past years have provided new possibilities to unravel this fascinating world.

References

1. Xie CQ, Shen FX, Yao MS (2011) A novel method for measuring the charge distribution of airborne microbes. Aerobiologia 27 (2):135–145. doi:10.1007/s10453-010-9183-x
2. Womack AM, Bohannan BJM, Green JL (2010) Biodiversity and biogeography of the atmosphere. Philos T R Soc B 365 (1558):3645–3653. doi:10.1098/rstb.2010.0283
3. Lighthart B (1997) The ecology of bacteria in the alfresco atmosphere. Fems Microbiol Ecol 23 (4):263–274. doi:10.1016/S0168-6496(97)00036-6
4. Lee S, Fuhrman JA (1987) Relationships between Biovolume and Biomass of Naturally Derived Marine Bacterioplankton. Applied and Environmental Microbiology 53 (6):1298–1303
5. Gorny RL, Dutkiewicz J, Krysinska-Traczyk E. (1999) Size distribution of bacterial and fungal bioaerosols in indoor air. Annals of Agricultural and Environmental Medicine 6 (2):105–113
6. Arslan D, Legendre M, Seltzer V, Abergel C, Claverie JM (2011) Distant Mimivirus relative with a larger genome highlights the fundamental features of Megaviridae. Proceedings of the National Academy of Sciences of the United States of America 108 (42):17486–17491. doi:10.1073/pnas.1110889108
7. La Scola B, Audic S, Robert C, Jungang L, de Lamballerie X, Drancourt M, Birtles R, Claverie JM, Raoult D (2003) A giant virus in amoebae. Science 299 (5615):2033–2033
8. Tang JW (2009) The effect of environmental parameters on the survival of airborne infectious agents. Journal of the Royal Society Interface 6:S737–S746
9. Taylor PE, Flagan RC, Valenta R, Glovsky MM (2002) Release of allergens as respirable aerosols: A link between grass pollen and asthma. Journal of Allergy and Clinical Immunology 109 (1):51–56. doi:10.1067/mai.2002.120759
10. Kuparinen A, Katul G, Nathan R, Schurr FM (2009) Increases in air temperature can promote wind-driven dispersal and spread of plants. P Roy Soc B-Biol Sci 276 (1670):3081–3087. doi:10.1098/rspb.2009.0693
11. Elbert W, Taylor PE, Andreae MO, Poschl U (2007) Contribution of fungi to primary biogenic aerosols in the atmosphere: wet and dry discharged spores, carbohydrates, and inorganic ions. Atmospheric Chemistry and Physics 7 (17):4569–4588

12. Reponen T, Hyvarinen A, Ruuskanen J, Raunemaa T, Nevalainen A (1994) Comparison of Concentrations and Size Distributions of Fungal Spores in Buildings with and without Mold Problems. Journal of Aerosol Science 25 (8):1595–1603. doi:10.1016/0021-8502(94)90227-5
13. Gorny RL, Reponen T, Willeke K, Schmechel D, Robine E, Boissier M, Grinshpun SA (2002) Fungal fragments as indoor air biocontaminants. Applied and Environmental Microbiology 68 (7):3522–3531. doi:10.1128/Aem.68.7.3522-3531.2002
14. Genitsaris S, Kormas KA, Moustaka-Gouni M (2011) Airborne algae and cyanobacteria: occurrence and related health effects. Front Biosci (Elite Ed) 3:772–787
15. Sharma NK, Rai AK, Singh S, Brown RM (2007) Airborne algae: Their present status and relevance. J Phycol 43 (4):615–627. doi:10.1111/j.1529-8817.2007.00373.x
16. Tegen I, Fung I (1994) Modeling of Mineral Dust in the Atmosphere—Sources, Transport, and Optical-Thickness. Journal of Geophysical Research-Atmospheres 99 (D11):22897–22914. doi:10.1029/94jd01928
17. Marshall WA (1996) Aerial dispersal of lichen soredia in the maritime Antarctic. New Phytol 134 (3):523–530. doi:10.1111/j.1469-8137.1996.tb04370.x
18. Despres VR, Huffman JA, Burrows SM, Hoose C, Safatov AS, Buryak G, Frohlich-Nowoisky J, Elbert W, Andreae MO, Poschl U, Jaenicke R (2012) Primary biological aerosol particles in the atmosphere: a review. Tellus Series B-Chemical and Physical Meteorology 64
19. Hamilton WD, Lenton TM (1998) Spora and Gaia: How microbes fly with their clouds. Ethology Ecology & Evolution 10 (1):1–16
20. Welsh DT, Viaroli P, Hamilton WD, Lenton TM (1999) Is DMSP synthesis in chlorophycean macro-algae linked to aerial dispersal? Ethology Ecology & Evolution 11 (3):265–278
21. Smith DJ, Griffin DW, Jaffe DA (2011) The High Life: Transport of Microbes in the Atmosphere. EOS, Transactions, American Geophysical Union 92 (30):249–256
22. Pöschl U (2005) Atmospheric aerosols: Composition, transformation, climate and health effects. Angew Chem Int Edit 44 (46):7520–7540. doi:10.1002/anie.200501122
23. Aller JY, Kuznetsova MR, Jahns CJ, Kemp PF (2005) The sea surface microlayer as a source of viral and bacterial enrichment in marine aerosols. Journal of Aerosol Science 36 (5–6):801–812. doi:10.1016/j.jaerosci.2004.10.012
24. Blanchar Dc, Syzdek L (1970) Mechanism for Water-to-Air Transfer and Concentration of Bacteria. Science 170 (3958):626-&. doi:10.1126/science.170.3958.626
25. Andreas EL (1998) A new sea spray generation function for wind speeds up to 32m s(-1). J Phys Oceanogr 28 (11):2175–2184. doi:10.1175/1520-0485(1998)028<2175:Anssgf>2.0.Co;2
26. Lee JS, Weon BM, Park SJ, Je JH, Fezzaa K, Lee WK (2011) Size limits the formation of liquid jets during bubble bursting. Nat Commun 2. doi:Artn 367, doi:10.1038/Ncomms1369
27. Imai M, Watanabe T, Hatta M, Das SC, Ozawa M, Shinya K, Zhong GX, Hanson A, Katsura H, Watanabe S, Li CJ, Kawakami E, Yamada S, Kiso M, Suzuki Y, Maher EA, Neumann G, Kawaoka Y (2012) Experimental adaptation of an influenza H5 HA confers respiratory droplet transmission to a reassortant H5 HA/H1N1 virus in ferrets. Nature 486 (7403):420-+. doi:10.1038/Nature10831
28. Burrows SM, Elbert W, Lawrence MG, Poschl U (2009) Bacteria in the global atmosphere—Part 1: Review and synthesis of literature data for different ecosystems. Atmospheric Chemistry and Physics 9 (23):9263–9280
29. Genitsaris S, Moustaka-Gouni M, Kormas KA (2011) Airborne microeukaryote colonists in experimental water containers: diversity, succession, life histories and established food webs. Aquatic Microbial Ecology 62 (2):139–U151
30. Chen PS, Tsai FT, Lin CK, Yang CY, Chan CC, Young CY, Lee CH (2010) Ambient Influenza and Avian Influenza Virus during Dust Storm Days and Background Days. Environmental Health Perspectives 118 (9):1211–1216. doi:10.1289/Ehp.0901782
31. Gloster J, Sellers RF, Donaldson AI (1982) Long-Distance Transport of Foot-and-Mouth-Disease Virus over the Sea. Vet Rec 110 (3):47–52
32. Imshenetsky AA, Lysenko SV, Kazakov GA (1978) Upper Boundary of Biosphere. Applied and Environmental Microbiology 35 (1):1–5
33. Haensch S, Bianucci R, Signoli M, Rajerison M, Schultz M, Kacki S, Vermunt M, Weston DA, Hurst D, Achtman M, Carniel E, Bramanti B (2010) Distinct Clones of Yersinia pes-

tis Caused the Black Death. Plos Pathog 6 (10). doi:ARTN e1001134, doi:10.1371/journal.ppat.1001134
34. Potter CW (2001) A history of influenza. J Appl Microbiol 91 (4):572–579. doi:10.1046/j.1365-2672.2001.01492.x
35. WHO (2004) The World Health Report 2004—Changing history. World Health Organization, Geneva
36. Lopez AD, Mathers CD, Ezzati M, Jamison DT, Murray CJL (2006) Global and regional burden of disease and risk factors, 2001: systematic analysis of population health data. Lancet 367 (9524):1747–1757
37. Douwes J, Eduard S, Thorne PS (2008) Bioaerosols. In: Heggenhougen HJ, Quah SR (eds) The encyclopaedia of public health, vol 1. Academic Press, San Diego, pp 287–297
38. Douwes J, Thorne P, Pearce N, Heederik D (2003) Bioaerosol health effects and exposure assessment: Progress and prospects. Annals of Occupational Hygiene 47 (3):187–200. doi:10.1093/annhyg/meg032
39. Klevens RM, Edwards JR, Richards CL, Horan TC, Gaynes RP, Pollock DA, Cardo DM (2007) Estimating health care-associated infections and deaths in US hospitals, 2002. Public Health Rep 122 (2):160–166
40. Liu AH, Leung DYM (2006) Renaissance of the hygiene hypothesis. Journal of Allergy and Clinical Immunology 117 (5):1063–1066. doi:10.1016/j.jaci.2006.03.027
41. Strachan DP (1989) Hay fever, hygiene, and household size. Bmj 299 (6710):1259–1260
42. Jaenicke R (2005) Abundance of cellular material and proteins in the atmosphere. Science 308 (5718):73–73
43. Deguillaume L, Leriche M, Amato P, Ariya PA, Delort AM, Poschl U, Chaumerliac N, Bauer H, Flossmann AI, Morris CE (2008) Microbiology and atmospheric processes: chemical interactions of primary biological aerosols. Biogeosciences 5 (4):1073–1084
44. IPCC (2007) Climate Change 2007—The Pysical Science Basis. Cambridge University Press, New York
45. Hoose C, Möhler O (2012) Heterogeneous ice nucleation on atmospheric aerosols: a review of results from laboratory experiments. Atmospheric Chemistry and Physics 12 (20):9817–9854. doi:10.5194/acp-12-9817-2012
46. Temkiv TS, Finster K, Hansen BM, Nielsen NW, Karlson UG (2012) The microbial diversity of a storm cloud as assessed by hailstones. Fems Microbiol Ecol 81 (3):684–695. doi:10.1111/j.1574-6941.2012.01402.x
47. Isard SA, Gage SH, Comtois P, Russo JM (2005) Principles of the atmospheric pathway for invasive species applied to soybean rust. Bioscience 55 (10):851–861. doi:10.1641/0006-3568 (2005) 055[0851:Potapf]2.0.Co;2

Chapter 4
Dispersion in the Atmosphere

Lennart Thaning and Leif Persson

The atmosphere is often divided into different layers, or spheres with intermediate pauses coupled with the vertical variation of temperature, as depicted in Fig. 4.1. Conditions in Fig. 4.1 represent the standard atmosphere according to ISO 2533:1975 (identical to the ICAO Standard Atmosphere up to 32 km).

As the standard atmosphere by definition reflects average conditions in this case both for the whole globe and over time, there are considerable deviations depending on season and latitude. For example, height to the tropopause in winter at high latitudes is only 7–8 km and in the tropics often 17–18 km, while the standard atmosphere states around 11 km.

Almost all "weather", as humans experience it, takes place in the troposphere.

When it comes to atmospheric pollution, human releases of gases and aerosols etc. often affect the lowest part of the troposphere, i.e. the planetary boundary layer (PBL). The PBL can be defined as the layer through which the "free" atmosphere, above the PBL, has contact with the surface of the earth. This contact is manifested through turbulent vertical fluxes of such features as temperature and horizontal momentum, and consequently the presence of turbulence is a main feature of the PBL.

Coupled to the characteristics of the turbulence and wind profile at different heights, the PBL can be divided into different sub-layers. In Fig. 4.2, an example of such sub-division is shown (also including the troposphere).

In the daytime, the height of the PBL is typically one tenth of that of the troposphere, but this ratio varies with season and time of day. Since the PBL is a turbulent layer, pollutants released close to ground will be more or less trapped in it for hours and perhaps days. So if we restrict our self to distances less than couple of hundreds of kilometres (in dispersion meteorology often called the local scale), the PBL is the layer of interest regarding dispersion processes.

Over land there is a clear tendency to a daily variation in the PBL structure. In night time, especially in situations with low wind speed and with low cloudiness and therefore an effective radiation cooling of the ground, a stable boundary layer

L. Persson (✉) · L. Thaning
Division of CBRN Defence and Security, FOI—Swedish Defence Research Agency, Cementvägen 20, 901 82, Umeå, Sweden
e-mail: leiper@foi.se

P. Jonsson et al. (eds.), *Bioaerosol Detection Technologies,* Integrated Analytical Systems, DOI 10.1007/978-1-4419-5582-1_4, © Springer-Verlag New York 2014

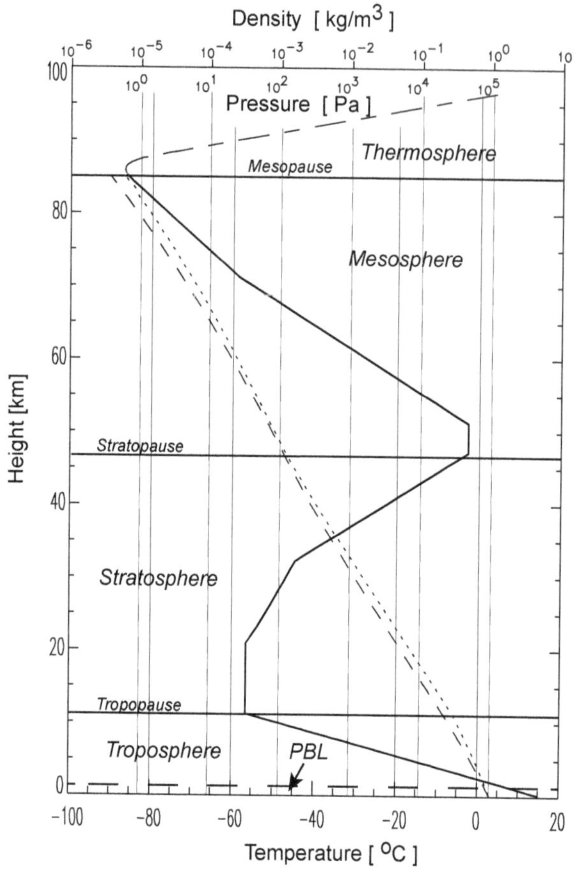

Fig. 4.1 Structure of the standard atmosphere with respect to the vertical temperature variation. In addition to the temperature (*full line*), the pressure (*dashed*) and the density (*dotted*) are shown. Scales for pressure and density are displayed at the top

(SBL) with weak turbulence will be formed. The height of SBL may, in extreme situations, be well below 100 m. However in day time, if the cloudiness and the wind speed are not too high, the ground will be heated from the solar radiation and a strongly turbulent, convective boundary layer (CBL) will be formed. Such conditions are often called unstable and typical boundary layer heights may become 1–2 km.

The daily variations described above are most obvious during summer, while in winter time stable conditions may prevail also in day time, especially on high latitudes. Other condition that may decrease or remove this daily variation is high wind speed. If the wind speed is more than around 3 m/s on 10 m height a neutral boundary layer, NBL (not to be confused with Nocturnal Boundary Layer for which this abbreviation is sometimes used) will be formed, regardless of the time of the day. Typical boundary layer heights in NBL are around 500–1000 m. Furthermore,

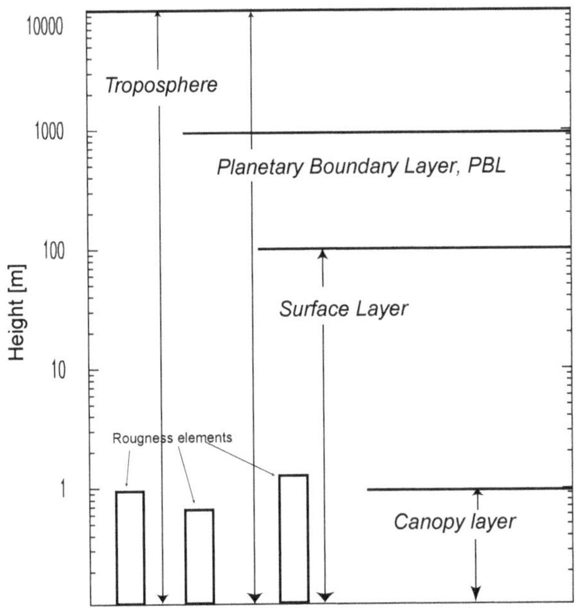

Fig. 4.2 Subdivision of the troposphere and the planetary boundary layer (PBL). The heights of the different layers are typical examples from which large deviations can be expected at a certain place and time of year and day

if cloudy conditions are prevailing, the cooling and heating of the ground will be counteracted resulting in neutral conditions also in low wind situations.

Note that the formation of different PBLs discussed above is valid only over land. Over sea, where radiation cooling and heating affects the surface temperature very slowly, the daily variations are replaced by seasonal variations. In fall and winter when the water is relatively warm compared to the land surface, unstable conditions will dominate while in spring and early summer stable conditions can be expected.

In the lower part of the PBL, approximately up to one tenth of the PBL height, a layer characterized by nearly constant vertical fluxes can be found. This layer which often is called the surface layer (SL) is, for obvious reasons, the most investigated and well-known part of the atmosphere both from a dynamical and an air quality point of view.

Turbulence is generated when the wind speed is retarded close to the ground. This retardation of wind speed and generation of turbulence are dependent of the structure of the ground or the roughness. For example, over a beach, the roughness elements are grains of sand resulting in a rather low generation of turbulence while over a forest with trees or a city with building, a high generation is expected. The layer in-between the roughness elements, which in some applications, for example, dispersion in an urban area, can be of high importance, and is called the canopy layer (CL).

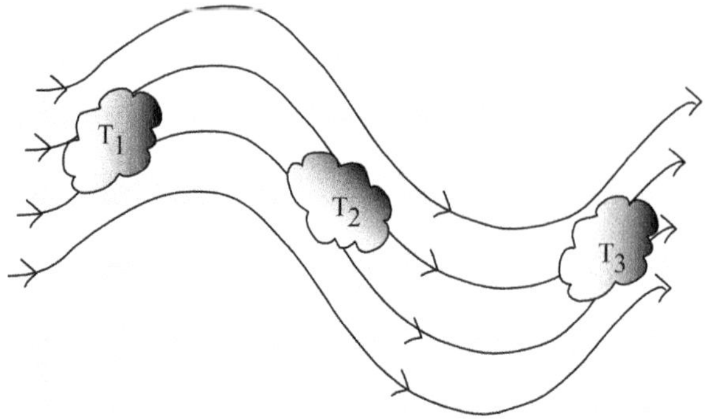

Fig. 4.3 Turbulent eddies much larger than the cloud transport the cloud. T_1–T_3 designates three instants of time, where $T_3 > T_2 > T_1$

Timescales in the Different Layers

The characteristic timescale for mixing of gases and aerosols (τ_m) is different in the different layers. In CL and SL, the vertical mixing is a dominating mechanism and τ_m is of the order of seconds (CL) and minutes (SL). In the PBL, where vertical mixing is still of high importance, τ_m typically is a couple of hours. In the troposphere, above the PBL, the turbulence and consequently the vertical turbulent mixing are very small. In fact, if turbulent mixing was the only available mechanism, τ_m would be of the order of a year. However, in the absence of turbulence, vertical transport due to up- and down-winds in, for example, synoptic weather systems (low and high pressure systems), plays an important role and τ_m becomes of the order of a month.

It may be interesting in this context to mention that the time scale for horizontal mixing in the troposphere is a couple of months within one hemisphere and a year between the hemispheres. Furthermore, the time scale for exchange between the troposphere and the lower part of the stratosphere is of the order of one year.

Dispersion Processes in PBL

When an aerosol or a gas is discharged to the atmosphere, it will be transported with the winds and at the same time dispersed (or diluted) due to the turbulent fluctuations. One way to describe these processes is to relate them to the size ratio between the turbulent eddies and the aerosol/gas cloud.

Turbulent eddies which are larger than the aerosol or gas cloud just transport the cloud, without dilution or deformation, as illustrated in Fig. 4.3.

Turbulent eddies, which have size of the same order of magnitude as the released cloud, will deform it and eddies with smaller dimension will dilute and enlarge the cloud due to turbulent diffusion, cf. Figs. 4.4 and 4.5.

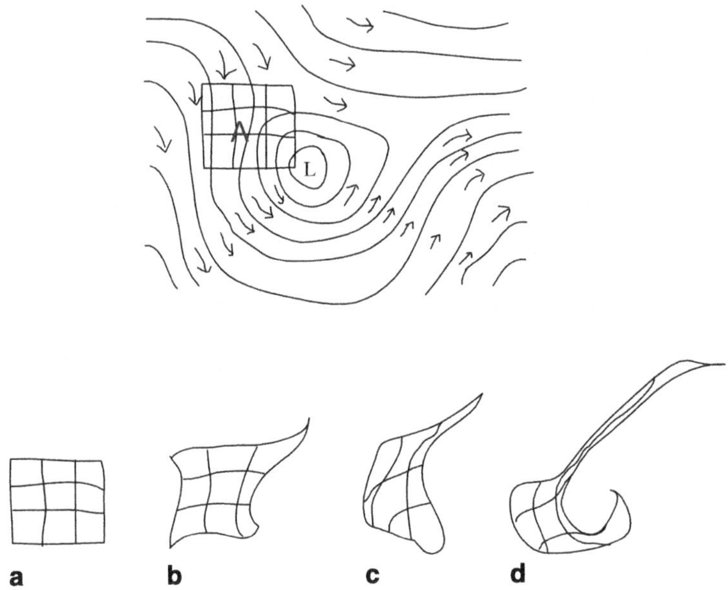

Fig. 4.4 Turbulent eddies of the same dimension as the cloud, deform the cloud. *a–d* designates different instants of time where $b>a$, $c>b$ etc

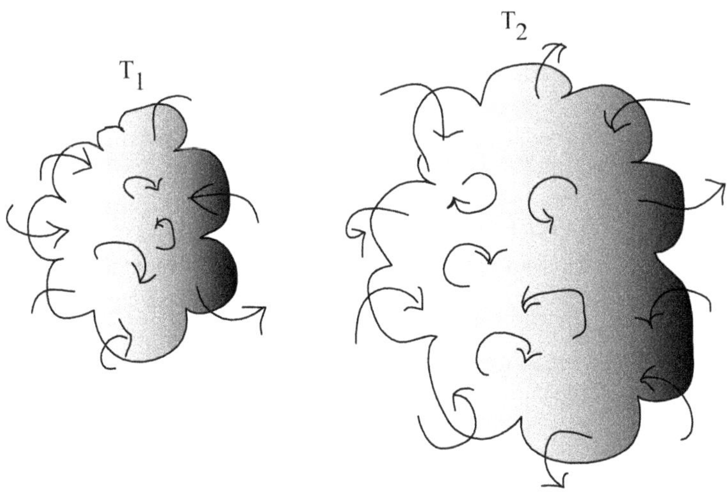

Fig. 4.5 Small turbulent eddies dilutes the cloud. T_1 and T_2 designates two instants of time, where $T_2>T_1$

Other Processes that Influences the Concentration

There are processes other than dispersion that can change the concentration in a cloud containing discharged aerosols or gas. In this context, biological decay is one example of such a mechanism. It is generally a function of time, humidity, solar radiation and temperature. Another example is deposition, either dry or wet. "Dry" deposition means that precipitation is not a part of the process while "wet" refers to removal of aerosols due to precipitation. In dry deposition settling or sedimentation, in addition to turbulent transport, is an important mechanism that can bring the aerosol particles very near to a deposition surface, from where processes like Brownian diffusion or impaction can finish the deposition. The settling is a function of size and form of the aerosol particles, particle density, air density and air temperature.

A process that may increase the concentration in the cloud is resuspension for example, when deposited B-particles are whirled up again by the wind or passing vehicles and thereby reintroduced to transport and dispersion in the atmosphere.

The interested reader can find detailed descriptions and discussions of atmospheric boundary layers and dispersion features in for example, Pasquill and Smith [7], Arya [1, 2], Stull [10], Garratt [5] and Kaimal and Finnigan [6].

Examples of Dispersion Model Results

In order to illustrate and provide the reader with some qualitative and quantitative understanding of dispersion processes on the local scale, some model calculations are shown below. The model used is a stochastic particle model developed at the Swedish Defense Research Agency, FOI. The version used for these illustration is based on horizontal homogeneity and steady meteorological conditions, and the variations of wind speed and wind direction with height make use of similarity profiles proposed by Zilintikevich et al. [14] for stable and neutral stratification (SBL and NBL) and Zililntikevich et al. [13] for unstable stratification (CBL), for further details the reader is referred to Schönfelt [8] and Sehlstedt [9]. One important feature for the model is that not only the SL but whole PBL is covered. Among other things this means that a wind veering with height, caused by the friction against ground and the rotation of the earth, is present. This turning with height, which may become very large during stable conditions, increases the predicted horizontal spreading of the plume.

Figures 4.6–4.8 depict the concentration, at 1.5 m above ground (henceforth referred to as ground level concentration) in three different weather situations but for the same release rate, which rather arbitrarily is chosen in order to give a maximum ground level concentration between 1 and 10. Furthermore, a continuous release of particles with diameter 10 μm on 10 m height is assumed.

The chosen weather situations correspond to the three different types of boundary layers discussed above: NBL, SBL and CBL (unstable stratification). In the

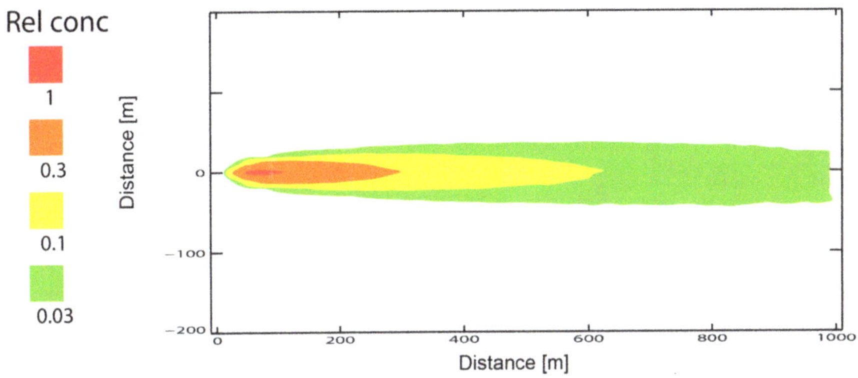

Fig. 4.6 Calculated average ground level concentrations in neutral stratification (NBL). Wind speed = 5 m/s at 10 m height. Note that there are somewhat different scales on the axis

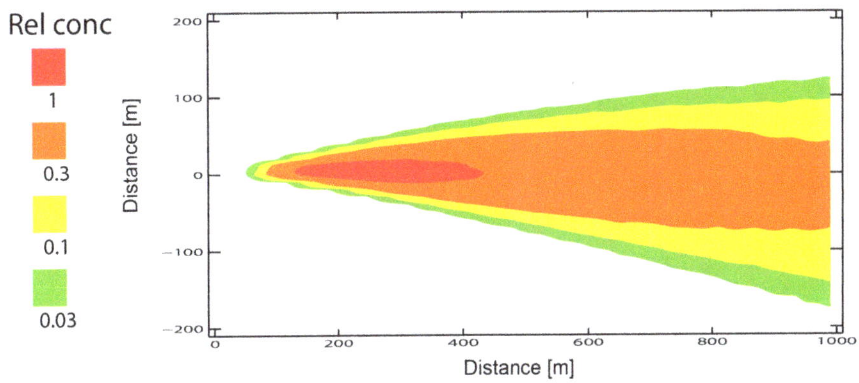

Fig. 4.7 Same as Fig. 4.6 but in SBL. Wind speed = 2 m/s

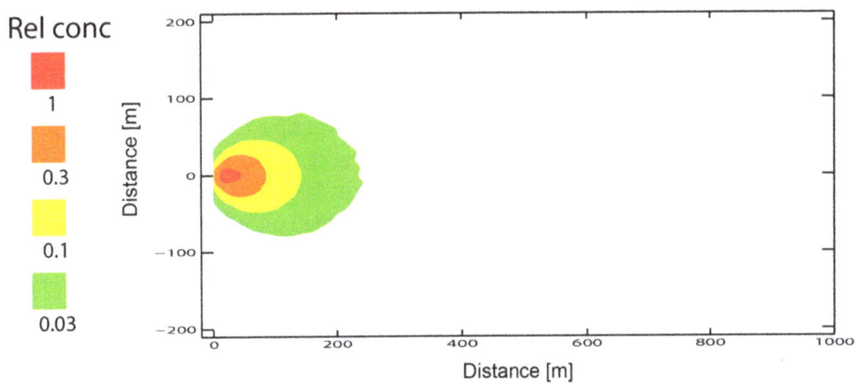

Fig. 4.8 Same as Fig. 4.6 but in CBL. Wind speed = 2 m/s

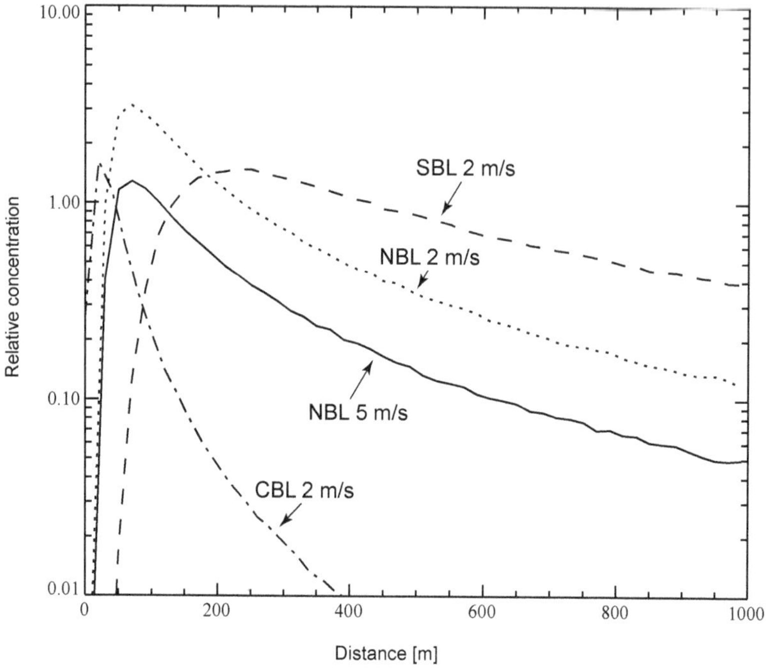

Fig. 4.9 Ground level concentrations in the center of the plumes for different stability conditions

NBL case a wind speed of 5 m/s at 10 m height is assumed, while for CBL and SBL the corresponding wind speeds are 2 m/s. In terms of the often used Pasquill classification, the PBL and CBL simulated in the model correspond to "F" and "A", i.e. extremely stable and unstable conditions. The NBL corresponds to Pasquill class "D".

The dispersion pattern for the NBL is shown in Fig. 4.6 and for SBL in Fig. 4.7. The model calculations clearly predict higher concentrations at longer distances in the SBL, compared to NBL. This feature is mainly a consequence of the smaller turbulence and dispersion in the vertical direction in SBL compared to NBL. The strong wind veering in SBL simulated in the model, actually results in a somewhat broader plume in SBL. However, there are situations, for example, in valleys where the ground geometry may prohibit a strong wind veering with height, and where consequently even higher ground level concentrations would occur in SBL. But there are also situations where relatively slow but large fluctuations in the wind direction would cause an even broader average plume and less average concentrations.

In the CBL, the high boundary layer and the high turbulence level result in fast dilution of e.g. a released organism and consequently relatively low ground level concentrations should be expected, as can be seen in Fig. 4.8.

In Fig. 4.9, the centerline concentrations close to ground are depicted for the same weather situations as shown in Figs. 4.6–4.8. Furthermore, calculations for

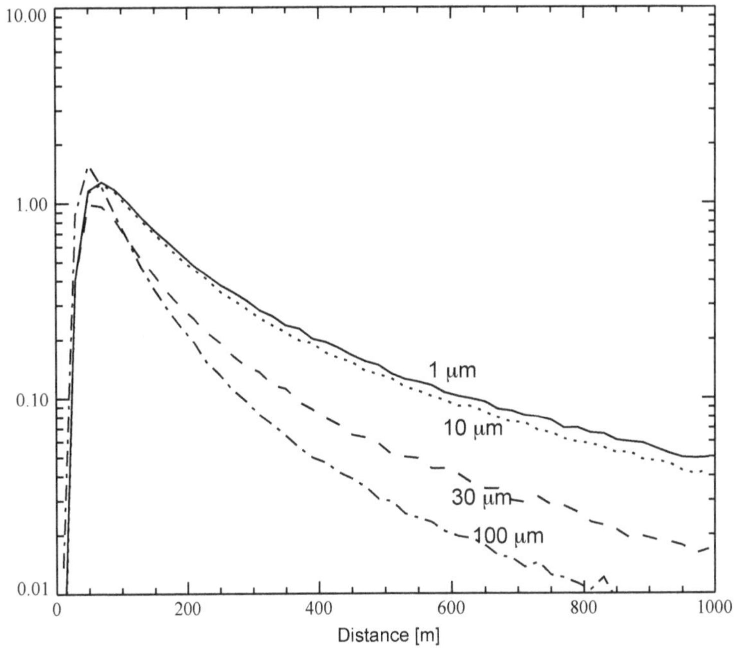

Fig. 4.10 Calculated influence of particle diameter. Ground level concentrations in plume center for particle diameters: 1, 10 ,30 and 100 µm. Stability NBL, 5 m/s

a NBL with wind speed 2 m/s at 10 m height has been added, in order to make it easier to compare the effects of different stratifications. It should be noted that the most common reason for neutral stratification is high wind speed, although NBL may occur also in low wind situations, for example, under cloudy conditions when radiation cooling or warming of ground is reduced.

As mentioned earlier the release height used in the simulations is 10 m, and as can be seen in Fig. 4.9, this causes rather different concentration patterns close to the source as compared to at some distance. Due to the high turbulence level and dispersion rate in the CBL, the ground level concentration at very short distances actually becomes higher than in all other simulated weather situations. In the SBL with the slowest dispersion rate, the maximum concentrations occurs at a longer distance, because it takes longer time plume to reach ground. If the source of the release is close to ground, this effect would not show up.

It may also be worth mentioning that concentration ratio for the two NBL is almost constant at all distances and approximately 2.5 m, which is the ration between the two assumed wind speeds (5 and 2 m/s). The main reason for the concentration differences in these two cases is the difference in initial dilution right at the source point, since the organisms released during a certain time interval are injected into a volume that is approximately proportional to the wind speed.

Dry deposition will decrease the concentrations, since organisms are removed from the plume. The impact on the concentration shown in Fig. 4.10 is due to the fact that an increase in particle diameter causes an increase in the settling velocity and consequently also in the deposition. In the simulations shown in Fig. 4.10, spherical particles with diameters 1, 10, 30 and 100 μm was assumed along with a particle density of 1000 kg/m^3. This corresponds to settling velocities 0.035, 0.3, 2.7 and 25 cm/s, respectively.[1] From Fig. 4.10, in which the results for the NBL—5 m/s case are shown, it can be seen that the difference between the 1 and 10 μm curves is almost negligible and also that for the largest particles (100 μm) the concentration has decreased not more than to almost one tenth of the concentration for 1 μm particles at 800 m. At first sight this may seem surprising since a settling velocity of 25 cm/s gives a downward transportation of about 40 m in 160 s which is the time for 800 m horizontal transportation with speed 5 m/s. But, there is also a turbulence induced vertical transport in the atmosphere, which occasionally can move the particles upwards and in the 5 m/s-case, these turbulent up winds (typically 0.5 m/s) are large enough to still keep some particles in the air at low altitudes at distance 800 m.

Another feature that is present in Fig 4.10 is that close to the source, the concentrations of 100 μm particles are larger than those of the smaller particles. This is due to the fact that the plume center, where the highest concentrations are expected, will descend to ground level rather fast (about 40 s for the 100 μm particles).

Wet deposition, in the simulations below defined as removal of organisms by precipitation falling through the plume, will also act to diminish the concentration. Under the assumption that the precipitation rate is constant over the area of interest, the process can be modelled as an exponential decrease of the concentration, i.e. $C = C_0 e^{-\Lambda t}$ where C_0 is the concentration at time t=0. The washout coefficient $\Lambda[s^{-1}]$ is a function of precipitation rate and particle size. Typical Λ-values for rain span from 5.E−05 (s^{-1}) for small particles and low rain rates (1 μm and 1 mm/h) to 5.E-3 for large particles and heavy rain (30 μm and 1 mm/h), according to Baklanov & Sørensen [4].

In Fig. 4.11, some model results for the NBL 5 m/s scenario are shown.

The ground level concentrations shown in Fig. 4.11 are calculated assuming a particle diameter 10 μm and precipitation rates 1, 10, 30 mm/h which correspond to Λ-values 2E-04, 2E-03 and 4E-03. The inverse of the washout coefficient, that can be interpreted as a typical decay time, which in our scenario become approx. (1.3 h, 8 and 4 min). The transport time through the area (length1000 m) with wind speed 5 m/s is around 3 min so we shouldn't expect any drastic concentration decay, which is also confirmed by the simulations demonstrated in Fig. 4.11. Actually, the curve for the 1 mm/h case is so close to the 0 mm/h curve that they are almost not distinguishable and also with 30 mm/h rain rate the simulations predict only around a 50% reduction.

Note that 30 mm/h is a very high precipitation rate which may occur in showers, while typical values coupled to the passage of a warm front in conjunction with mid latitude low pressure systems are 0.5–1 mm/h.

---

[1] The settling velocities are calculated under the assumption of normal temperature and pressure. However, it may be pointed out that settling velocities are rather insensitive to the meteorological variables.

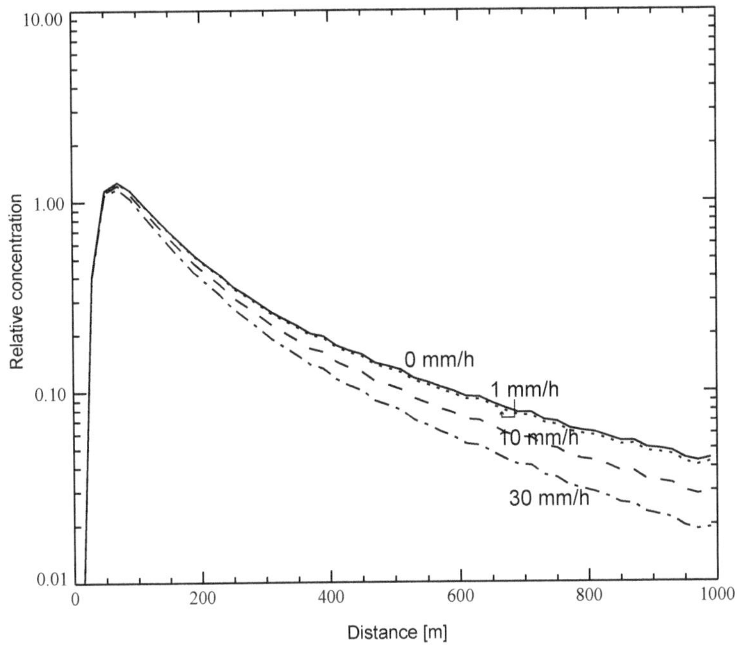

Fig. 4.11 Calculated effect of precipitations. Ground level concentration in plume center for precipitations rates: 0, 1, 10 and 30 mm/s. Particle diameter = 10 μm. Stability NBL, 5 m/s

As in the case with settling velocity and dry deposition, the effect of wet deposition is a function of time, so with a slower wind speed larger effect can be expected at the same distance from the source.

The curves exhibited in Fig. 4.11 can also be used to show the effects of biological decay if the Λ-values are transformed into e.g. decimal reduction times, which often are used for biological decay. The curves may then be interpreted as the ground level concentrations for a release of biological organisms with decimal reduction times: 0, 3.2 h, 20 and 10 min respectively.

All the examples above are based on model calculations with a model that strictly is valid only for rather simple conditions, i.e. no horizontal variations in the mean wind and steady state, i.e. no meteorological variation in time. Still the effect of many important dispersion processes can be demonstrated with this type of model. The limitations of meteorological models are discussed in the next section.

On Modelling in Meteorology

The atmosphere is an immensely complex system, with phenomena and interactions on a wide range of temporal and spatial scales. Our knowledge about the atmosphere is gradually developed by an incremental scientific process, where hy-

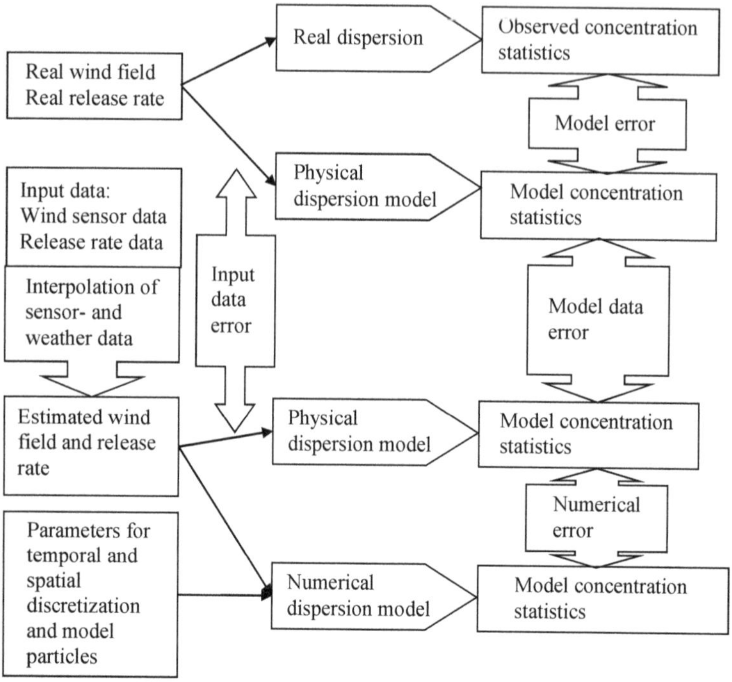

Fig. 4.12 Flow of information and errors in the modelling process

potheses are formulated in terms of certain physical and chemical quantities of interest, and physicochemical laws are formulated in a mathematical language, often approximated by a numerical method (algorithm) implemented on a computer. The hypotheses are tested against usually indirect and sparse observations of the quantities of interest in field experiments. Parts of the physicochemical models may be confirmed by controlled laboratory experiments.

Any feasible model is an abstraction with a lot of phenomena and influencing factors left out.

These factors cause seemingly random fluctuations in repeated measurements and observations. There is usually a payoff between model complexity ("size" of parameter space) and the statistical uncertainty in the model. A larger parameter space may give more freedom to explain the observed data. On the other hand, the available experimental data are limited, so when more parameters are used, less amount of data are available for each set of parameter values. This gives a larger statistical uncertainty in the statistical analysis for a model with more parameters. Also, parameter estimation in a larger parameter space is usually more difficult, so larger uncertainty in the parameters may degrade the better modeling capabilities of a more complex model. Model parameters are often distributed, and averaged over large spatial domains, while measured parameters are necessarily measured as averages over small domains, which may give values not representative for the larger area. For example, the local wind direction may be different from the average wind direction of the area due to effects of the local topography.

Because of the sparseness of the data, datasets are usually suitable for some models but not for others; in other words, field experiments and models must be carefully adapted to each other.

Models may be thought of on a scale between two extremes, phenomenological (or knowledge-based) models and behavioural models [11]. Phenomenological models are based on physical laws, usually in the form of conservation equations (for energy, momentum, constituent mass, etc.), whilst behavioural models consists of a pure data fit, with no underlying physical theory. In meteorology, a mixture of these extremes are often used, where large-scale phenomena are modelled by physicochemical conservation equations, and the influence of smaller scale phenomena on these larger scales are given by behavioural models, so called *subgrid* models.

To discuss the errors in a model, we may distinguish between the "ideal" physicochemical model in the mathematical form of a partial differential equation (field model) or stochastic differential equation (particle model), and the numerical algorithm computing approximations of the field model. We call the former a "physical model" and the latter a "numerical model".

We can identify three sources of error in the total prediction or estimation error (difference between observations and numerical model result) as follows:

- Model error, caused by the physical model being an abstraction of the real dispersion process. This error cannot be handled within the formal "model world".
- Model data error, caused by errors in the input data, propagated through the physical model.
- Numerical error, caused by the numerical discretisation of the physical model and the finite numerical precision of the computer. This includes sampling errors of Monte Carlo methods like Lagrangian stochastic particle models [12].

See the Fig. 4.12. Recommendations on the evaluation atmospheric dispersion models can be found in (ASTM [3]).

References

1. Arya S (1988) Introduction to Micrometeorology. Academic Press, San Diego
2. Arya S (1999) Air Pollution Meteorology and Dispersion. Oxford University Press, New York
3. ASTM Standard Guide for the Statistical Evaluation of Atmospheric Dispersion Model Performance. ASTM Standard D 6589-00
4. Baklanov A, Sørensen J (2001) Parameterisation of radionuclide deposition in atmospheric long-range transport modelling. Phys. Chem. Earth (B) 26:787–799
5. Garratt J (1992) The atmospheric boundary layer. Cambridge University Press, Cambridge
6. Kaimal J, Finnigan JJ (1994) Atmospheric Boundary Layer Flows—Their Structure and measurements. Oxford University Press, New York
7. Pasquill F, Smith F (1983) Atmospheric Diffusion. Ellis Horwood, (John Wiley & Sons) Chichester

8. Schönfeldt F (1997) A Langevin equation dispersion model for the stably stratified planetary boundary layer. FOA-R-97-00523-862-SE
9. Sehlstedt S (2000) A Langevin equation dispersion model for the unstably stratified boundary layer. FOA-R-00-01408-862-SE
10. Stull R (1988) An Introduction to Boundary Layer Meteorology. Kluwer Academic Publishers, Dordrecht
11. Walter É, Pronzato L (1997) Identification of Parametric Models from Experimental Data. Springer, Berlin
12. Wilson J, and Sawford B (1996) Review of Lagrangian Stochastic Models for Trajectories in the Turbulent Atmosphere. Boundary-Layer Meteorology 78:191–210
13. Zilintikevich S, Fedorovich E, Shabalova M (1992) Numerical model of a non-steady atmospheric planetary boundary layer, based on similarity theory. Boundary-Layer-Meteorology 59:387–411
14. Zilintikevich S, Johansson P-E, Mironov D, Baklanov A (1998) A similarity-theory model for wind profile and resistance law in stably stratified planetary boundary layers. Journal of Wind Engineering and Industrial Aerodynamics 74–76:209–218

Chapter 5
Aerosol Sampling and Transport

Jorma Keskinen and Marko Marjamäki

Introduction

The instruments described in the next section of this book aim at detecting biological particles suspended in air. This chapter describes the art and components of sampling the aerosol and transporting the particles to the actual detection unit, while keeping them airborne. Depending on the detection principle, later stages may require transferring the particles into another medium such as a liquid. Figure 5.1 shows the main aerosol components of a point detector.

In normal aerosol measurement, the objective is to take a representative sample of the ambient air and to transport it to the detection unit while keeping the aerosol concentration and size distribution unchanged. This is accomplished by avoiding both oversampling and undersampling and by avoiding particle losses during sample transport. For the present purpose, the sensitivity of the whole instrument for the biologically harmful particles is the main priority. Therefore, undersampling or loss of particles outside the interesting size range is not as harmful as normally. As particles lost during the transport contaminate the instrument interior and may be later re-entrained, the losses should anyhow be minimized. Unwanted particles such as those too large in diameter should be removed from the air stream in a well-defined manner. To increase instrument sensitivity, the transport system usually features a concentrator for coarse particles.

The design of the sampling, transport, and concentrator phases is based on the kinetics of aerosol particles. We start by taking a look at this subject, followed by more detailed description of aerosol sampling, aerosol transport, coarse particle concentration, and finally particle transfer to the detection volume.

J. Keskinen (✉)
Department of Physics, Tampere University of Technology,
P.O. Box 692, FI 33101, Tampere, Finland
e-mail: jorma.keskinen@tut.fi

M. Marjamäki
Nuclear & Thermal Power, Fortum Power and Heat Oy,
P.O. Box 100, FI 00048, Fortum, Finland
e-mail: marko.marjamaki@fortum.com

P. Jonsson et al. (eds.), *Bioaerosol Detection Technologies*, Integrated Analytical Systems,
DOI 10.1007/978-1-4419-5582-1_5, © Springer-Verlag New York 2014

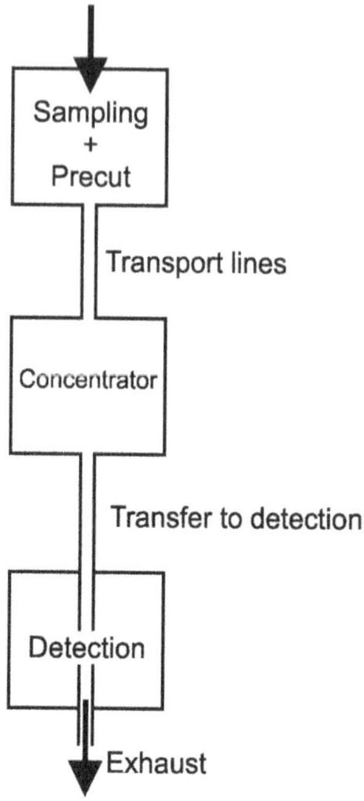

Fig. 5.1 Sampling and transport system and its components

Aerodynamics of Bioaerosol Particles

To be measured by the detection system, the particles usually need to be transferred from the ambient into the detection volume of the instrument. To this end, a basic thing needed is aerosol flow in a flow channel, such as a pipe with circular cross section. Fluid dynamics in general is complicated, but pipe flows with different volume flow rate (Q), pipe diameter (W), and different fluid properties can be compared using a dimensionless quantity called Reynolds number, Re:

$$\mathrm{Re} = \frac{\rho_g W U}{\eta} = \frac{\rho_g 4Q}{\pi \eta W} \tag{5.1}$$

where U is the average fluid flow velocity $4Q/\pi W^2$, ρ_g is the fluid density, and η is the fluid viscosity (see, e.g.[12]). From here on, we limit ourselves to air in normal conditions.

At low Reynolds numbers, the flow in the tube is unidirectional, formed of coaxial *laminae,* and is called laminar. At higher Reynolds numbers, the flow forms turbulent eddies that behave statistically. The flow is laminar at least up to Re=2000, while over 4000 it is turbulent. In between, both cases are possible. For fine particle transport, laminar flow is not essential, but for the coarse bioaerosol particles turbulence causes unwanted deposition onto the tube walls. Figure 5.2 illustrates laminar aerosol flow.

As discussed, numerous phenomena affect particle deposition and could be used for particle classification. For bioaerosol measurement, however, it is useful to limit

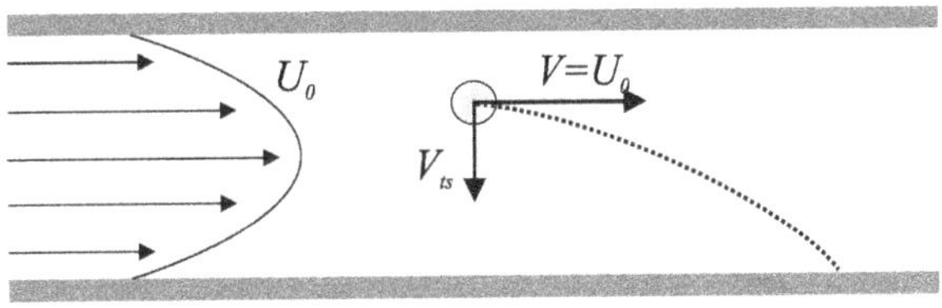

Fig. 5.2 Laminar aerosol flow in a horizontal pipe. Settling under gravitational force

to gravity and inertial effects. These cover the most important size classification and loss mechanisms in the size range of interest. Bioaerosol particles, as any other objects, obey Newton's second law, the basic equation of motion:

$$\mathbf{a} = \frac{\mathbf{F}}{m_p} \tag{5.2}$$

where **a** is the acceleration (vector) of the particle, m_p is the mass of the particle, and **F** is the vector sum of all the forces on the particle. This means that if the net force on the particle is zero, it will continue with constant velocity. To change the velocity or direction of the particle, a force must be exerted on it. This tendency to maintain the state of motion is often called the inertia of the particles.

If the particle moves relative to the gas (air), the air exerts a drag force (air resistance) on the particle, depending on the relative velocity. For particles of interest here, a particularly useful approximation for this drag force $\mathbf{F}_D$ is given by the Stokes law, here somewhat modified (e.g. [18]):

$$\mathbf{F}_D = -\frac{3\pi\eta d_p \mathbf{V}}{C} \tag{5.3}$$

where η is the air viscosity, d_p is the particle diameter, and **V** is the velocity of the particle relative to the air. Stokes law strictly holds only for spherical particles much larger than the mean free path of the air and for incompressible laminar flow around the particle. However, using modifications, this law can be applied to most cases involving the bioaerosol particles. The most frequent modification is to include a correction factor C (Slip correction) for particles that are comparable to the free path of air, as has been done in Eq. (5.3). This factor is within 15% of unity for particles larger than 1 µm in diameter. Any other complications we neglect and use Eq. (5.3).

It turns out that under varying external forces, the particles rapidly go through an acceleration phase (Eq. (5.2)). After that, the particle reaches a velocity that produces a drag force equal put opposite to the external force, and the particle will travel with constant velocity. As an example, drag force rapidly balances the gravitational force $\mathbf{F}_G$ on the particle so that the total force on the particle is zero: $\mathbf{F}_D = -\mathbf{F}_G$. As a result, the particles fall down with a constant velocity called the terminal settling velocity. For the absolute values, the force balance can be written as

$$\frac{3\pi\eta d_p V_{ts}}{C} = m_p g = \frac{\pi}{6}\rho_p d_p^3 g$$

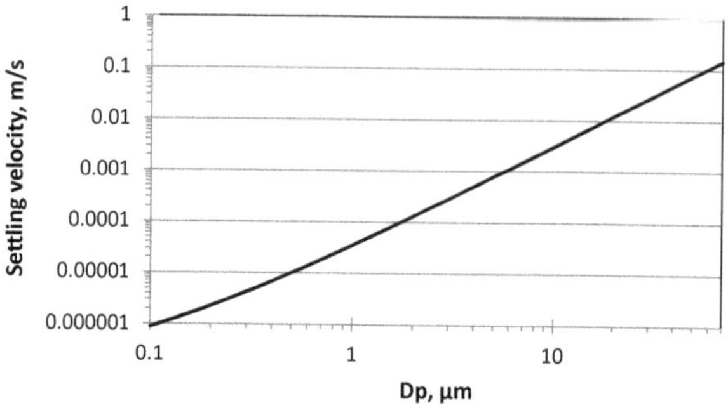

Fig. 5.3 Particle settling velocity as a function of aerodynamic particle diameter

where g is the acceleration of gravity, and ρ_p is the density of the particle. Solving for the settling velocity gives

$$V_{ts} = \frac{\rho_p g d_p^2 C}{18\eta} \tag{5.4}$$

There is square-law dependence between the settling velocity and particle diameter. Figure 5.3 shows the settling velocity as a function of particle diameter for unit density spheres. Real particles have varying shapes and varying density, and therefore the settling velocity cannot be read from Fig. 5.3 for the diameter. However, we do the opposite and define particle diameter through settling velocity. Measuring the settling velocity for a real particle of undefined shape and density, we calculate the diameter of a unit density sphere that would settle with the same velocity. We call this diameter the *aerodynamic diameter* of the real particle.

As an example, under normal conditions in air, unit density spheres with diameters of 1, 10, and 100 µm settle with velocity values of 0.035, 3.1, and 250 mm/s, respectively. For a three-dimensional flow system, one can use Eq. (5.4) to predict the vertical movement of the particles, as illustrated in Fig. 5.2. A practical consequence of this is that the residence time in horizontal flow lines must be kept less than a second to avoid settling of 10 µm particles to the bottom of the line. Another thing to observe is that 100 µm particles fall one meter in 4 s. Therefore, particles of this size in air are always fresh, that is, they must be the result of a recent generation or dispersion process.

Let us consider a case illustrating the effect of the inertia of the particle. Let the particles travel with an air flow with a velocity of U. Then assume that the air flow is stopped abruptly at time $t=0$. Immediately after this the particle travels with velocity U relative to the air, but the air drag force slows it down. We now look how far the particle travels before stopping down. Treating this as a one-dimensional case, we can write the equation of motion as:

$$a = \frac{F_D}{m_p} = -\frac{3\pi\eta d_p V/C}{\frac{\pi}{6}\rho_p d_p^3} = -\frac{18\eta V}{\rho_p d_p^2}$$

where V is the momentary velocity of the particle, with an initial value of U. Note that the drag force and acceleration are into the opposite direction of the velocity. We next use the definition of acceleration and velocity and write $a=-dV/dt=-d^2s/dt^2$, where s is the momentary distance travelled by the particle. Separating the variables and integrating twice gives the distance travelled by the particle at a given time. The total distance S travelled is obtained by setting the end time infinite ($t=\infty$):

$$S = U\frac{\rho_p d_p^2}{18\eta} = U\tau \qquad (5.5)$$

where we have used a new quantity, the particle relaxation time t:

$$\tau = \frac{\rho_p d_p^2}{18\eta} \qquad (5.6)$$

If the air properties are unchanged, relaxation time only depends on particle density and particle diameter. Eq. (5.5) can also be used to calculate the distance S that the particle travels in its original direction when the air flow turns abruptly. If there is a wall in the original distance closer than the distance S, the particles will hit the wall and be lost from the air flow. This kind of deposition phenomena is usually called inertial deposition and takes place for instance in flow line contractions and elbows and bends. For our unit density example particles of 1, 10, and 100 µm, the relaxation time values are 3 µs, 5 µs, 0.31 ms, and 31 ms, respectively. For an initial velocity of 10 m/s, this gives stopping distances of 0.035, 3.1, and 310 mm, respectively. As a first approximation, we could compare these with the radius of curvature in flow line bends. The numbers show that the inertial losses in normal flow lines can be avoided for 1 µm particles but have to be taken seriously already for 10 µm particles. More detailed tools for calculating the losses in flow lines are presented in the section "Transport of Particles".

Inertial deposition is also used for particle size classification in devices called impactors. Figure 5.4 shows a cross-sectional view of a rectangular nozzle (slit nozzle) impactor. The sample aerosol enters the impactor through an acceleration nozzle (width W) where the flow velocity is increased to a jet velocity value U. After the nozzle, the flow is forced to make a sharp turn in front of a solid wall called impaction plate. Small particles follow the gas streamline, but larger ones impact onto the plate.

Consider the flow line $W/4$ halfway between the flow centerline and the nozzle wall, as shown in Fig. 5.4. Particles entering on this flow line and with stopping distance larger than $W/4$ impact on the plate. For this flow line, the condition for impaction is $S \geq W/4$.

In general, the inertial deposition is characterized by a dimensionless number called the Stokes number (Stk), defined simply as the ratio of particle stopping distance to a geometric dimension of the flow system. For impactors, it is customary to use half of the nozzle width W (for slit nozzle) or half of the diameter W (circular nozzle) as the geometric dimension. Therefore

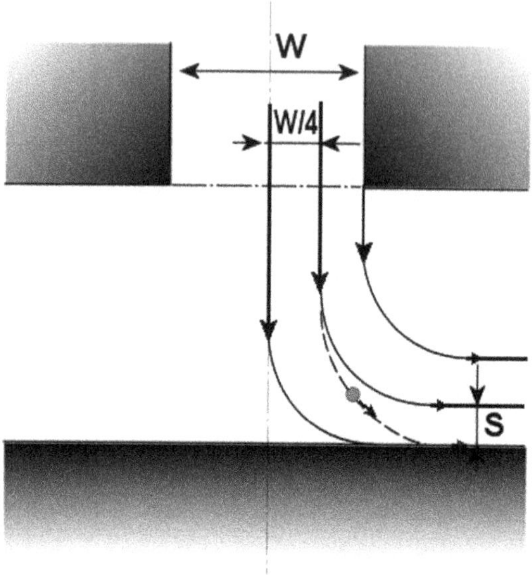

Fig. 5.4 Cross-sectional view of an impactor

$$Stk = \frac{S}{W/2} = \frac{2\tau U}{W} = \frac{\rho_p d_p^2 UC}{9\eta W} \tag{5.7}$$

For our example flow line, particles having $Stk \geq 0.5$ impact onto the plate. To impact, particles entering closer to the nozzle wall would need to have a longer stopping distance (higher Stokes number), whereas particles closer to the nozzle symmetry plane would need to have shorter stopping distance (lower Stokes number). As described by John [24], we therefore expect that $Stk = 0.5$ corresponds to the particle diameter with 50% collection efficiency (d_{p50}), and write $Stk_{50} = 0.5$. For circular nozzle impactors the flow pattern gets closer to the impaction plate, and Stk_{50} is approximately 0.25. Sometimes the square root of the Stokes number is used, as it is proportional to particle diameter. We therefore expect $\sqrt{Stk_{50}} = 0.7$ for rectangular slit nozzle and $\sqrt{Stk_{50}} = 0.5$ for a circular nozzle. Plotting the fraction of particles collected as a function of particle diameter or $\sqrt{Stk}$ produces an impactor curve, as shown in Fig. 5.5.

The flow Reynolds number somewhat affects the flow pattern and thus also the critical Stokes number. If the impactor geometry is kept similar and the flow is similar enough (Reynolds number ≈ constant), the $\sqrt{Stk_{50}}$-value is constant for different cut diameters. This is demonstrated in Fig. 5.5. Therefore, Eqs. (5.1) and (5.7) are used together in designing impactors for varying flow rates and different d_{p50} values. To reach lower particle diameters, smaller nozzle diameters and low pressure can be used. To maintain a high flow rate with small nozzle dimension, either multiple circular nozzles or long slits can be used. Inertial impactors are widely applied to classify and collect particles from a few nanometers to tens of micrometers. Impactors are often used as preseparation devices to prevent particles larger than desired entering the instrument. Marple et al. [33] provide a good review and basic guide to impactor characteristics, design, and use.

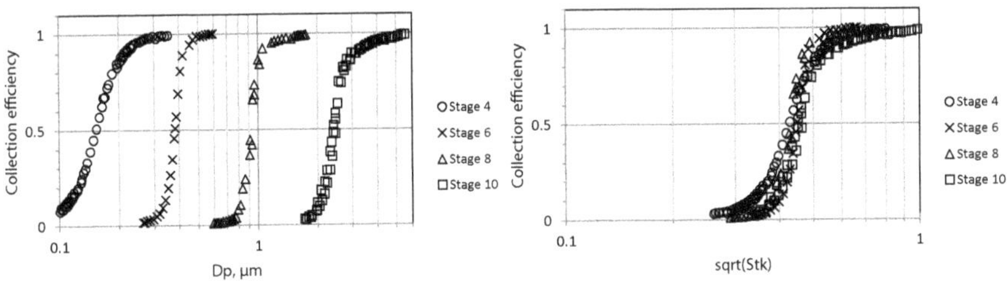

Fig. 5.5. Collection efficiency curves for four impactors as a function of particle diameter (*left*) and square root of Stokes number (*right*). (Data from Marjamäki et al. [31])

Sampling of Particles

For point detectors, the measurement begins at sampling the aerosol. This may take place in outdoors or inside a building, vehicle, or vessel. In the latter cases, sampling can be done either from the essentially still air or from an air flow in a ventilation duct. In sampling, the aerosol is extracted from its environment with a sample inlet or probe. Generally, bioaerosol sampling utilizes same techniques and principles that are used for nonbiological particles. In aerosol sampling, it is usually advantageous to sample the aerosol with as little bias as possible. However, in bioaerosol detection it may be beneficial to have some kind of precut in the sampling system to remove the largest particles from the sample for example to prevent clogging and contamination of the sampling and measurement system. In addition to precut, particle concentrators are often used and integrated in the sampling system for point detectors to enrich the concentration of particles within the size range of interest.

The efficiency of sample extraction depends on the inlet design and particle size. Particles must be influenced by the sample gas flow to be drawn into the inlet without being deposited into the inlet. This sampling efficiency, P_s, is the fraction of the particles that are drawn and transported through the inlet. The sampling efficiency depends on the particle size and it is defined as:

$$P_s = P_{as} P_{tr}$$

where P_{as} = aspiration efficiency and P_{tr} = transmission efficiency of the inlet. Aspiration efficiency is the concentration of particles entering the inlet divided by the concentration of particles in the environment where the sample was drawn from. Transmission efficiency is the fraction of the aspirated particles that are transmitted through the inlet to the rest of the sampling system.

Sampling from a Flowing Gas

There are two common ways to draw sample from the flowing gas: thin-walled sample probe and blunt sampler. Blunt samplers include a range of sampler inlets with varying shapes and dimensions where the inlet is small compared to the overall sampler dimensions. For example thick-walled sample probe can be thought as a blunt sampler. Blunt samplers have several disadvantages: particle deposition on the face of the sampler, particle bounce and difficulty of having a representative sample for larger particles [3]. Therefore we concentrate on thin-walled sample probes. The book by Vincent [45] is recommended for the reader interested in blunt samplers.

Sampling efficiency for a thin-walled sample probe is a product of the aspiration efficiency and the transport efficiency. Best efficiency is gained with isoaxial and isokinetic sampling. Aspiration efficiency for isoaxial sampling can be calculated with the following equation [1]:

$$P_{as} = 1 + \left(\frac{U_0}{U} - 1\right)\left(1 - \frac{1}{1 + k\,Stk}\right) \tag{5.8}$$

where

$$k = 2 + 0.617\left(\frac{U_0}{U}\right)^{-1}$$

In the equation, Stokes number is calculated using the diameter of the sampling probe and the ambient gas flow velocity, U_0. U is the sampling velocity. This equation can be used when $0.05 \le Stk \le 2.03$ and $0.05 \le U_0/U \le 5.6$. For isokinetic sampling ($U = U_0$), the efficiency reduces to unity. For sub-isokinetic sampling ($U < U_0$) the aspiration efficiency is larger than unity and for super-isokinetic ($U > U_0$) it is less than unity.

Transmission losses of the probe are mainly caused by particle inertia. For isoaxial probes, gravitational losses in the probe can be considered as a deposition from a tube flow and can be estimated with the equations shown in chapter "Transport of particles". However, inertial losses are important for the sampling efficiency. Liu et al. [22] state that for sub-isokinetic sampling, some of the particles having velocity vectors directed toward the walls deposit on the probe walls. They give the transmission efficiency as

$$P_{tr,i} = \frac{1 + \left(\frac{U_0}{U} - 1\right) \Big/ \left(1 + \frac{2.66}{Stk^{2/3}}\right)}{1 + \left(\frac{U_0}{U} - 1\right) \Big/ \left(1 + \frac{0.418}{Stk}\right)} \tag{5.9}$$

Equation is valid for $0.01 \le Stk \le 100$ and $1 \le U_0/U \le 10$. For super-isokinetic sampling, Hangal and Willeke [15] say that turbulent deposition occurs in a vena contracta that is formed in the probe inlet. They give the inertial transmission efficiency for super-isokinetic sampling as

$$P_{tr,i} = \exp\left(-75 I_v^2\right) \tag{5.10}$$

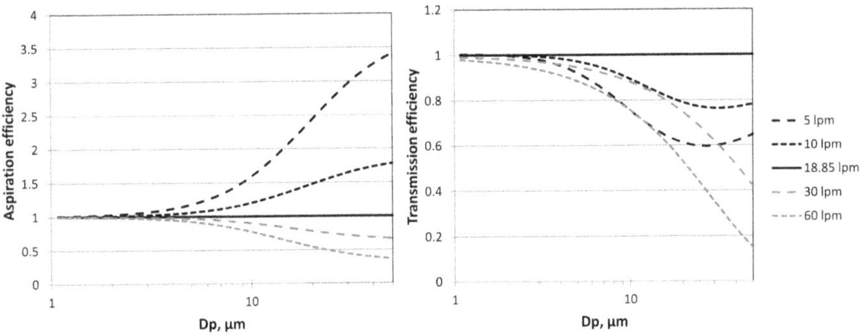

Fig. 5.6 Aspiration efficiency (*left*) and transmission efficiency (*right*) as a function of particle size. Calculated assuming sampling with a thin-walled probe with an inner diameter of 10 mm from the 4 m/s flow. Flow rate of 18.85 lpm corresponds to isokinetic conditions ($U = U_0 = 4$ m/s)

where

$$I_v = 0.09\left(Stk\frac{U - U_0}{U_0}\right)^{0.3}$$

for $0.02 \leq Stk \leq 4$ and $0.25 \leq U_0/U \leq 1$.

Figure 5.6 illustrates the aspiration and transmission efficiencies of the thin-walled probe. Isokinetic conditions are essential especially for particle sizes over 10 µm.

In the case of nonisoaxial sampling, additional deposition caused by impaction into the inside probe wall can occur. Therefore it is not recommended. The case of non-isoaxial sampling has been studied by Durham and Lundgren [7] and Hangal and Willeke [17] and they present equations that can be used to calculate the aspiration and transport efficiencies for such geometries.

Sampling in Calm Air

When sampling the aerosol from the ambient air, entry of particles into the sampling inlet or probe can be affected by gravitational settling, inertia of particles, geometry and orientation of the inlet and the strength and direction of ambient wind [6]. Sampling from calm air efficiently requires the sampling gas velocity to be low enough to allow particles to adapt to the sampling gas flow. At the same time, sampling gas velocity needs to be high enough not to allow gravitational settling to deposit particles significantly. Davies [6] developed criterion for representative sampling through a tube in arbitrary orientation. Brockmann [3] formulates the Davies criterion as two conditions that need to be true:

$$Stk_i \leq 0.016$$

$$\frac{V_{ts}}{U} \leq 0.04$$

where the Stokes number, Stk_l, is calculated with the average sampling velocity, U, and the inlet diameter, d. The presented criterion is stringent and applies to any nozzle orientation. For vertical sampling Brockmann [3] gives conditions that give greater than 95 % sampling efficiency:

$$\frac{U^2}{gd} \leq 1$$

or

$$\frac{4Q}{\pi d^{2.5}\sqrt{g}} \leq 1$$

where Q is the volumetric flow rate of the sample, d is the inlet diameter and g is the acceleration due to gravity.

For ambient bioaerosol sampling care should be taken to protect the measurement setup from the weather effects like rain or wind speed and other possible disturbing factors like flying insects. For this purpose similar techniques and inlets that are used for ambient PM collection could be used also for bioaerosol point detection. There are numerous different designs and manufacturers for these inlets. Especially the PM10 inlets designs such as described by Liu and Pui [28] or European standard EN-12341 [9] are useful also for bioaerosol sampling.

Transport of Particles

After the aerosol is sampled from its environment, it needs to be transported to the measurement instruments. The transport of the aerosol is usually performed using sample lines. During the transport, particles will get deposited inside the sample lines. Transport efficiency for the system in question is a function of the particle size and is defined as the fraction of the particles that are not lost during the transit through the system. The transport efficiency can be defined separately for all of the different deposition mechanisms and different flow elements of the sampling system. Then the total transport efficiency can be calculated as a product of the all of these transport efficiencies.

There are several mechanisms that can cause deposition of particles into the sample lines: gravitation, diffusion, inertia of the particles, electrical effects, thermophoresis and diffusiophoresis [3]. Diffusion is an important effect for fine particles less than 0.5 μm in diameter. Electrical forces can cause deposition if the particles are charged. In the case of bioaerosols particles could get charged during the dissemination process. The best way to reduce the electrical deposition is to have electrically conducting and grounded sampling lines. Thermophoresis is driven by a temperature difference and therefore it is not a common mechanism for normal ambient sampling. Diffusiophoresis requires nonuniform gas mixture where diffusing

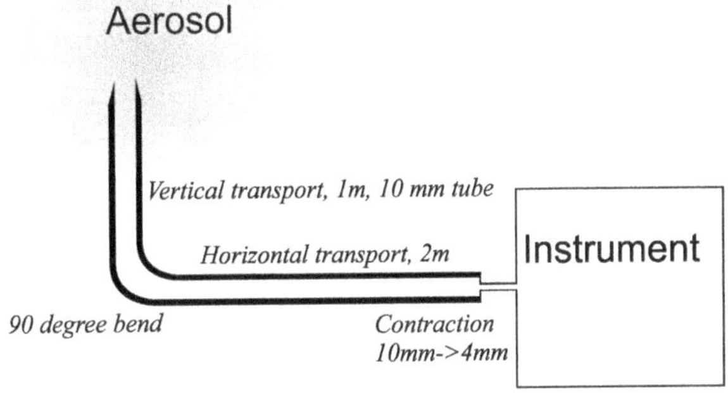

Fig. 5.7 Studied sampling and transport system

gas molecules cause the transport of the particles to the direction of the diffusion of the heavier molecules. In bioaerosol sampling such phenomena should be very rare. The most important mechanisms for coarser particles, like bioaerosols, are gravitational settling and inertial deposition. Gravitational settling occurs in horizontal sections of the sample line. Inertial deposition occurs in the sample line bends and contractions. Both of these mechanisms are heavily influenced by the configuration of the sample line, that is, the amount of bends, contractions and horizontal sections. Flow variables, like laminar or turbulent flow, flow speed and residence time, also affect the particle deposition. Generally short residence times are preferred, not only to reduce particle deposition but also to hold back coagulation and possible chemical reactions and evaporation/condensation.As an example we will study and calculate the sampling and transport efficiency for the system presented in Fig. 5.7. Different transport efficiencies for the system will be calculated in the following sub-chapters and in the end the total transport efficiency for the system is shown.

Gravitational Settling

Gravitational settling causes particle deposition in the nonvertical sampling lines. Fuchs [11] and Thomas [44] independently solved the deposition by gravitational settling in a horizontal tube with circular cross section assuming laminar parabolic flow profile. The transport efficiency is given by:

$$P_{t,g} = 1 - \frac{2}{\pi}\left[2\varepsilon\sqrt{1-\varepsilon^{2/3}} - e^{1/3}\sqrt{1-\varepsilon^{2/3}} + \arcsin\left(\varepsilon^{1/3}\right)\right] \tag{5.11}$$

$$\varepsilon = \frac{3}{4}Z = \frac{3}{4}\frac{LV_{ts}}{dU}$$

where Z = gravitational deposition parameter, L = length of the tube, V_{ts} = particle terminal settling velocity, d = diameter of the tube and U = flow velocity.

Heyder and Gebhart [16] used experimental data to modify the results of Fuchs and Thomas to take into account inclination in the sample line. Their equation can be used to calculate the transport efficiency for inclined (angle θ) circular tube:

$$P_{t,g} = 1 - \frac{2}{\pi}\left[2k\sqrt{1-k^{2/3}} - k^{1/3}\sqrt{1-k^{2/3}} + \arcsin\left(k^{1/3}\right)\right] \tag{5.12}$$

$$k = \varepsilon \cos\theta = \frac{3}{4}\frac{LV_{ts}}{dU}\cos\theta$$

The equation is valid for:

$$\frac{V_{ts}\sin\theta}{U} \ll 1 \tag{5.13}$$

For turbulent flow conditions, the correlations originally presented by Schwendiman and co-workers need to be used [3]. For horizontal tube transport efficiency is:

$$P_{t,g} = \exp\left(-\frac{4Z}{\pi}\right) = \exp\left(-\frac{dLV_{ts}}{Q}\right) \tag{5.14}$$

where Q is the volumetric flow rate. For the inclined tube equation is:

$$P_{t,g} = \exp\left(-\frac{4Z\cos\theta}{\pi}\right) = \exp\left(-\frac{dLV_{ts}\cos\theta}{Q}\right) \tag{5.15}$$

For the use of Eq. (5.15) the condition of Eq. (5.13) must also be fulfilled. Equation (5.13) states that the axial settling velocity is small compared to the flow velocity.

Figure 5.8 shows the gravitational transmission efficiency in a tube with a 10 mm inner diameter with different flow rates and tube lengths. Particles less than 1 μm are transmitted regardless of the flow rate and length. For particles larger than 10 μm, high flow rates or transport lines with very short horizontal sections are needed. However, high flow rates cause turbulence and inertial losses as will be seen later.

Turbulent Inertial Deposition

Coarse particles are inertially deposited due to the eddies in turbulent flow. Coarse particles have high inertia and they cannot follow the flow streamlines and are therefore deposited to the tube walls. Transport efficiency in a tube with turbulent deposition can be calculated as [3]:

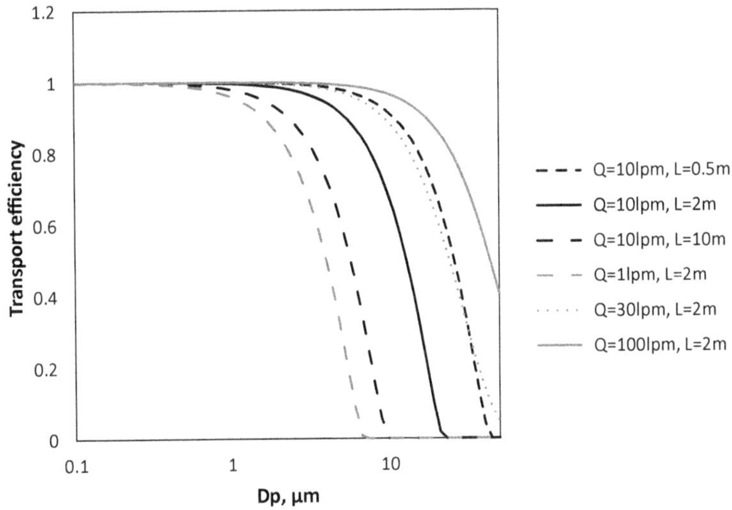

Fig. 5.8 Gravitational settling in a tube with a 10 mm inner diameter. Transport efficiency shown as a function of particle size. For flow rates of 1 lpm and 10 lpm flow is laminar, for flow rates of 30 lpm and 100 lpm flow is turbulent

$$P_{t,t} = \exp\left(-\frac{\pi d L V_t}{Q}\right) \tag{5.16}$$

where V_t is the turbulent deposition velocity. The turbulent deposition velocity can be calculated as [25]:

$$V_t = \frac{V_+ U}{5.03\mathrm{Re}^{1/8}}$$

where Re = flow Reynolds number and V_+ is the dimensionless deposition velocity:

$$V_+ = 6 \cdot 10^{-4} \tau_+^2 + 2 \cdot 10^{-8}\,\mathrm{Re}$$

where τ_+ is the dimensionless particle relaxation time defined as:

$$\tau_+ = 0.0395\, Stk \mathrm{Re}^{3/4}.$$

where Stk = particle Stokes number calculated with the tube diameter and average flow velocity. Equation (5.9) includes the combined effect of turbulent inertial deposition and diffusion. Equation can be used when τ_+ is smaller than 12.9. When τ_+ is larger 12.9, V_+ can be regarded as constant value of 0.1.

The next figure shows the turbulent deposition efficiency for a 3 m long tube with a 10 mm inner diameter with different flow rates. At high flow rates the deposition is severe for particles larger than 2 µm (Fig. 5.9).

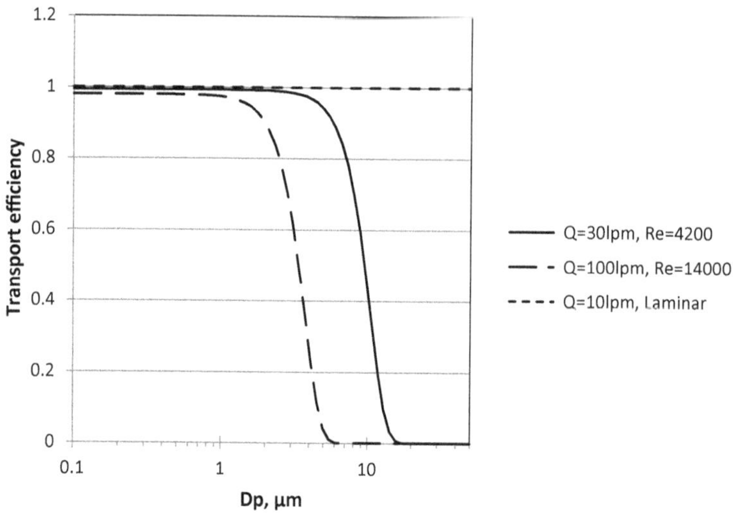

Fig. 5.9 Turbulent inertial deposition in a tube with a 10 mm inner diameter and a length of 3 m. Transport efficiency shown as a function of particle size. At 10 lpm the flow is laminar

Inertial Deposition in a Bend

In a bend the gas flow is diverted and, due to their inertia, particles may deviate from the flow streamline and deposit on the wall. Pui et al. [39] studied the deposition in a 90° bend for a Reynolds number of 1000 and curvature ratios of 5.6 and 5.7. The curvature ratio is defined as the curvature radius of the bend divided by the tube radius. Tube diameters ranged from 4 to 8.5 mm. A fit to the data gives the following equation [3]:

$$P_{b,i} = \left[1 + \left(\frac{Stk}{0.171}\right)^{0.452\frac{Stk}{0.171}+2.242}\right]^{-\frac{2}{\pi}\phi} \tag{5.17}$$

where ϕ is the angle of curvature in degrees. Equation can be used as a guide to predict the transmission efficiency in a bend for laminar flow.

For turbulent flow, Pui et al. [39] found that the deposition does not depend on the Reynolds number of the flow. Brockmann [3] gives the following equation to estimate the transmission efficiency in the case of turbulent flow:

$$\eta_{b,i} = \exp\left(-2.823\,Stk\,\phi\right) \tag{5.18}$$

To reduce losses in bends, the curvature ratio should be kept as high as possible, preferably over 4 [3]. Particle Stokes numbers should also be kept small. Transmission

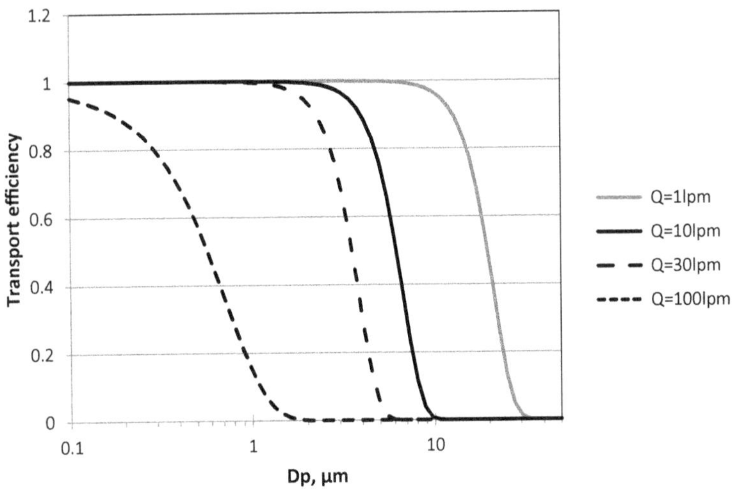

Fig. 5.10 Transport efficiency of a 90° bend shown as a function of particle size. Flow rates of 1 lpm and 10 lpm are laminar, other flow rates are turbulent

efficiency, calculated with the presented equations, over a 90° bend is shown in Fig. 5.10. The bend is in a tube with 10 mm inner diameter. Even as low as 10 lpm flow rate cuts off particles over 10 μm.

Inertial Deposition in Contractions

Contractions in the sampling tube cause flow streamlines to change direction. If particle inertia is large enough, the particle cannot follow the streamline and can get deposited onto the tube wall just before the contraction. Contractions occur when tube diameter is changed but they can also occur in valves, orifices, tee connectors, etc. To calculate the transport efficiency, the formulation of Muyshondt et al. [34] can be used:

$$P_{c,i} = 1 - \frac{1}{\left[1 + \left\{\frac{Stk\left[1 - \left(\frac{A_0}{A_i}\right)\right]}{3.14\exp(-0.0185\theta)}\right\}\right]^{-1.24}} \tag{5.19}$$

where θ is the contraction half-angle in degrees (90° corresponds to blunt contraction), A_0 = cross sectional area after the contraction, A_i = cross sectional area before the contraction.

Transmission efficiency for the case of blunt contraction from 10 to 4 mm tube is shown in Fig. 5.11 for different flow rates.

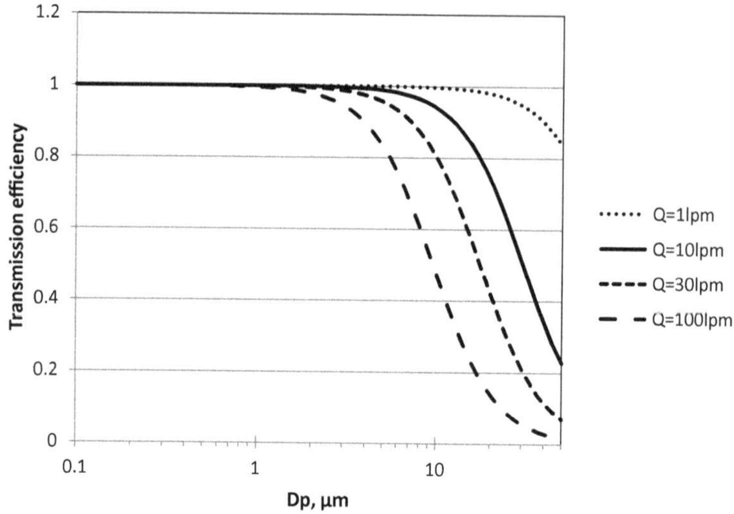

Fig. 5.11 Transport efficiency of a blunt contraction from 10 to 4 mm tube shown as a function of particle size

General Guidelines

Figure 5.12 shows the total transport efficiency for the example case. The total transport efficiency is the product of all of the relevant transport efficiencies. The lowest losses are for the volumetric flow rate of 10 lpm. Even for this flow rate, the transport efficiency drops rapidly as particle diameter increases to 5 µm. For 10 µm it is practically zero. For the 1 lpm case the losses are dominated by gravitational settling. For higher flow rates the losses are dominated by the inertial deposition in the bend. For the 100 lpm case the inertial losses caused by the turbulent flow are also significant.

Sampling lines should be as short as possible, especially the amount of horizontal sections should be kept at minimum. To minimize the electrostatic effects sampling lines should be manufactured from a conducting material. Metal tubing is preferred, but for example Tygon tubing is an acceptable substitute. Teflon is especially poor because of its electrical properties and it should be avoided for aerosol transport tubes. Tube contractions and also enlargements should be avoided. Connectors should be chosen so that their inner diameter matches as well as possible the inner diameter of the tubing. If valves need to be used, ball valves are preferred since in general they offer less disturbance to the flow.

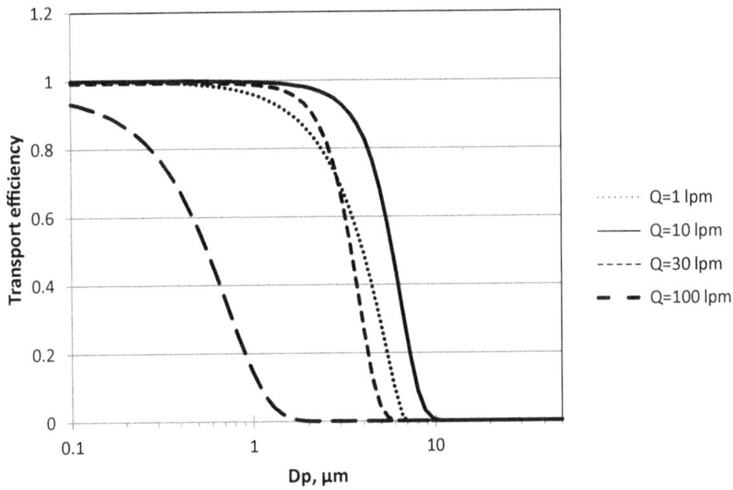

Fig. 5.12 Overall transport efficiency of the studied case

Coarse Particle Concentrator

The number of biological particles entering the actual bioaerosol detection volume is proportional to the volumetric aerosol flow rate entering the detection volume. However, it is not always possible to increase the flow rate to increase the instrument sensitivity. For instance, to keep the detection volume small for high illumination intensity, the flow rate of fluorescent detection units is typically limited to 1 L/min [20, 21, 38].

Bioaerosol particles are relatively large, that is, larger than 0.5 to 1 μm in diameter. Both in remote and urban surroundings, the number size distribution is usually dominated by smaller particles, typically less than 0.1 μm in diameter (see, e.g. [42]). The detection system may respond also to these smaller particles. As an example, fine combustion aerosol particles such as diesel soot, may be fluorescent causing unwanted fluorescent background signal.

To increase the overall sensitivity of the detector to biological particles and to minimize the effect of the fine background particles, it is customary to concentrate the sample for coarse particles. The most typical concentrator is a virtual impactor. The virtual impactor relies on the same principle of inertial classification as the conventional impactor described earlier. The operation of a virtual impactor is shown schematically in Fig. 5.13. The sample aerosol flow (Q_0) enters the virtual impactor through a high speed nozzle. The aerosol jet exiting the nozzle is directed at a collection probe. A major part of the flow (major flow, Q_{major}) makes a sharp turn to a side passage. Because of their inertia, large particles cross the streamlines and enter the collection probe. This leads to an increased concentration of large particles in the minor part of the flow (Q_1) passing through the collection probe.

The virtual impactor in its present form was first described by Conner in 1966 [5]. Early theoretical and experimental studies on the design and performance of the virtual impactor have been published Marple and Chien [32], Forney et al. [10], [4], and by Loo and Cork [30]. The last one is still a very good practical guide to virtual impactor design for coarse particles. The virtual impactor has been widely used as a dichotomous sampler to get size classified filter samples of coarse and fine particles [8]. The fact that the concentration of coarse particles is increased in the minor flow makes it possible to use the device as a concentrator for real-time instruments. Probably the first ones to use the virtual impactor as a concentrator for a real-time particle measurement instrument were Keskinen et al. [23]. They connected the virtual impactor to the input of an optical particle counter to increase the instrument sensitivity for coarse particles. Since then, the same principle has been applied in a number of studies (e.g. [27, 36, 46]. See also Ho [19] for a recent review on the use of virtual impactors in biological aerosol detection.

The increase in particle concentration in the minor flow can be characterized by a concentration factor $K(d_p)$

$$K(d_p) = \frac{C_1}{C_0}$$

where C_1 is the concentration in the minor flow and C_0 is the concentration in the sample inlet. The concentration factor depends on the aerodynamic particle diameter in the same fashion as the collection efficiency curve of a normal impactor. Figure 5.13 shows a typical concentration factor curve for a virtual impactor. The minimum of a $K(d_p)$-curve is one, as the concentration of small particles is unchanged. The maximum theoretical value for $K(d_p)$ corresponds to the case when all the coarse particles in the inlet enter the minor flow:

$$K_{\max} = \frac{Q_0}{Q_1} = \left(\frac{Q_1}{Q_0}\right)^{-1}$$

where Q_0 and Q_1 are the sample volumetric flow rate and the minor flow rate, respectively. The diameter where $K(d_p)$ reaches 50% of $K_{\max}$ is often called the cut-point diameter of the virtual impactor. Equation (5.7) can be used in dimensioning the virtual impactors as well. However, the Stk_{50} value is highly dependent on the flow ratio (e.g. [32]). The decrease in $K(d_p)$ for particle diameters over 5 µm in Fig. 5.13 is a real effect caused by particle losses onto the impactor walls.

In principle, the concentration factor $K_{\max}$ can be increased infinitely by decreasing the flow ratio Q_1/Q_0. In practice, the particles losses increase as the flow ratio decreases (e.g. [30]). Therefore, the minimum flow ratio used is usually 0.05, corresponding to a maximum concentration factor of 20. It is, however, possible to use several virtual impactor stages in cascade (e.g. [35]). The concentration factor of such a multistage system is the product of the factors of the individual stages. This

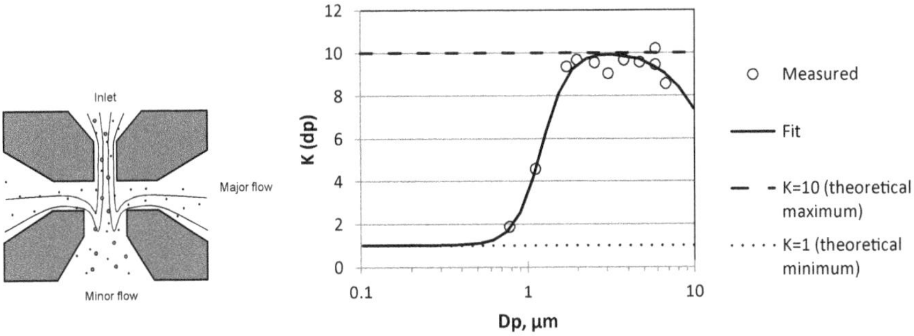

Fig. 5.13 Cross section of a virtual impactor (*left*). Concentration factor (*right*, see text). (Data from Rostedt et al. [41])

way, it is possible to achieve nominal overall concentration factor in the order of 1000 (e.g. [13]).

To achieve good sensitivity, one often wishes to sample as high a flow rate as possible. To scale a virtual impactor for higher flow rates, one should keep the Stokes number constant and allow only moderate change for Reynolds number in the nozzle. An analysis of Eqs. (5.1) and (5.7) shows that keeping the Stokes number constant while increasing the flow rate leads to increased jet velocity. The increase in jet velocity leads to higher pressure drop over the nozzle, which leads to higher pumping power consumption. It is possible to use several circular nozzles [40], but this leads to a rather cumbersome construction. The obvious choice is to use rectangular slit nozzle (see, e.g. [43]). For highest flow rates, either the slit length can be maximized [14] or one can use a multi-slit arrangement [2].

Virtual impactors designed for cutpoints in the 2.5 µm range have proven to be rather straightforward to design and reliable to operate. To achieve lower cutpoint, one should aim for the same Stokes number with a smaller particle diameter (d_p) in Eq. (5.7). It is obvious that the nozzle diameter/width W should be decreased. It is possible to reach submicrometer [43] and even nanometer range cutpoints [26]. However, the alignment of the acceleration and sampling nozzles becomes critical. Alignment may be maintained by using micromachining techniques [37]. However, for small dimensions also the contamination and plugging of the nozzles becomes an issue (e.g. [40]).

It seems that the virtual impactors used in commercial concentrator units have been conservatively designed. There seem to be no units claiming sub-micrometer cutpoint, and the individual concentration factors are usually less than 10. The available tests still report rather high losses: the maximum concentration factor was 80% of the nominal for the best one, while staying less than 50% for the rest [13, 22]. It is interesting to note that the highest measured concentration factor was achieved with a multi-stage construction with low individual concentration factors (less than 6).

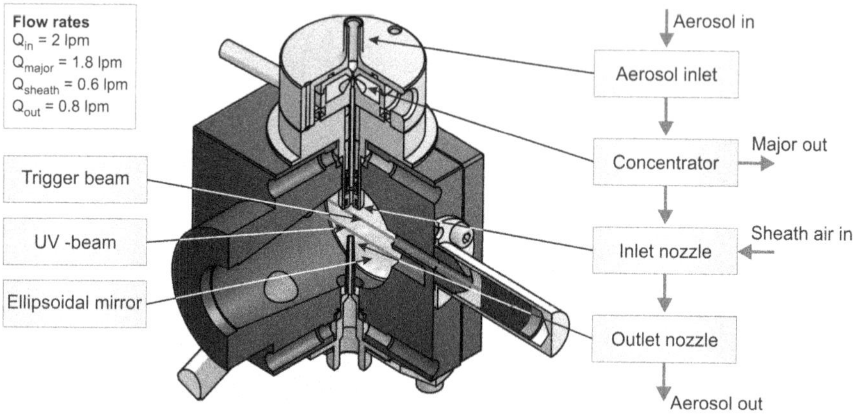

Fig. 5.14 Example of aerosol transport system integrated into fluorescence detection optics [41]

Transfer of Particles into Detection Volume

The final stage of particle transport is the transfer of the particles into the volume where they are detected or characterized. This may involve collecting the particles onto a surface or transferring them into a liquid. For UV-LIF detection of bio-aerosol particles, the particles need to be illuminated with the UV-light and the fluorescence light need to be collected. This should happen in a well defined volume called the detection volume. As it is unwise to try to detect the fluorescence through windows, a free jet of the aerosols is formed. This jet is directed into an optical chamber. The overall detection volume is the intersection of the aerosol jet, the excitation light beam, and the detection optics detection volume. To obtain the highest sensitivity and to detect all the particles, the jet diameter (or width) should equal the width of the uniformly illuminated part of the light beam. Another goal is to minimize the contamination of the detection system, and hence to minimize particle deposition onto the surfaces. This is usually accomplished by surrounding the aerosol jet with a layer of clean air, called sheath air. The details of the flow system for different detection schemes usually need to be checked with CFD and tested experimentally. Figure 5.14 shows an optical chamber designed for fluorescence detection. Note that a concentrator stage is integrated into the system [41].

References

1. Belyaev SP, Levin LM (1974) Techniques for collecting of representative aerosol samples. J Aerosol Sci 5:325–338
2. Bergman W, Shinn, Lochner R et al (2005) High air flow, low pressure drop, bio-aerosol collector using a multi-slit virtual impactor. J Aerosol Sci 36:619–638

3. Brockmann JE (2001) Sampling and Transport of Aerosols. In: Baron PA, Willeke K (ed) Aerosol Measurement: Principles, Techniques, and Applications, 2nd edn. Wiley, New York
4. Chen BT, Yeh HC (1987) An improved virtual impactor: Design and performance. J Aerosol Sci 18:203–214
5. Conner WD (1966) An inertial-type particle separator for collecting large samples. JAPCA J Air Waste Ma 16:35–38
6. Davies CN (1968) The entry of aerosols into sampling tubes and heads. Br J Appl Phys (J Phys D: Appl Phys) 1:921–932
7. Durham MD, Lundgren DA (1980) Evaluation of aerosol aspiration efficiency as a function of stokes number, velocity ratio and nozzle angle. J Aerosol Sci 11:179–188
8. Dzubay TG, Stevens RK (1975) Ambient air analysis with dichotomous sampler and X-ray fluorescence spectrometer. Environ Sci Technol 9:663–668
9. European Committee for Standardization (1998) EN 12341:1998 Air quality—Determination of the PM 10 fraction of suspended particulate matter—Reference method and field test procedure to demonstrate reference equivalence of measurement methods. Brussels
10. Forney LJ, Ravenhall DG, Lee SS (1982) Experimental and theoretical study of a two-dimensional virtual impactor. Environ Sci Technol 16:492–497
11. Fuchs NA (1964) The Mechanics of Aerosols. Pergamon Press, Oxford
12. Granger RA (1995) Fluid Mechanics. Dover Publications, New York
13. Haglund JS, Chandra S, McFarland AR (2002) Evaluation of a high volume aerosol concentrator. Aerosol Sci Technol 36:690–696
14. Haglund JS, McFarland AR (2004) A circumferential slot virtual impactor. Aerosol Sci Technol 38:664–674
15. Hangal S, Willeke K. (1990) Aspiration efficiency: Unified model for all forward sampling angles. Environ Sci Technol 24:688–691.
16. Heyder J, Gebhart J (1977) Gravitational deposition of particles from laminar aerosol flow through inclined circular tubes. J Aerosol Sci 8:289–295
17. Hangal S, Willeke K (1990) Overall efficiency of tubular inlets sampling at 0–90 degrees from horizontal aerosol flows. Atmos Environ A-Gen 24A:2379–2386.
18. Hinds WC (1999) Aerosol Technology: Properties, Behavior, and Measurement of Airborne Particles, 2nd edn. Wiley, New York
19. Ho J (2012) Use of Virtual Impactor (VI) Technology in Biological Aerosol Detection. Kona Powder Part J 29:16–26
20. Ho J, Stanley NJ, Kuehn TH (2011) Feasibility of using real-time optical methods for detecting the presence of viable bacteria aerosols at low concentrations in clean room environments. Aerobiologia 27:163–172
21. Kaye PH, Stanley WR, Hirst E et al (2005) Single particle multichannel bio-aerosol fluorescence sensor. Opt Express 13:3583–3593
22. Kesavan J, Bottiger JR, McFarland AR (2008) Bioaerosol concentrator performance: comparative tests with viable and with solid and liquid nonviable particles. J Appl Microbiol 104:285–295
23. Keskinen J, Lehtimäki M, Janka K (1987) Virtual impactor as an accessory to optical particle counters. Aerosol Sci Technol 6:79–83
24. John W (1999) A simple derivation of the cutpoint of an impactor. J. Aerosol Sci. 30:1317–1320
25. Lee KW, Gieseke JA (1994) Deposition of particles in turbulent pipe flows. J Aerosol Sci 25:699–709
26. Lee P, Chen D-R, Pui DYH (2003) Experimental study of a nanoparticle virtual impactor. J Nanopart Res 5:269–280
27. Liebhaber FB, Lehtimäki M, Willeke K (1991) Low-cost virtual impactor for large-particle amplification in optical particle counters. Aerosol Sci Technol. 15:208–213
28. Liu BYH, Pui DYH (1981) Aerosol sampling inlets and inhalable particles. Atmos Environ 15:589–600

29. Liu BYH, Zhang ZQ, Kuehn TH (1989) A numerical study of inertial errors in anisokinetic sampling. J Aerosol Sci 20:367–380
30. Loo BW, Cork CP (1988) Development of High Efficiency Virtual Impactors. Aerosol Sci Technol 9:167–176
31. Marjamäki M, Keskinen J, Chen D-R et al (2000) Performance evaluation of electrical low pressure impactor (ELPI). J Aerosol Sci 31:249–261
32. Marple VA and Chien CM (1980) Virtual impactors: a theoretical study. Environ Sci Technol 14:976–984
33. Marple VA, Olson BA, Rubow KL (2001) Inertial, Gravitational, Centrifugal, and Thermal Collection Techniques. In: Baron PA, Willeke K (ed) Aerosol Measurement: Principles, Techniques, and Applications, 2nd edn. Wiley, New York
34. Muyshondt A, McFarland AR, Anand NK (1996) Deposition of aerosol particles in contraction fittings. Aerosol Sci Technol 24:205–216
35. Novick VJ, Alvarez JL (1987) Design of a multistage virtual impactor. Aerosol Sci Technol 6:63–70
36. Pan Y-L, Hartings J, Pinnick RG et al (2003) Single-particle fluorescence spectrometer for ambient aerosols. Aerosol Sci Technol 37:628–639
37. Park D, Kim Y-H, Woo Park C et al (2009) New bio-aerosol collector using a micromachined virtual impactor. J Aerosol Sci 40:415–422
38. Pinnick RG, Hill SC, Nachman P, Pendleton JD, Fernandez GL, Mayo MW, Bruno JG (1995). Fluorescence particle counter for detecting airborne bacteria and other biological particles. Aerosol Sci Technol 23:653–664
39. Pui DYH, Romay-Novas F, Liu BYH (1987) Experimental study of particle deposition in bends of circular cross section. Aerosol Sci Technol 7:301–315
40. Romay FJ, Roberts, Marple VA et al (2002) A high-performance aerosol concentrator for biological agent detection. Aerosol Sci Technol 36:217–226
41. Rostedt A, Putkiranta M, Marjamaìki M et al (2006) Optical chamber design for aerosol particle fluorescent measurement. Proceedings of SPIE—The International Society for Optical Engineering 6398, art. no. 63980G
42. Seinfeld JH, Pandis SN (1998) Atmospheric Chemistry and Physics: From Air Pollution to Climate Change. Wiley, New York
43. Sioutas C, Koutrakis P, Burton RM (1994) Development of a low cutpoint slit virtual impactor for sampling ambient fine particles. J Aerosol Sci 25:1321–1330
44. Thomas JW (1958) Gravity settling of particles in a horizontal tube. J Air Pollut Control Assoc 8:32–34
45. Vincent JH (2007) Aerosol Sampling—Science, Standards, Instrumentation and Applications. Wiley, Chichester
46. Wu JJ, Cooper DW, Miller RJ (1989) Virtual impactor aerosol concentrator for cleanroom monitoring. J Environ Sci 32:52–56

Part II
Principles and Technologies for Bioaerosol Detection

Chapter 6
Light Scattering and Particle Charge Techniques for the Detection of Biological Warfare Agents

James M. Clark

Background

To be fully effective a biological point detection system must be capable of reliably detecting the presence of a potentially dangerous aerosol immediately it reaches the detection site. This first detection can then be succeeded by sequential analysis techniques that confirm that it is biological, deliberately generated and finally identify the agent being used.

In many ways the first step of immediate detection can be the most difficult to achieve. The particulate components of the atmosphere vary enormously in both space and time. The components that make up this atmospheric aerosol include mineral dusts, marine salts, a very wide variety of industrial pollutants, and many biological components: both primary and secondary. The relative proportions of the components are generally totally unpredictable and will often show very large variations at a single site in what might appear to be very similar meteorological conditions. It is against this varying and unpredictable background that a potentially very small pathogenic challenge must be recognized.

Until quite recently the only technique available for analyzing the biological components of the atmospheric aerosol was to collect them by impaction or filtration and analyze them by microscopy or, in the case of bacteria and viruses, growing the collected material in nutrient or cellular media appropriate to the species being assayed. Clearly this class of techniques cannot be used for immediate detection and techniques needed to be developed that enabled analyses to be carried out in the airborne state and so eliminate the inevitably time consuming element of particle collection.

The measurements that is possible to make on microscopic airborne particles are very limited. However, the presence of particles does affect any radiation passing through the medium and this can be employed to measure at least some characteristics of an aerosol. The early measurement techniques simply used the attenuation of light

J. M. Clark (✉)
Pitanwerth, Above Hedges, Pitton, Salisbury SP5 1DS, UK
e-mail: dclark@biral.com

P. Jonsson et al. (eds.), *Bioaerosol Detection Technologies*, Integrated Analytical Systems,
DOI 10.1007/978-1-4419-5582-1_6, © Springer-Verlag New York 2014

passing through an aerosol to provide a relative concentration level. This gives a reasonable estimate of concentration when the particle number density is high but becomes difficult to use when concentrations are lower and the attenuation caused by the particles is of the same order as fluctuations in the light source.

The next stage was to measure the light scattered by the particles outside of the envelope of the interrogating beam. This permitted much more accurate measurements on low concentrations of particles as the scattered light could be viewed against a dark background. In this measurement configuration, while the variability of the light source may still cause some inaccuracy in the absolute value, only the presence of aerosol particles in the measurement space will cause light to be scattered.

Although the measurement of the light scattered by a volume of aerosol gives an indication of the quantity of particulate matter in the measurement space it says little about the number or size of the particles without prior knowledge of one of these parameters. In order to measure the size and number concentration of particles in an aerosol with unknown characteristics it is necessary to make light scattering measurements on the individual particles. The development of high intensity, controllable light sources together with high sensitivity light sensors made this a realizable target. The growing requirement for clean room facilities in manufacturing and the need to measure the particulate characteristics of the air in industrial environments for occupational health monitoring gave rise to the progressive development of single particle counting and sizing instruments based on light scattering.

The theoretical explanation [12] of the scattering of light by isolated particles suspended in a fluid predicted accurately the angular pattern of the scattered light from spheres and cylinders. This assisted in the design of particle sizing instruments by predicting, for instance, the optimum angle at which to place the sensor relative to the light beam to minimize the effect of the material optical constants on the size measurement. However, although many advances in the theoretical treatment of light scattering were made [2], the practical applications of these advances were very limited until techniques were being sought to recognize biological agent material in the aerosol phase.

Of course light scattering is not the only means by which airborne particles can be characterized. Other techniques for particle sizing such as measuring the acceleration of particles to determine their aerodynamic diameter are now very widely used for many applications. While this sizing technique does not have any particular merit in terms of differentiating potential biological agents from other components of the atmospheric aerosol, it was one of the first techniques used to warn of major changes to the atmospheric aerosol in the earliest biological detection systems.

Techniques developed to measure other properties of aerosol particles do have potential. One such measurable property is the charge carried by aerosol particles [4], which is of interest in a number of fields, primarily those concerned with particle dynamics. However, charge can also be used to determine the approximate age of an aerosol which can be employed to recognize a component in a mixed aerosol that has been relatively recently generated.

The Approach to Real Time Detection

The difficulty in recognizing the presence of a biological agent in the highly variable atmospheric aerosol is that it does not have a unique signature that is measurable in the aerosol phase. Deliberately generated pathogenic bacteria and viruses are effectively identical chemically, and biologically very similar, to the many natural bacteria and viruses that are present in the atmosphere. They can be reliably distinguished only by the conformation of proteins on their surfaces and by their DNA/RNA. Currently there are no techniques that enable these subtle characteristics to be measured in airborne suspension. They can of course be used to identify species once the particles have been collected into liquid suspension.

Effective real time detection is however possible by using other indicators. The components of the atmospheric aerosol, and their concentration, vary hugely in both space and time, and it is generally not possible to predict in advance the pattern of such variation. However, it is often possible to establish some recognizable patterns at a particular location, over an extended measurement period, particularly when the aerosol characterization is linked to meteorological measurements. An effective detector must therefore be able to learn the patterns in the location where it is sited and compare changes to these patterns with the likely characteristics of a biological agent release. In order to recognize significant changes to the atmospheric aerosol it is important to measure as many of the characteristics of the aerosol particles as possible.

The "natural" variations in the characteristics of the atmospheric aerosol will invariably be due to a combination of factors. These can include:

- Meteorological variations, both local and upwind, including wind speed, wind direction, temperature and precipitation
- Time of day
- Season
- Local activity
- Upwind activity at various distances
- Periodic pollutant introduction
- Combustion sources

The changes that occur due to these factors will generally have characteristic periodicity that may range from a few seconds to months. These can be divided into those with variations shorter than the monitoring period and those of the same order or longer. The first of these can largely be classified as cyclical variations to existing sources and the second as the introduction of new sources. Elements of the second class may eventually join the first as the monitoring period increases.

To recognize an attack with a biological agent it is necessary to be able to both distinguish it from the cyclical changes and to find characteristics that differentiate it from other new sources, both natural and pollutant. In order to do this it is necessary to measure as many distinguishing characteristics as is practicable.

Options for Particle Characterization

Before this field began to be explored for biological agent detection the only measurements made directly on particles in airborne suspension were the number count and size. Both of these can be measured from the light they scatter when passing through an illuminated measuring space. However, although these are certainly the easiest characteristics to extract, the pattern of light scattered by a particle also depends and the shape of the particle, its homogenicity, and the optical constants of the material or materials of which it is made [2]. Separation and measurement of these variables are possible only to a limited extent, particularly for particles where all the characteristics are unknown. However, a shape characteristic for particles can be extracted from the scattered light pattern independent of the other parameters involved.

Shape has been used as a means of describing particles since the early days of aerosol science. Until recently this was always achieved by the collection of particles on to a surface for microscopic examination. A range of techniques have evolved to quantify the shapes found which most frequently have the objective of converting the observed shape to an equivalent sphere in order, for example, to estimate mass or predict aerodynamic properties. Highly individual and reproducible shapes such as are seen in some pollens and spores, have also been used directly for species recognition.

The early attempts to measure shape simply extended the established technique by acquiring a microscopic image of particles in airborne suspension and applying automated image analysis techniques. This produced some effective, though rather complex, instruments. However, for the purpose of biodetection, the potential value of this technique is very limited. As a great deal of data has to be collected to produce an image only a very limited throughput of single particles can be achieved. So when the target is only a very small proportion of the total aerosol it is likely to take an unacceptably lengthy period to acquire sufficient data for reliable detection. A second problem is that the size of the respirable particles likely to be deployed is sufficiently close to the wavelength of the illuminating light that shape resolution is lost.

Light scattering can also be used to determine the complex refractive index of particles but only under highly constrained conditions. These require the particle to have a regular geometry, generally a sphere, and to be homogeneous. These requirements are very unlikely properties of biological agents as large scale generation techniques generally produce irregular particulate composites.

The generation of aerosols using disruptive techniques invariably results in charge separation, producing particles that are quite highly charged [7]. Where the particles are generated from a liquid suspension the evaporation of the liquid from the droplet to leave a solid residue will significantly increase the charge to size ratio. These charges will gradually decay by interaction with gaseous ions and will eventually reach an equilibrium level. This is a relatively slow process and measuring nonequilibrium charges on particles gives a measure of the time since they were generated.

Of course charge cannot be directly deduced from light scattering and techniques using the dynamics of the particles in an electrical field must be employed. However, the scattering of light by the particle may still be used to determine its position in the field [4].

Light Scattering

The scattering of light by airborne particles is a phenomenon that has been extensively studied since the late nineteenth century. It is of major importance in the atmosphere, giving rise to a number of visual effects including the blue color of the sky and the red color of sunsets.

Scattering is the term given to the dispersion pattern from a beam of light following interaction with an object that is either smaller or of the same size order as the wavelength of the light. Once the object becomes much larger than the wavelength of the illuminating light the dispersion the object causes can be described by the reflection, refraction and diffraction of classical geometric optics.

For particles that are much smaller than the wavelength of the illuminating light (<50 nm for visible wavelengths) the intensity of the scattered light is given by the theory developed by Rayleigh [2]. For these particles, and indeed gas molecules, the incident wave creates an oscillating dipole that re-radiates the energy absorbed in all directions. For unpolarized incident radiation the intensity of the pattern of the radiated light is given by

$$I(\theta) = \frac{I_0 \pi^4 d^6}{8R^2 \lambda^4} \left(\frac{m^2 - 1}{m^2 + 2} \right)^2 (1 + cos^2 \theta), \tag{6.1}$$

where $I(\theta)$ is the total scattered intensity at distance R from the centre of the particle and over the angle θ. I_0 is the intensity of the illuminating light, d the diameter of the particle, λ the wavelength of the light and m the refractive index.

The scattered intensity is a very strong function of the size of the particle but is independent of the particle shape, being effectively a point scatterer relative to the wavelength of the light. The strong inverse relationship of $I(\theta)$ with λ means that shorter wavelengths are scattered much more effectively than longer wavelengths, which is why the sky appears blue.

For particles whose size is of the same order as the wavelength of the illuminating light the intensity pattern of the scattered light field can only be determined by solving Maxwell's electromagnetic field equations to give the disturbance to the propagation of the field caused by the presence of a particle [12]. The equations were solved by Mie for spheres and cylinders to give an intensity value $I(\theta)$ for the scattered field at a distance R from the particle center:

$$I(\theta) = \frac{I_0\, \lambda^2 (i_1 + i_2)}{8\pi^2 R^2}. \tag{6.2}$$

The apparent simplicity of this equation is deceptive as i_1 and i_2, the polarised intensity functions, are complex series that include the geometric parameters of the particle as well as its optical constants. The equation was solved by Mie for spherical and cylindrical geometries but the analytical solution technique is complex involving the use of Bessel and Hankel functions.

The relationship between the incident and scattered fields can be expressed in terms of the Mueller matrix

$$\begin{bmatrix} I_s \\ Q_s \\ U_s \\ V_s \end{bmatrix} = \frac{1}{k^2 r^2} \begin{bmatrix} S_{11} & S_{12} & S_{13} & S_{14} \\ S_{21} & S_{22} & S_{23} & S_{24} \\ S_{31} & S_{32} & S_{33} & S_{34} \\ S_{41} & S_{42} & S_{43} & S_{44} \end{bmatrix} \begin{bmatrix} I_i \\ Q_i \\ U_i \\ V_i \end{bmatrix}, \tag{6.3}$$

where I, Q, U and V are the components of the Stokes vector, describing the incident and scattered fields. The 4×4 scattering matrix is the operator which transforms the incident vector to scattered vector.

For a homogeneous sphere the relationship simplifies to

$$\begin{bmatrix} I_s \\ Q_s \\ U_s \\ V_s \end{bmatrix} = \frac{1}{k^2 r^2} \begin{bmatrix} S_{11} & S_{12} & 0 & 0 \\ S_{12} & S_{11} & 0 & 0 \\ 0 & 0 & S_{33} & S_{34} \\ 0 & 0 & -S_{34} & S_{33} \end{bmatrix} \begin{bmatrix} I_i \\ Q_i \\ U_i \\ V_i \end{bmatrix}. \tag{6.4}$$

The four nonzero elements remaining in the scattering matrix are functions of elements of the amplitude scattering matrix

$$\begin{bmatrix} E_{pa_s} \\ E_{pe_s} \end{bmatrix} = \frac{e^{ik(r-z)}}{-ikr} \begin{bmatrix} S_2 & S_3 \\ S_4 & S_1 \end{bmatrix} \begin{bmatrix} E_{pa_i} \\ E_{pe_i} \end{bmatrix}, \tag{6.5}$$

where k is the wave number and r and z define the position in the scattered field. The Stokes parameters are functions the polarisation components E_{pa} and E_{pe}.

The amplitude scattering elements S_i, (where i=1, 2, 3, 4) are functions of the scattering angles θ and φ. θ is the angle in the plane of the incident light and φ is the azimuthal angle. It is from the variability of the scattered intensity with φ that a shape factor for particles can be inferred from the scattered light pattern.

Particle Shape from Scattered Light

The theoretical treatment of light scattering is restricted to shapes that are describable geometrically. Analytical solutions are restricted to spheres, cylinders and shells of those shapes. For other readily describable shapes such as cubes, cuboids, ellipsoids, an approximate model of the scattered light pattern can be predicted by numerical modeling techniques. However, most aerosol particles, whether natural

Fig. 6.1 Some examples of particles likely to be found in the atmosphere

or artificially generated, have very complex shapes. This may be because they have been formed by rupturing of crystalline or amorphous solids, crystallization from solution or agglomeration. Figure 6.1 shows some typical examples of particles collected from the atmosphere.

The shape of such particles is a complex concept and is not susceptible to description by any single parameter [13]. Shapes that cannot be described in terms of simple geometric parameters can be approximated digitally but this requires a complex data set. Such techniques are impractical when attempting to acquire data rapidly on a very large number of particles. For this a simple parameter is required, analogous to the shape factor that defines an "equivalent sphere" in particle/fluid dynamics.

A parameter that can be extracted from the pattern of light scattered by single particles is the asymmetry [9]. The intensity of the light scattered at any angle around the axis of the incident beam is a function of the size, shape, and optical constants of the particle as well as the wavelength and polarization of the illuminating beam. In most practical applications it is not possible to invert the matrix of these parameters to extract information on the particle shape. However, in addition to the angular variability in scattered intensity in the plane of the optical axis θ, there is also variability in the 2π azimuthal angular range surrounding the optical axis φ [14].

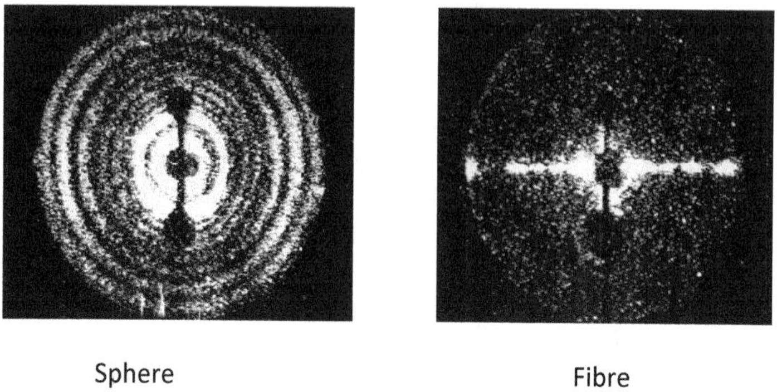

Fig. 6.2 Azimuthal angular pattern of light scattered by a sphere and a fibre

The angular variability of the intensity of the scattered light around the azimuthal angles correlates with the geometric asymmetry of the particle scattering the light. This is illustrated in Fig. 6.2 showing images of the light scattered by a sphere and a fiber [6]. The concentric rings of variable intensity in the light scattered by the sphere are the cyclical variations in intensity seen in the angular pattern in the plane of the optical axis, typical of spherical particles. However the important aspect is that spheres scatter at constant intensity around the azimuthal angles. The fibre, on the other hand, scatters in a maximally asymmetric pattern, the angle of which is a function of the particle's orientation in the measurement space. Other particle geometries produce a range of different scattered light patterns that are typical of that class of shapes (Fig. 6.3).

Measurement Technique

In order to obtain the most complete measure of a particle's shape characteristics it is desirable to measure scattering around the complete azimuthal angular range over as wide an axial angle as possible. The technique developed to do this [10] uses an ellipsoidal mirror to capture the scattered light. If the measurement space is placed at the primary focus of the ellipsoid then the mirror's geometry ensures that all light originating there passes through the secondary focus of the ellipsoid. If a spatial filter in the form of a small aperture is placed at the secondary focus this ensures that only light that originates at the primary focus passes through the filter to the light intensity measurement system beyond. If the particle measurement space is sufficiently tightly constrained and there is only a single particle in it then the only light reaching the intensity measurement plane is that scattered by the particle.

Figure 6.4 illustrates the optical arrangement required for the measurement. To obtain a symmetric, forward scattering profile the illuminating beam must be delivered along the axis of the mirror and focused to coincide with the primary focus of the mirror. In the developed instruments this is achieved by mounting a small plane mirror on an optical flat sealed to the ellipsoidal mirror. The beam from a

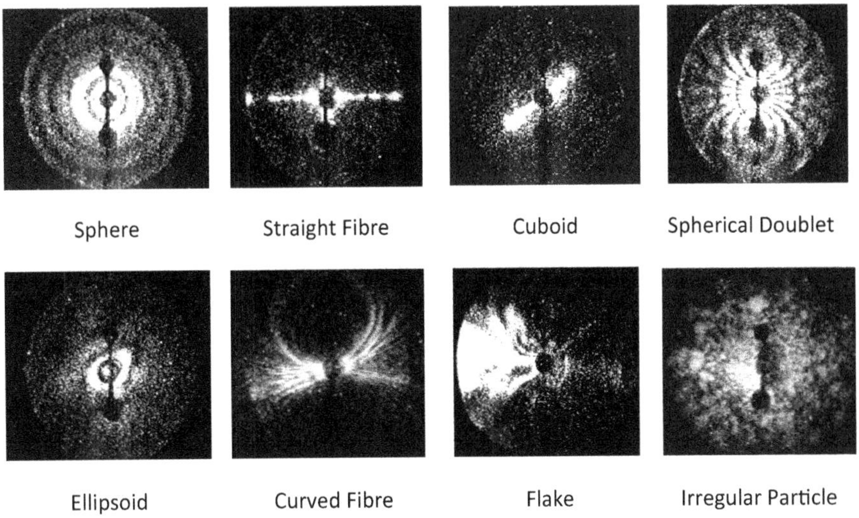

Fig. 6.3 Azimuthal angular pattern of light scattered by a range of particle geometries

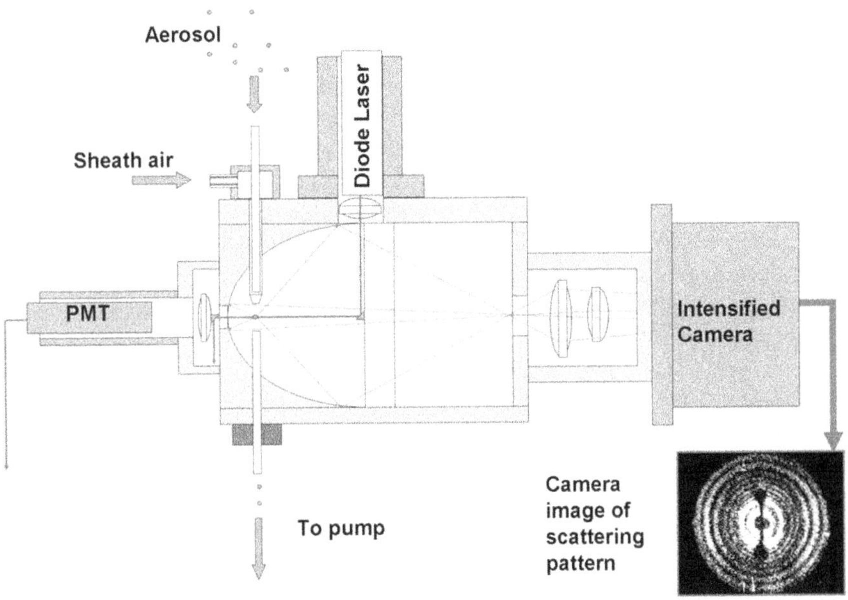

Fig. 6.4 An optical arrangement for the measurement of the scattered light pattern of a particle

diode laser, focused to the appropriate length, is directed to the beam-turning mirror through a small aperture perpendicular to the optical axis. Beyond the coincident focus of the beam and the mirror, the diverging beam passes through an axial hole in the mirror and immediately into a beam dump and power measurement device. The hole in the mirror is slightly larger than is required for the exit of the beam so that the intensity of the narrow, forward-scattered light can be measured separately. This is valuable as the intensity measured to about 25° from the optical axis gives the best measure of particle size, independent of the particle shape and optical constants.

The measurement of shape relies on relative values within the scattered field and so is largely independent of illumination intensity. However, the measurement of particle size relies on absolute intensity values. Total laser power varies with time and so it is necessary to continuously monitor it and apply an appropriate correction to the scattered value measured. In addition to this variation in total intensity there is also a variation across the beam, with the variation generally approximating to a Gaussian function. Ideally particles would be directed exactly through the center of the beam so that they would always be exposed to the maximum incident intensity. However, the degree of control over particle trajectory that can be achieved is limited and so another approach is required. The technique that has been adopted is to focus the beam asymmetrically, using a cylindrical lens. This broadens the Gaussian curve so that variation in intensity is minimal within the envelope of fluidic containment of the particle trajectory.

Aerosol is introduced into the measurement space through a fine tube, typically at a flow rate of about 1 l/min. To contain the particle trajectory within a very narrow envelope two techniques may be adopted. The tube diameter, position and aerosol flow rate are adjusted so that the *vena contracta* of the unconstrained flow is coincident with the measurement space. Additionally a higher velocity flow of clean filtered air is introduced concentrically around the sample stream and the shear forces between converging flows further constrains the trajectories of particles contained in the sample flow. The clean air envelope has a further function in that it minimizes the potential for particle losses from the sample stream and consequently greatly reduces the potential for particulate contamination of optical surfaces. Figure 6.5 shows a chamber designed to make these measurements.

The rate at which, in principle, the shape and size of particles can be measured is constrained by the need to have only one particle in the measurement space at any time. Of course at any concentration there will be some possibility of coincidence but when this occurs very infrequently it can be recognized from measurement of the residence time in the beam. Data for which the residence time is outside the limit for a single particle can be discarded from the analysis of particle shape and size without significantly changing the distribution of these parameters. For the size of measurement space readily achievable with standard optics, particle rates of up to 60,000 particles per second can be achieved before coincidence in the space becomes unacceptably high. However, the actual number of particles that can be measured depends on the resolution required, as the limiting step is the data acquisition and processing.

Data Acquisition in Relation to Shape Resolution

The images of the scattered light pattern from particles with various shapes, shown in Figs. 6.2 and 6.3, were produced using an intensified, high resolution CCD camera. The data acquisition and processing load required to produce such images is little different to that required for microscope images of particles. The rate at

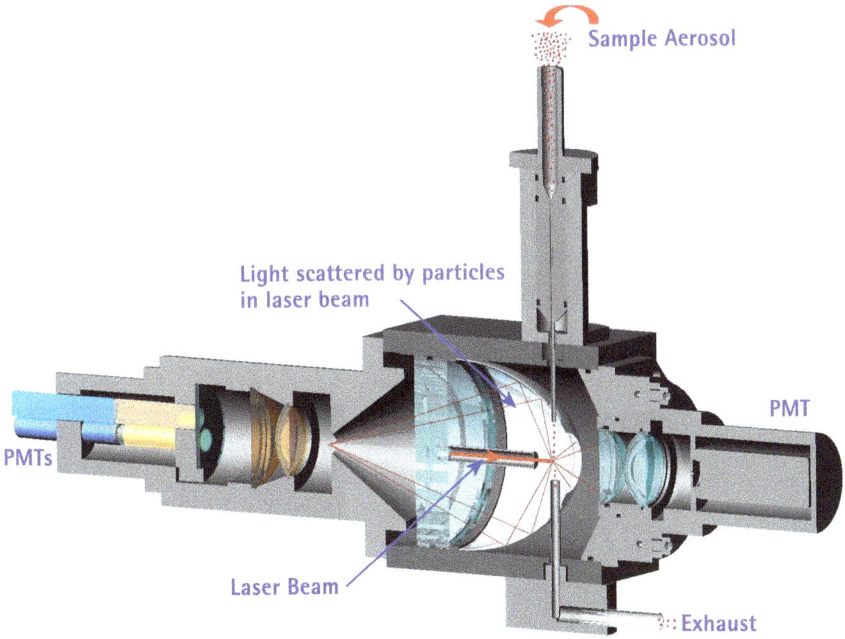

Fig. 6.5 Scattering chamber for a developed particle shape measuring instrument

which particles can be processed is therefore quite slow and so, as with microscopy, the technique is unsuitable as a means of detecting low concentrations of biological materials in the atmospheric aerosol.

In an exercise to examine how some shape information could be acquired as fast as possible the intensified CCD camera was replaced with three photomultiplier tubes (PMTs) placed at equally spaced intervals around the optical axis [9]. Figure 6.6 shows the image of the scattered light pattern from a sphere and a fiber transposed on to the field of view of the PMTs. It can be seen that the sphere gives an equal intensity across all three while the fiber is highly asymmetric, scattering predominantly into only one of the PMTs. Particles with other, nonspherical shapes scatter different proportions of the total scattered light to each of the PMTs and these proportions can be used to assign the particle an asymmetry factor. However, as the ratio of intensities also depends on the orientation of the particle as it passes through the measurement space, the asymmetry factor for any specific nonspherical shape is a distribution, not a single number. For particles with characterisable geometric shapes such as cuboids, flakes or fibers, this gives rise to highly characteristic distributions.

In the work that originally examined how a shape factor for particles could be acquired rapidly it was not expected that the comparison of three intensity fields would provide sufficient discrimination. For applications requiring the detailed characterization of particle shape this proved to be true and considerable work has gone into developing custom-designed detection arrays to achieve different levels

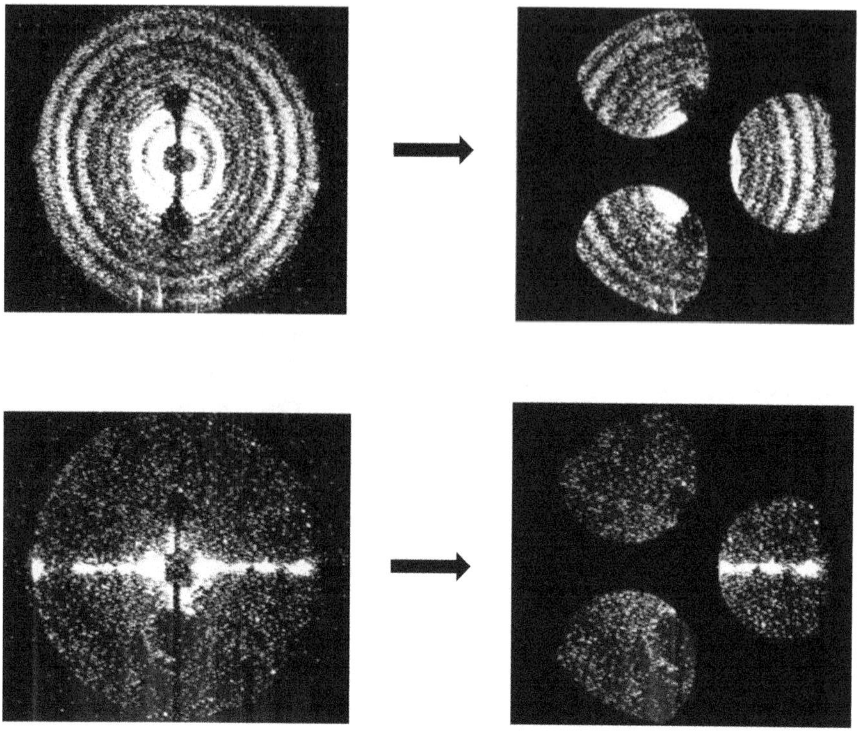

Fig. 6.6 Transposition of light scattering patterns to a 3-detector arrangement

of shape characterization appropriate to a range of research applications. Figure 6.7 illustrates the type of resolution that can be achieved with a larger number of detector elements arranged to meet specific characterization objectives.

For the characterization of a new source of particles in the atmospheric aerosol for the purpose of differentiating deliberately generated biological agents from natural and pollutant sources detailed shape information has proved to be of little value. This is because the particles of interest have highly variable and nonspecific shapes and so there are no target characteristics around which a specific array can be built. In addition to which, the 3-detector system proved to be far more sensitive than expected and is, for example, easily able to differentiate ellipsoids with 2:1 and 3:1 aspect ratios (Fig. 6.8).

Data Display for Particle Shape and Size

Although data displays have limited value in terms of recognizing significant changes in the atmospheric aerosol they do have considerable value to observers with some expertise, particularly during trials and exercises.

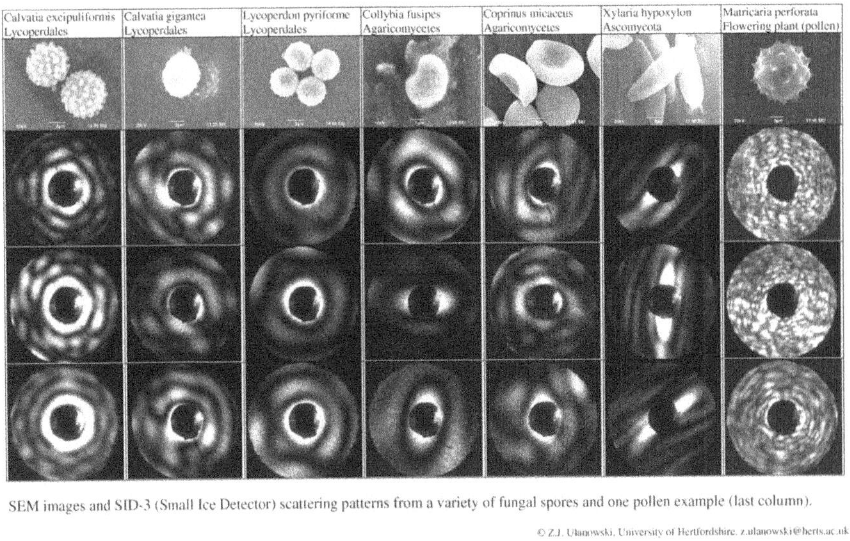

Fig. 6.7 Examples of more detailed scattering patterns with an array of detectors

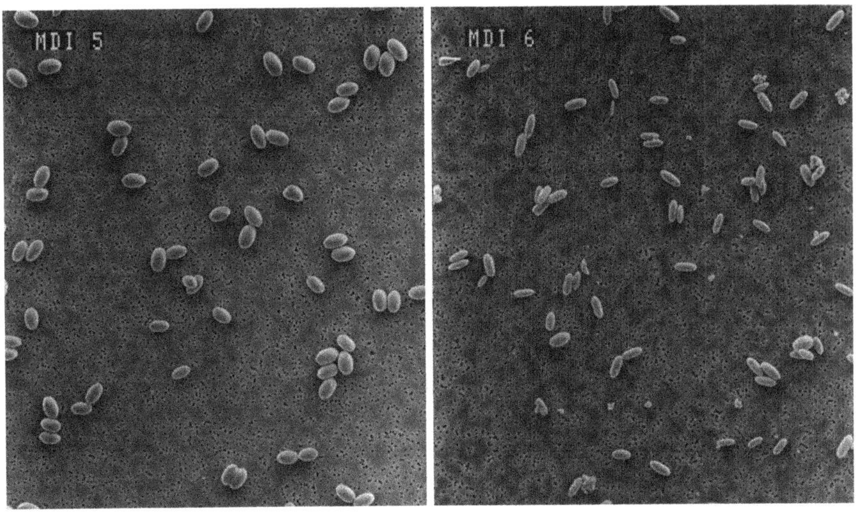

Fig. 6.8 Ellipsoids 2:1 and 3:1 aspect ratios

One means of displaying shape is to mimic the 3-detector configuration with a triangular centroid plot, as shown in Fig. 6.9. In this the position of each data point is determined by the relative intensity values at the three detectors, labeled E1, E2, and E3. If all the detectors see the same intensity, as happens with spherical particles, then the data point will appear at the center of the plot. The greater the vari-

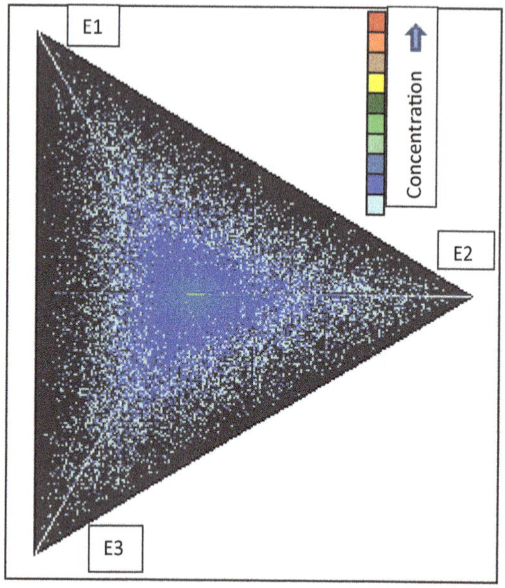

Fig. 6.9 Centroid plot of particle shape data

ance between the values the closer the point will be to the periphery of the plot and the position around the periphery will be a function of the orientation of the particle in the measurement space. For most particles with amorphous shapes the degree of asymmetry is indicated by the density of data points at various distances from the center of the plot as illustrated in Fig. 6.9. However, particles with specific shapes or combinations of shapes exhibit patterns that are often recognizable. Figure 6.10 illustrates this with the pattern displayed by partially agglomerated, mixed length fibers.

In mixed aerosols it is often possible to enhance the visual impression of changes by filtering out as much of the background as possible. This can be done using the shape factor to eliminate particles that are close to spherical or, alternatively, those with an asymmetry of greater than a specific value. Similarly filtering can be carried out on the size measurement so that the shape of particles only within a particular size band is displayed.

Shape and size can also be displayed together to give a complete view of the measured parameters of an aerosol and how these may be changing with time (Fig. 6.11).

Measurement of Particle Charge

The object of measuring particle charge in the context of biological agent detection is to determine the age of components of the atmospheric aerosol. Particles in the size range 1–5 μm, the most likely range for biological weapons, which are part of the background aerosol, have lifetimes in the atmosphere that typically range from

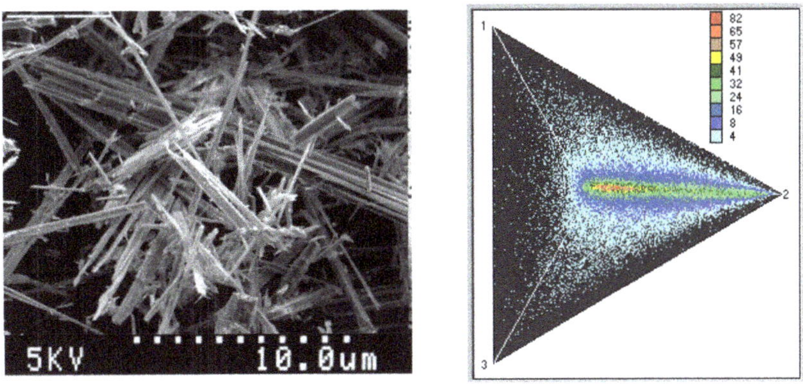

Fig. 6.10 Data pattern for fibres

Fig. 6.11 Contour Plot of Shape and Size

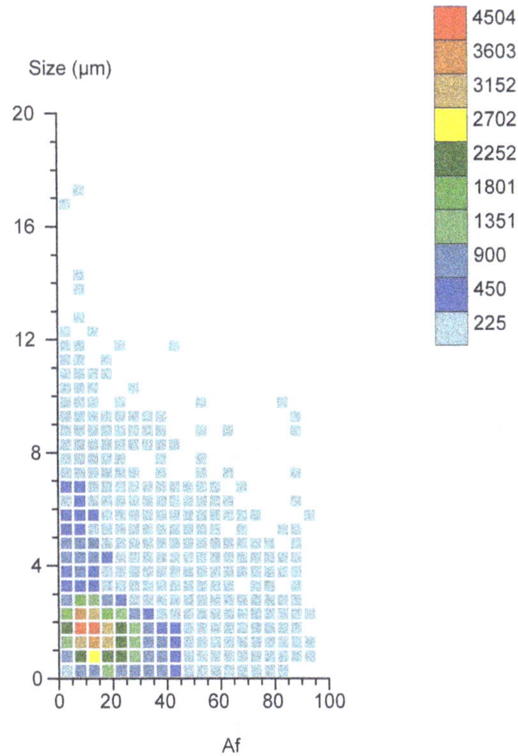

many hours to weeks and so are very likely to be in equilibrium with the atmospheric population of gaseous ions.

When aerosols are generated by disruptive processes, whether dispersion of a dry powder or the atomization of a liquid suspension, then charge separation and tribocharging invariably lead to the generation of an aerosol with highly charged particles [15]. When such particles are disseminated into the atmosphere they

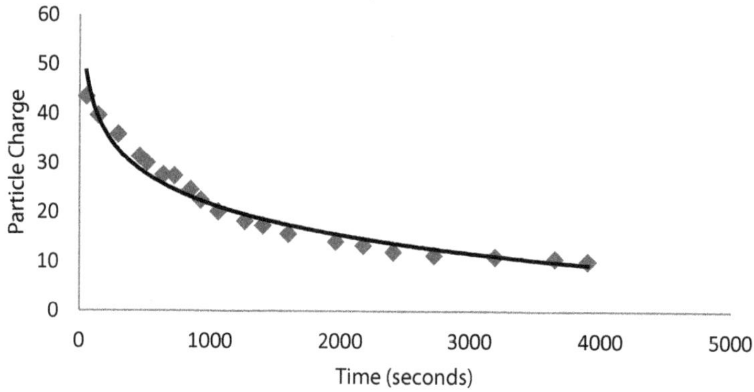

Fig. 6.12 An example of particle charge decay following spray generation

interact with the gaseous ions present and this gradually reduces the net charge until it reaches an equilibrium distribution of charge, often referred to as the Boltzmann distribution. This is the same distribution of charge as will be found on the long-term components of the background aerosol.

The rate at which charge is lost will depend on the concentration of ions in the atmosphere [7]. This can vary significantly due to meteorological events such as thunderstorms or at sites close to power lines, but generally the ion concentration remains fairly constant and predictable. Measuring the rate at which aerosol particles lose their charge is therefore a powerful means of determining how recently the aerosol was generated. A typical rate of decay for particles of approximately 1 μm is shown in Fig. 6.12.

Knowledge of the nonequilibrium charge, and so time since generation, of particles cannot directly differentiate between biological and nonbiological material. The value of this technique to nonspecific detection is that, when combined with knowledge of the wind speed and direction, it can pin-point quite accurately the origin of a new source of particles that has been generated by disruptive techniques.

As a typical half-life for charged particles in a normal atmosphere is of the order 20 min (Fig. 6.12) it is feasible to discriminate recently generated material from the background for at least an hour after generation. As not all particles are the same size, and larger particles naturally carry more charge, it is necessary to use a technique that measures the charge density on particles.

Measurement Techniques

The charge on particles can be deduced from their mobility in a known electrical field. So if a charged particle is in an electrical field it will achieve a terminal velocity determined by equating the electrical force to the Stokes drag.

$$V_{TE} = \frac{n_p e E C}{3\pi \eta D_p}, \tag{6.6}$$

where V_{TE} is the terminal velocity of the particle, n_p is the number of unit charges on the particle, e is the unit charge, E is the field strength, C is the Cunningham slip factor, η is the viscosity of the gas and D_p is the diameter of the particle.

It is possible to measure the velocity of a particle in an electrical field but, as can be seen from Eq. (6.6) there are two unknowns: n_p and D_p and so to find the total charge on an unknown particle it is necessary to measure the size as well as the velocity [4]. The inverse of this principle is used in instruments such as differential mobility analyzers (DMAs) where sub-micrometer particles are sized on the basis of their velocity in an electrical field. The success of this technique is predicated on the assumption that the maximum charge on a particle will be a single unit charge. The technique is therefore limited to particles with the minimum equilibrium charge and to particles in the nanometer size range.

For larger particles with multiple charges using velocity measurements requires the simultaneous measurement of particle size. No fieldable instrument has been developed for use in biodetection, although the principle has been demonstrated using phase doppler anemometry. This technique combines laser doppler anemometry, which is widely used in the analysis of fluid flows, with a measure of the phase shift of the light scattered by the particles in the fluid which can be calibrated to give a particle size parameter. This technique has two major limitations however: the particle shape must be close to spherical to give a reliable correlation between phase shift and size and the technology is too complex and expensive for a fieldable system.

Other techniques are available that could, realistically, be developed into a fieldable system. A variant of the Differential Mobility Analysis technique can be envisaged in which the sampled particles are separated by size or have their size measured by a separate technique as part of the analysis. Another possibility is to measure the charge collected by an electrode on to which the particle is deposited. This is the principle used in the electrical low pressure impactor in which the particle sizes are separated by a cascade impactor and the number of particles are counted at each stage by very sensitive current measurement. This technique could in principle be adapted and significantly simplified, as the low pressure required to collect nanometer size particles would not be required for this application.

Data Analysis Techniques

Although techniques for displaying data can be very useful to experts in the field, particularly during tests and trials, for operational deployment automated analysis is always required. Such an autonomous system must analyze the data and issue appropriate warnings without any manual intervention. To achieve this, a wide variety of techniques can be used, ranging from simple statistical measures of variation to the most sophisticated artificial intelligence techniques.

Statistical Analysis

Simple statistical analysis of collected data has been used with some success in trials to differentiate releases of simulants from natural variations in the atmospheric aerosol. The objective of such analyses is to differentiate significant changes in measured parameters from the natural variability found in the background data by statistical matching of populations. This technique uses libraries containing the expected characteristics of releases such as the medians, means and variance, of the measured parameters and compares data sets from the unknown samples with the contents of the library.

A range of statistical tests such as the χ^2 test, can be used to obtain the best match and the probability that the unknown data come from the same population as the best match in the library. The test relies on statistical measures that are valid only if the distribution of the values is normal or has been transformed to normal. Parameters such as standard deviation only have true meaning when they relate a distribution that can be described mathematically. They may still have a value in describing the characteristics of imperfect distributions but when they are used in formal statistical tests the outcome may be unreliable. Other statistics are available and are used in specific contexts but χ^2 is almost certainly the most appropriate in this context.

The great advantage of such simple techniques is that they are very easy to implement and require minimal processing power. They are also unambiguous and relatively easy to understand. However, such techniques do have some significant disadvantages for the automated detection of biological agents. Firstly, they assume that a library can be compiled that contains all the elements that a detection and characterization system may encounter. As most applications involve making measurements against the almost infinitely variable atmospheric background that can never be truly the case. In this application the target of the detection is also likely to have characteristics that are difficult to accurately predict. A further problem is that they match the measured parameters separately and so, for example, if particle shape showed a significant match but size did not, then some other process must decide whether or not to issue a warning.

Knowledge Based Techniques

Knowledge based techniques belong to a class of data analysis techniques often referred to as Artificial Intelligence (AI). They are composite techniques that can make use of all classes of information, by constructing an empirical rule base and testing how well data fit the rules. The knowledge may be statistical, qualitative or heuristic. Where this wins over the straightforward statistical approach is in the flexibility that can be applied to the development of the rule base. For example: when statistics such as means and standard deviations are used these can be applied much more readily to poorly defined distributions as the calculated parameters can be used as indicators rather as inputs to formal statistical tests. The ability to use a range of data types within the rule-base enables a much more flexible analysis technique that can be adapted to a wide range of varying circumstances.

Examples of the type of logic and qualitative rules appropriate to such an analysis might include:

- Was the time at which significant changes occurred in different measured parameters coincident?
- Were the changes relationally similar to data derived from experiments and trials?
- Was the time related to the inception of cyclical changes in the background (e.g., diurnal)?
- Was it coincident with a weather change (wind direction, precipitation)?
- How long did the changes persist—and was the period of the change different to observed periodicity in the background?
- Were the conditions for the release of biological agents favorable (wind speed and direction, atmospheric stability, UV light levels).

Other, nonmeasurement based rules such as threat levels can also be include in the analysis.

Once the rules have been established and quantified decision levels determined, they are placed in a hierarchy that leads to a decision point on whether or not an alarm should be signaled. If required, levels of certainty can be associated with the alarm, although these will be necessarily qualitative.

Learning Machines

Learning machines are a class of AI technique that are appropriate to analysis situations where exact rules governing a relationship cannot be established but a significant quantity of data demonstrating the relationship is available. The established data is used to train adaptable algorithms to recognize the relationships in new data.

Artificial Neural Networks

Artificial neural networks (ANNs) were the first class of techniques to create “intelligent” computers that were capable of learning from experience. Algorithms were developed that changed the weighting associated with tracks through the system in response to learning data where the end class was assigned. So if it was required to design a system that sorted two differentiable items X and Y it would be necessary to make characterizing measurements on a wide range of examples of both. Data files on each would be passed through the system with it configured so that the system always directed it to the correct answer and the algorithm would modify the weightings appropriately. Once this “supervised learning” process had been completed then unknown, mixed data could be fed through the system and it would correctly classify each item as either X or Y.

ANNs are particularly appropriate for classification applications, where large quantities of representative data are available but where no direct analytical or statistical relationship between the descriptive parameters and the result can be established. This makes aerosol particle characterization an ideal candidate for this type of analysis as most direct measurement techniques are capable of generating data on many thousands of particles per second.

Many variations on the early neural network techniques have been developed and these are now used for a very wide range of applications. However, considerable care needs to be exercised when choosing an appropriate technique as all are not appropriate to every potential application. Because data is used to alter the algorithm weightings, aspects of the data that may be ignored in other forms of analysis may be taken account of when modifying the weightings and so give rise to unexpected effects. For instance, in a simple and frequently used form of ANN, the weighting applied is proportional to the quantity of training data used. So if more data on X than Y is used in training then the classification of unknown data will be biased to select X over Y. This form of bias can be very subtle and so introduce errors that would not be readily recognized.

An advanced form of neural network that generates multi-dimensional data "maps" from the learning process has been used very successfully with real time nonspecific detection technology. It has the great advantage that maps for any target can be added to or removed from the network without the need to re-train the others. Classification of data from an unknown source is carried out by a process known as learning vector quantization. This assigns each multi-parameter data package to the map that gives the best fit. This technique has been employed for a number of years in military bio-detection systems without any reports of systematic misclassification.

One limitation with all ANN techniques is that it is very difficult to classify data as an unknown because data from an unknown is, by definition, unavailable. This is a problem when analyzing atmospheric data because there will almost always be unknown components. Techniques are available that can be used to overcome this problem, at least to some extent. However it is a limitation that must be borne in mind whenever these techniques are being considered for automated detection.

A further difficulty that often occurs with ANN techniques is when too many parameters, or problem dimensions, are included in the analysis. For the limited number of parameters used to describe the size and shape of particles in nonspecific biodetection the dimensionality of the analysis has been well within the limits of the technique. However, as it becomes feasible to include real-time generic classification of particles, for example by measuring particle fluorescence characteristics, then simultaneous analysis of all the data acquired may be beyond the capability of ANNs.

Support Vector Machines

Support vector machines (SVMs) are learning techniques that use a particular class of mathematical algorithms known as Kernels. They are beginning to replace neural networks in a wide range of difficult classification and pattern recognition applications. They have the potential to overcome many of the problems that have been encountered in the application of neural networks, including parameter weighting in the input vector, high dimensionality (many measured parameters) and a problem, often associated with complex data sets, called overfitting.

In relatively simple terms, the technique uses linear learning machines (linear discriminators) combined in a "kernel" function to separate classes, in high dimensional feature space, by linear hyperplanes. The mathematics of the technique is outside of the scope of this overview but the concept may be easier to understand than it first appears.

It may be helpful to consider initially a simple linear regression where the objective is to draw the most representative line through noisy data. This is generally achieved by computing the minimum of the sum of the squared deviations from the regression line (least squares). If this principle is extended to as many dimensions as the data have descriptive features (f) then data can be separated in f-dimensional space by hyperplanes with dimensions of f-1.

A great advantage of kernel based learning machines is that there is no requirement to directly define the relationship between characterization variables as the kernel can be designed so that this becomes implicit. This overcomes what is known as "the curse of dimensionality" which is often encountered when the number of characterization parameters in ANN analysis is increased.

Future Trends and Developments

There are many ways in which this class of techniques may be developed and improved in the future. Some of these will be incremental improvements made feasible by developments to the technologies that they use and these are likely to lead to smaller and lower-cost instruments. Others will come from new measurement techniques that are constantly being developed for characterizing airborne particles. The latter class is likely to emerge from testing programs at the national defense institutes where new techniques, often developed for other purposes, can be shown to be effective in detecting simulated events.

Higher Resolution Measurements

Many projects have demonstrated that the size and shape of airborne particles can be measured with considerably greater resolution than has been achieved in the technologies that have been deployed. The important question is: do higher resolution measurements give more certain detection? The answer is almost certainly not, though this has not been fully explored, largely because a significant improvement is thought so unlikely.

Biological weapons are likely to employ fairly crude techniques to generate aerosols. Liquid suspensions may be sprayed using devices similar to those used in agriculture; powders may be dispersed using air jets or explosives. The object is to generate inhalable particles that can remain in suspension for long periods. Developing techniques to deliver particles with highly specific and reproducible characteristics is technically difficult and would achieve very little in terms of effective dose delivery. High resolution measurements are therefore of little value as they are not likely to increase discrimination between agent particles and those in the background aerosol.

Measurement of Other Parameters

It is theoretically possible to deduce other particle parameters such as optical constants, homogeneity and density from light scattering patterns, particularly when combined with other measurements such as particle dynamics. However such measurements may be restricted to particular particle geometries and may, under some conditions, be ambiguous. They are also very complex measurements to make and are likely to require very complex and so expensive instruments. As far as the author is aware there is no research being undertaken that will add new measurable parameters to this class of techniques within the foreseeable future.

Networking

Networking arrays of this class of technologies does have the potential to deliver significant improvements in performance, in terms of reducing the potential for both false positive and false negative alarms. In scenarios where the agent is being generated covertly at some distance upwind of the target area it will form a widespread cloud. If this passes over an array of this class of detector they will alarm in real time and, if they are networked and connected to wind measurement systems, the correlated responses will give high confidence that a cloud of material has been detected. Uncorrelated responses from individual sensors will be much more likely to relate to transient local sources that are more likely to be interferents.

While some work is known to have been carried out on detector networks there is little published data on what may be gained over the deployment of isolated detectors. A factor inhibiting such trials is the cost of procuring large numbers of detectors together with the expense and difficulty of running the large scale field trials that would be necessary to demonstrate the utility of such networks.

Low Cost Sensors

The reduction in component costs and technical advances are combining to make it more feasible to design and manufacture relatively low cost sensors. Such sensors make the testing and deployment of networks, as described earlier, a much more affordable proposition. Having arrays of large numbers of sensors can deliver a range of significant advantages. These are point sensors and can only respond to material that reaches the sampling point. The more sensors there are in a given area the less likely it is that a cloud of agent could escape detection and so increasing numbers reduces the potential for false negatives. A large number of sensors linked to wind measurement will give a comprehensive picture of cloud cohesion and dynamics and so significantly reduce the potential for false positives due to local variability and interferent sources. Failure of a sensor is also much less of an issue if there are many rather than a few and this feeds back into simpler, and so cheaper, designs. Effective data reduction and communication from large arrays of low-cost sensors in a military environment is a potential problem area that may require further study.

Charge Measurement

There is currently no technology available to measure the charge on particles that would be suitable for deployment in a military environment. However there are a number of techniques that are used in the laboratory or for monitoring in other environments and one of these could almost certainly be adapted to meet the military monitoring requirements. If such an instrument were combined with size and shape measurement in a multi-parameter monitoring system it could deliver a very high level of confidence that a cloud was deliberately generated, originated from a single source and, with meteorological data, pinpoint the likely source.

Data Analysis

Possibly the greatest potential for advance in this field of detection is in the use of artificial intelligence to analyze the data. Significant progress has already been made in this area but the techniques are being developed very rapidly for many other applications and these can be adapted to improve the confidence in detection. These techniques are almost ideally suited to learning machines as they generate vast quantities of relatively simple data from which adaptable algorithms can learn patterns. There is already the potential to develop analysis systems that continuously learn patterns in the variation in the background aerosol where they are sited and so can continuously improve the discrimination they offer.

References

1. Aptowicz, K.B., Pinnick, R.G., Hill, S.C., Pan, Y.L. and Chang, R.K., 2006, Optical scattering patterns from single urban aerosol particles at Adelphi, Maryland, USA: a classification relating to particle morphologies, *J Geophys Res* 111, D12212.

2. Bohren, C.F. and Huffman, D.R., 1983, Absorption and scattering of light by small particles, Wiley, New York.
3. Dick, W.D., Ziemann, P.J., Huang, P.F. and McMurray, P.H., 1998, Optical shape fraction measurement of submicrometre laboratory and atmospheric aerosols, *Meas Sci Technol* 9(2):183–196.
4. Flagan, R.C., 1998, History of electrical aerosol measurements, *Aerosol Sci Tech* 5, 1–2.
5. Gucker, F.T. and Rowell, R.L., 1960, The angular variation of light scattered from single dioctyl phthalate aerosol droplets, *Discuss. Faraday Soc* 30(185).
6. Hirst, E., Kaye, P.H. and Guppy, J.R., 1994, Light scattering from non-spherical airborne particles: experimental and theoretical comparisons, *Appl Optics* 33(30): 7180–7186.
7. Hoppel, W.A. and Frick, G.M., 1986, Ion-aerosol attachment coefficients and the steady state charge distribution on aerosols in a bipolar environment, *Aerosol Sci Tech* 5:1–21.
8. Kaye, P.H., Eyles, N.A., Ludlow, I.K. and Clark, J.M. 1991, An instrument for the classification of airborne particles on the basis of shape, size and count frequency. *Atmos Environ* 25(3–4): 645–654.
9. Kaye, P.H., Hirst, E., Ludlow, I.K., Clark, J.M. and Micheli, F., 1992, Airborne particle shape and size classification from spatial light scattering profiles. *J Aerosol Sci* 23(6): 597–611.
10. Kaye, P.H., Alexander-Buckley, K., Hirst, E. and Saunders, S., 1996, A real-time monitoring system for airborne particle shape and size analysis. *J. Geophys Res-Atmos* 101 (D14): 19215–19221.
11. Kerker, M., 1997, Light scattering instrumentation for aerosol studies: An historical overview, *Aerosol Sci Tech* 27: 522–540.
12. Mie, G., 1908, Considerations on the optics of turbid media, especially colloidal metal sols *Ann.d. Physik* 25: 377–442.
13. Pan, Y.L., Aptowicz, K.B., Chang, R.K., Hart, M. and Eversole, J.D. 2003, Characterising and monitoring respiratory aerosols by light scattering, *Opt Lett* 28(8): 589–591.
14. Sachweh, B.A., Dick, W.D. and McMurray, P.H., 1995, Distinguishing between spherical and non-spherical particles by measuring the variability in azimuthal light scattering. *Aerosol Sci Tech* 23: 373–391.
15. Whitby, K.T. and Liu, B.Y.H., 1966, The electrical behaviour of aerosols, in Aerosol Science, editor Davies C.N., Academic Press.

Chapter 7
Bioaerosol Detection with Fluorescence Spectroscopy

Per Jonsson and Fredrik Kullander

Introduction

Fluorescence spectroscopy is probably the most explored technology for direct bioaerosol detection. The technology is often called laser-induced fluorescence (LIF) since lasers are normally used for the excitation. However, there are examples of other excitation sources, for example, light-emitting diodes and lamps. Fluorescence spectroscopy has been popular since its sensitivity enables instant spectral detection of down to single micron-sized particles. The fluorescence spectroscopy probes inherent properties of biological compounds. Therefore it can be used for detection of unknown types of harmful organisms or toxins even without a pre-measured library in contrasts to most other technologies. The main disadvantage of fluorescence spectroscopy is the relatively low specificity. Many harmless naturally occurring particles, including nonbiological particles, also fluoresce which may cause a nonacceptable false-alarm rate. The low specificity is due to the generally broad characteristics of both the absorption and the emission spectra with few distinct features that can be used for classification.

In the scientific literature much of the work has been focused on how to make the technology more specific, for example, by changing the excitation wavelength, or by resolving the emission either in the spectral or temporal domain. The use of dual excitation wavelengths is one way to probe different molecules of interests. Despite the fact that the fluorescence spectra are broad, a higher spectral resolution can be used for better classification. Measurement of the fluorescence lifetime may also be explored for classification. The scientific test systems exploring these extra features are often quite complex. Therefore another parallel trend has been to develop more compact and affordable systems with some sacrifice on the ability of classification. Today a number of commercial systems exist.

P. Jonsson (✉) · F. Kullander
Division of Sensor and EW Systems, FOI—Swedish Defence Research Agency,
Olaus Magnus väg 42, 581 11 Linköping, Sweden
e-mail: per.jonsson@foi.se

P. Jonsson et al. (eds.), *Bioaerosol Detection Technologies*, Integrated Analytical Systems,
DOI 10.1007/978-1-4419-5582-1_7,

Fluorescence is often combined with other technologies. Almost all systems use elastic scattering as an additional source of information about particle size or even shape as described in the previous chapter. Some experimental systems combine fluorescence with other complementary technologies. There are examples of combinations with atomic emission spectroscopy, for example, flame photometric detectors (FPDs) or laser-induced breakdown spectroscopy (LIBS), which gives information about elementary composition of the particles. Thermal absorption (IR or THz) or Raman spectroscopy can be used for identification of different molecules by detecting their vibrational levels. Fluorescence has even been used in combination with mass spectroscopy. The details of these complementary technologies can be found in other chapters of this book.

The next section of this chapter explains the fundamentals of the fluorescence followed by a section with fluorescence from biomolecules in focus. A large number of research groups and companies have developed fluorescence-based bioaerosol detection systems. The research performed with these systems are presented. Some of the considerations when designing a detection system are pointed out in the subsequent section. Some concluding remarks complete this chapter. Fluorescence-based standoff detection will be covered later in this book and the focus in this chapter will be on fundamentals and point-detection systems.

Fundamentals of Fluorescence

Interaction between light and matter is one of the fundamental processes in nature. In the field of spectroscopy, the interaction can be used to gain information about internal structure and dynamics of atoms and molecules, for example, electronic levels and vibrations in molecules. Spectroscopy can also be used to investigate interactions and dynamics between molecules, for example, chemical reactions. This overview is mainly focused on the internal structure and dynamics of molecules.

The electrons in atoms and molecules have certain discrete energy levels that are allowed according to quantum mechanical theory. In normal environmental conditions the majority of the molecules reside in the so-called ground state, where the total energy of the molecule (sum of electronic energies) is as low as possible. If the molecule gains enough energy, the electrons can be excited to a higher level. Optical absorption, that is, absorption of photons, is one way for the molecule to gain this energy. The molecule can relax down from the excited state to the ground state by emitting energy, for example, by emitting photons also called *luminescence*. A detailed picture of electronic energy levels can be illustrated in a so-called Jablonski diagram as shown in Fig. 7.1.

Every electronic state is degenerated due to vibrational modes between atoms in the molecule leading to a band of vibrational sublevels (states). At room temperature, the distinct sublevels, as shown in the figure, become more of a continuous band. These sublevels cause the broad spectrum, both in absorption and in fluorescence. Relaxation from a vibrational state down to the lowest vibrational

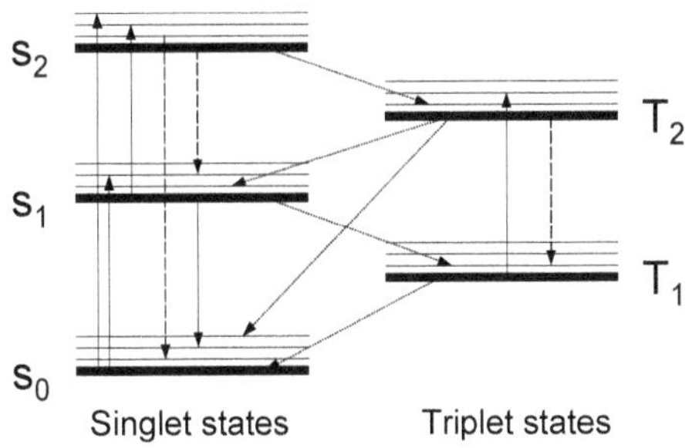

Fig. 7.1 Jablonski diagram of a generic molecule. The *bold horizontal lines* represent the electronic states with appurtenant vibrational sublevels. The *arrows* show the most probable transitions. The *solid lines* represent excitation possibilities (pointing *upward*) and fluorescence (pointing *downward*), the *dashed lines* internal conversion and dotted lines intersystem crossing

state in an electronic level is very fast, in the order of picoseconds. The singlet state S_0 is the ground state for the molecule. The molecule has many excited singlet states (more than the two shown in the figure). The long wavelength limit of the absorption corresponds to the energy difference between the first excited state S_1 and the ground state S_0. The relaxation from S_1 to S_0 is normally in the order of nanoseconds while relaxation between higher order excited states is much faster, in the picosecond range.

For every excited singlet state, there exists a triplet state of somewhat lower energy. The transition between a singlet and a triplet state is spin-forbidden and therefore has a low probability. The relaxation between different triplet states is as fast as relaxations between the corresponding singlet states. However, a molecule in the lowest triplet state has a very slow transition back to the ground state. It can vary from microseconds to seconds depending on the surrounding and therefore the first excited triplet state T_1 is often called metastable.

By historical reasons luminescence caused by relaxation from S_1 to S_0 is called *fluorescence* while luminescence caused by relaxation from T_1 to S_0 is called *phosphorescence*.

Organic molecules exhibiting efficient fluorescence commonly contain conjugated structures (alternating double bonds or aromatic ring structures). The molecules with conjugated bonds have an energy difference between S_1 and S_0 that corresponds to the energy of photons in a wavelength range approximately from 200 nm to 1 μm. The absorption is normally well below 200 nm for organic molecules without these conjugated bonds. This corresponds to an energy that is higher than the dissociation energy of most chemical bonds, which sets a short wavelength limit for efficient fluorescence. The light rather destroys the molecule than excites an electron. The upper limit at 1 μm is determined by the thermal excitation. Molecules with small difference between the energy levels can be excited thermally and therefore degrades due to the biradical character of the metastable triplet state [1].

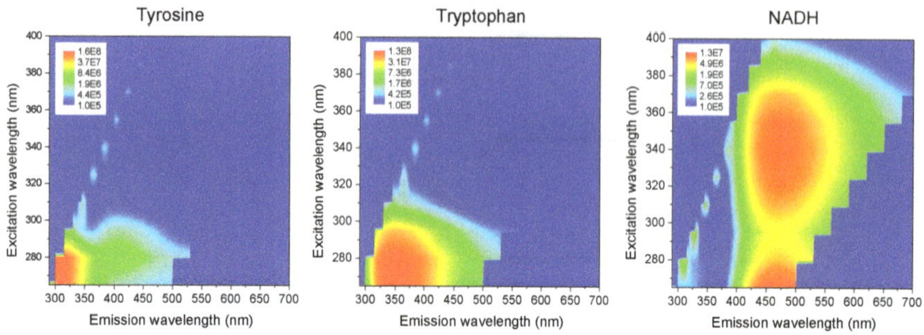

Fig. 7.2 The excitation-emission spectra of tyrosine, tryptophan and the reduced form of intracellular nicotinamide adenine dinucleotide (*NADH*; previously known as diphosphopyridine nucleotide or pyridine nucleotide). The *first two* are amino acids and NADH is a molecule involved in the metabolism of organisms. The features to the *left* in the graphs are due to insufficient subtraction of the water Raman scattering

Normally, the emitted photons in the fluorescence process have a longer wavelength (lower energy) than that of the absorbed photons due to the internal energy conversion process. Basically the molecules excited to S_1 relax from higher vibration levels down to the lowest vibrational level by dissipating heat before emitting a photon. Figure 7.2 shows the excitation-emission spectra of two amino acids and a molecule involved in the metabolism of organisms. These molecules are of particular interest for bioaerosol detection as we will see subsequently.

Fluorescence spectroscopy is mainly giving information about the electrical levels of the molecule, while the vibrational levels mainly broaden both the absorption and emission spectra. In thermal absorption spectroscopy (IR or THz) and Raman spectroscopy the same vibrational levels can be probed. For comparably small molecules the vibrational levels can provide a relatively specific "fingerprint" of the molecule. This is used for example in chemical detection. Bioaerosol particles normally contain many molecules and they are larger with a much more complex vibrational state space making it difficult to extract useful vibrational spectroscopic information. In addition, the strength of the measurable optical signal in vibrational spectroscopy is substantially weaker than the fluorescence signal. This is the reason why fluorescence is established in bioaerosol detection while vibrational spectroscopy still is an emerging technology with potential of better specificity.

Much of the information about fluorescence is contained in the fluorescence cross-section term $\sigma^F(\lambda, \lambda_l)$. This is a fictitious area usually measured in centimeter square that relates the excitation intensity $I_{exc}(\lambda_l)$ at the wavelength λ_l with the resulting emitted power at the wavelength λ as

$$P^F(\lambda, \lambda_l) = I_{exc}(\lambda_l)\sigma^F(\lambda, \lambda_l). \tag{7.1}$$

Following the nomenclature of Measures [2], the fluorescence cross section per unit wavelength interval can be expressed as

$$\sigma^F(\lambda, \lambda_l) = \sigma^A(\lambda_l)F(\lambda, \lambda_l), \tag{7.2}$$

where $\sigma^A(\lambda_l)$ is the absorption cross section at the exciting wavelength λ_l. $F(\lambda, \lambda_l)$ is the quantum yield factor and can be expressed as

$$F(\lambda, \lambda_l) = Q^F L^F(\lambda). \tag{7.3}$$

The quantum yield factor is the probability that a photon of wavelength λ is emitted for each photon absorbed at the excitation wavelength λ_l, while the quenching factor Q^F is the probability that a photon of any wavelength is emitted for each photon absorbed. Hence the fluorescence profile function $L^F(\lambda)$ is the fraction of the total fluorescence emitted into the unit wavelength interval centered around λ, and can be seen as the normalized fluorescence spectrum since

$$\int L^F(\lambda) d\lambda = 1. \tag{7.4}$$

The total fluorescence cross section (integrated over all emission wavelengths) accordingly becomes

$$\sigma^F(\lambda_l) = \sigma^A(\lambda_l) Q^F. \tag{7.5}$$

In many textbooks the factor Q^F is called the fluorescence quantum yield or the quantum efficiency of fluorescence. The factor can be expressed in terms of rates or lifetimes as

$$Q^F = \frac{\Gamma_R}{\Gamma_R + \Gamma_Q} = \frac{\tau}{\tau_{\text{rad}}}, \tag{7.6}$$

where Γ_R and Γ_Q are the radiating and quenching rates, respectively. This theory was developed for remote sensing of the atmosphere, where the main reason for nonradiative decay of an excited state is quenching by collisions between molecules in a dense gas media. In a bioaerosol the molecules are bound in particles with a low probability of collision quenching. However, if Γ_Q includes the total rate of nonradiative decay Eq. (7.6) is valid. The temporal expression of the quenching factor is simply the ratio between the *observed* lifetime or decay of the excited population τ and the radiative lifetime or decay of the excited state τ_{rad}, since $1/\tau = 1/\tau_{\text{rad}} + \Gamma_Q$.

In most cases the fluorescence can be assumed to be isotropic, that is, the probability of emission is equal in all direction. In this case, the differential cross section per unit wavelength interval is

$$\frac{d\sigma^F(\lambda, \lambda_l)}{d\Omega} = \frac{1}{4\pi} \sigma^A(\lambda_l) Q^F L^F(\lambda) \tag{7.7}$$

and the total fluorescence differential cross section is

$$\frac{d\sigma^F(\lambda_l)}{d\Omega} = \frac{1}{4\pi} \sigma^A(\lambda_l) Q^F. \tag{7.8}$$

Table 7.1 Maximal absorption and emission wavelengths for different biological compounds [3, 4]

Substance	λ_{abs}, nm	λ_{em}, nm
Phenylalanine	257	282
Tyrosine	275	303
Tryptophan	280	348
DPA	325	415
NADH	340	467
FAD	462	540

DPA dipicolinic acid, *NADH* reduced nicotinamide adenine dinucleotide, *FAD* flavin adenine dinucleotide

The above-mentioned expressions include most of the information that can be extracted by fluorescence spectroscopy. In practice however a theoretical prediction is difficult since bioaerosol particles can contain quite complex mixtures of molecules with parameters that are dependent of the environmental conditions. Normally it is not possible to calculate neither the absorption nor the fluorescence cross sections.

Fluorescence from Biological Molecules

Many molecules of biological origin have conjugated bonds and are therefore often good fluorophores (see above). This intrinsic fluorescence, or autofluorescence, is useful for biological detection. The most important molecules in bioaerosol detection are the aromatic amino acids, for example, phenylalanine, tyrosine and tryptophan, and some molecules related to metabolic processes in microorganisms, for example, reduced nicotinamide adenine dinucleotide (NADH) and flavin adenine dinucleotide (FAD). The excitation-emission spectra of three of them are shown in Fig. 7.2.

Amino acids are the building block of all biological organisms. NADH and FAD are substances originating from the metabolisms in organisms and they are as such a potential sign of if the organisms are alive. The presence of the intrinsic fluorescence in these important biological molecules has made fluorescence spectroscopy an important tool in biochemistry. An overview of this research area can be found in the book of Lakowicz [3].

The efficiency of the fluorescence is directly proportional to the absorption cross section. Therefore it is important that the excitation wavelength matches the peak of the absorption to achieve efficient fluorescence. The quantum yield (Q^F) depends on the molecule and its surrounding environment. The quantum yield normally is quite high for the molecule mentioned above, for example, about 15% for tryptophan in water at normal room temperature [3], and each microorganism contains many of those molecules.

Table 7.1 shows the optimal absorption wavelengths above 250 nm and the corresponding peak emission wavelengths for the most important molecules in bioaerosol detection. All molecules except for dipicolinic acid (DPA) are mentioned above. DPA is produced during spore formation and is important for biological warfare agents (BWAs) such as anthrax [4].

As we have seen, material of biological origin contains a variety of different molecules that do fluoresce. Different microorganisms will have different relationship of these molecules (and additional molecules) which will affect both the absorption and the emission spectra. Several groups have measured the fluorescence cross section of interesting biological particles [5–12]. It is not easy to compare the results for several reasons. The methods differ in the studies in many respects. The preparation of samples (solution, wet aerosol, dry aerosol, size etc.) is quite different. Also the excitation wavelengths are different and the emission is measured in various wavelength bands. In general the total fluorescence cross section $\sigma^F(\lambda_l)$ from Eq. (7.5) varies from 10^{-10} to about 10^{-12} cm^2 for particles in the micron size. If excluding excitation wavelengths of less than 250 nm, the maximum cross section for most of these particles is found at an excitation wavelength around 280 nm. A second local maximum can be found around 340 nm. It has been noticed that particles in dry samples have a significantly lower cross section than particles of the same species in liquid wet solutions [5, 10, 12]. Washing of BG spores (a common BWA simulant) has been shown to result in a lower cross section [11]. Over the years the species *Bacillus globigii* has taxonomically changed name from *Bacillus subtilis* var *niger* and to the current *Bacillus atrophaeus*. The acronym BG of the original name is still the most common to use.

During the last 10 years FOI has measured the excitation-emission spectra for biological molecules. The systematic measurements were made on liquid samples. Figure 7.3 shows some examples of bacteria in water solution. Other groups have performed similar measurements and a few of them have been published with similar results [13–16]. It is seen that the emission spectra are changing with excitation wavelength. At short excitations wavelength (of less than 300 nm) the emission around 350 nm dominates. At longer excitation wavelengths the emission around 450 nm dominates. The excitation wavelengths that give the maximal emission for the two regions are around 280 and 340 nm, respectively. This corresponds well with the spectra of tryptophan and NADH (see Fig. 7.2 and Table 7.1). It is also clear that the emission is stronger when exciting at approximately 280 nm compared to longer wavelengths.

The conclusions from these measurements are similar to those of Hill et al. presented in a report from Army Research Laboratory (ARL) in USA [16]. To achieve fluorescence from flavins excitation of less than about 530 nm is necessary. NADH and amino acid fluorescence can be excited with wavelength of less than about 420 and 315 nm, respectively. Shorter excitation wavelength can always generate emission from all molecules with longer excitation wavelength limits, since the excited electrons may decay nonradiatively down to these emission bands before emission. Excitation at 300 nm can generate emission from all the molecules amino acids NADH and flavins, while excitation at 400 nm can only generate emission from NADH and flavins.

As pointed out in the ARL report, the excitation wavelength should be chosen so that it generates fluorescence from the different types of biological particles while it should not unnecessarily excite interfering fluorescence from nonbiological particles. The excitation wavelength should also be chosen so that the fluorescence spectra contain as much differences between biological and nonbiological parti-

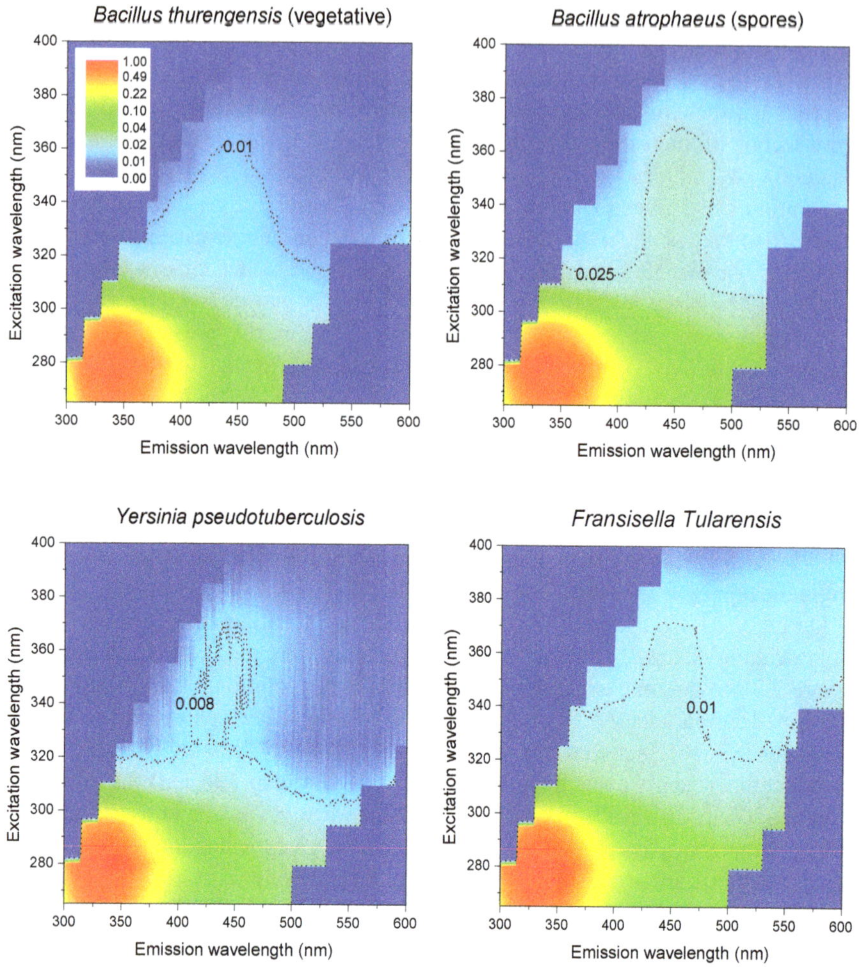

Fig. 7.3 Examples of normalized excitation-emission spectra of bacteria in water solution. The two examples on the *top* are *Bacillus thurengensis* (BT) and *Bacillus atrophaeus* (BG) that are commonly used as simulants for sporulating BWAs such as anthrax. The *bottom graphs* show the spectra of *Yersinia pseudotuberculosis* and *Fransisella tularensis*

cles as possible. Some organisms of interest do have low or no fluorescence from NADH or flavins, for example, some viruses and protein toxins, and an excitation wavelength of less than 315 nm is necessary. On the other hand, a short excitation wavelength tends to generate more fluorescence from potential interferents such as clay minerals. Several systems employ dual excitation wavelengths for better classification [10, 17–19].

Often, in many prototype and commercial systems, only just the presence of fluorescence in a certain broad band (typically about 100 nm) is measured. However, the emission spectra contain information that could be used for classification. Several groups have explored this possibility both for point detection [12, 20, 21] and standoff detection [22–25]. There is a debate which excitation wavelengths give the best information. Hill et al. have shown that some types of bioaerosol particles are easier to distinguish at 266 nm, whereas 355 nm is preferable for others [26]. The same group has been able to classify the background aerosol at different locations into ten different clusters when exciting at 263 nm [27, 28]. We and others have seen that there are more features in the spectra when exciting at less than 300 nm [29, 30]. Others mean that excitation of the amino acids, that is, at less than 300 nm, only will generate a signal from dead material while excitation at longer wavelength will indicate that the bacteria are alive, see, for example, the chapters of Ho and Simard et al. in this book. The researchers arguing for the short wavelength excitation mean that most of the fluorescence when exciting at wavelengths of 340 nm and above originates from growth media and culturing by-products and not from NADH [31]. The debate is not yet settled and as mentioned above several groups use multiple excitation wavelengths to take advantage of the two main excitation regions.

The emission spectra have been shown to be a robust signature for detection [32] (see also the chapter of Simard et al. in this book). However, an important complication is that the emission is highly dependent on the preparation of the bioaerosol, the way of dispersion and where the organism is in its life cycle. Washing of the microorganisms will radically change the spectra as can be seen in Fig. 7.4. Turex is a nonpure technical preparation of *B. thuringensis* spores so the great reduction in emission, especially at longer wavelengths, is not surprising but even for most of the BG spores the change is significant. A change of the emission spectra due to washing has been seen by many other groups [26, 33–37]. Another effect seen in clean samples is that bacteria of different species show little or no difference in their spectra when exciting at less than 300 nm [6, 7].

Dry samples and samples in water solution does not only have different quantum efficiencies as mentioned above, but also different emission spectra [12, 16, 38]. The spectra from bacterial spores can also change when exposed to ozone [39]. In addition, the life cycle state of microorganisms may have an impact on the emission spectra. Figure 7.5 shows the change of spectra for *Bacillus atrophaeus* kept in a water solution free from nutrition (growth media) for a few days in 26 °C. When kept in 4 °C the measured spectra were nearly constant. The difference is believed to arise from a faster spore formation in the warmer water since the bacteria keep trying to metabolite, but suffer from starvation and form spores. Other groups have seen similar differences in spectra of *Bacillus atrophaeus* when they are in their vegetative phase compared to when they appear as spores [9, 26].

Fluorescence decay time (so-called lifetime) can also be used to improve the discrimination of different biological compounds [3, 13, 40–44]. Petroleum-based products for examples have a strong fluorescence that could cause false-alarms in fluorescence-based trigger systems. However, the fluorescence from these products generally has longer lifetimes than the fluorescence signals from bioaerosol particles, which makes the classification fairly easy in the time domain. It is difficult to utilize this information in direct aerosol detection, but one group has published field trial results with a system measuring the fluorescence decay time [19].

In conclusion, detection of biological molecules with fluorescence is a powerful tool. The fluorescence signal is strong which allows for very sensitive and fast detection. The main drawback of fluorescence is the relatively low specificity. There are ways to increase the classification capability by utilizing multiple wavelengths,

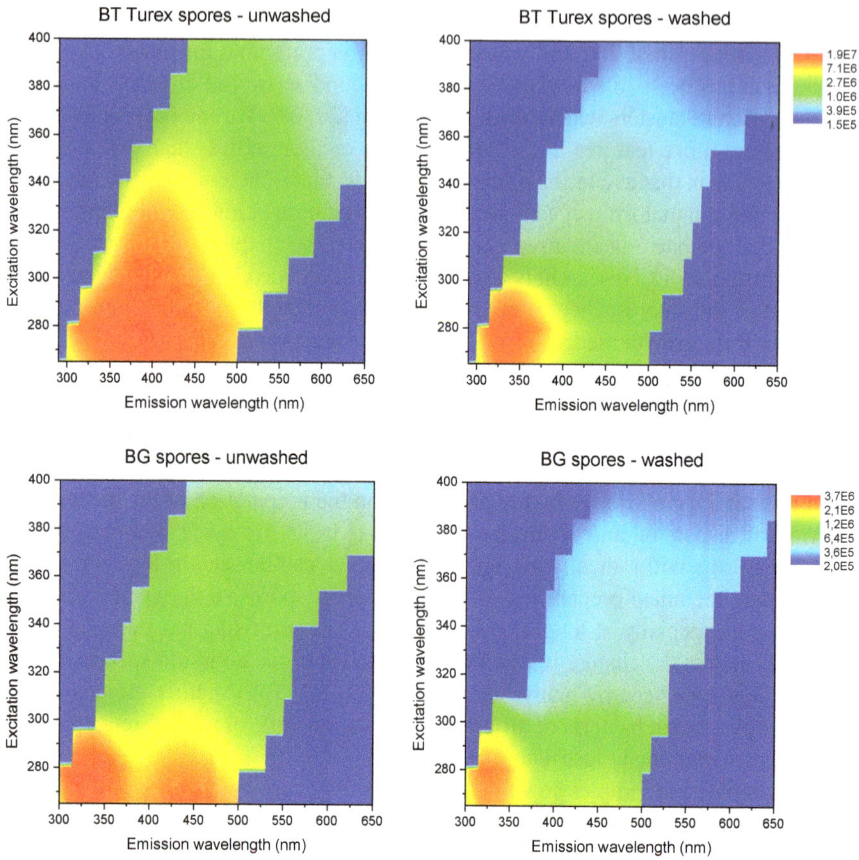

Fig. 7.4 Effect of washing in excitation-emission spectra of bacteria spores in water solution. The *top* and *bottom* graphs show *Bacillus thurengensis* (BT) and *Bacillus atrophaeus* (BG) spores, respectively. The *left* and *right* graphs show the spectra before washing and after washing, respectively

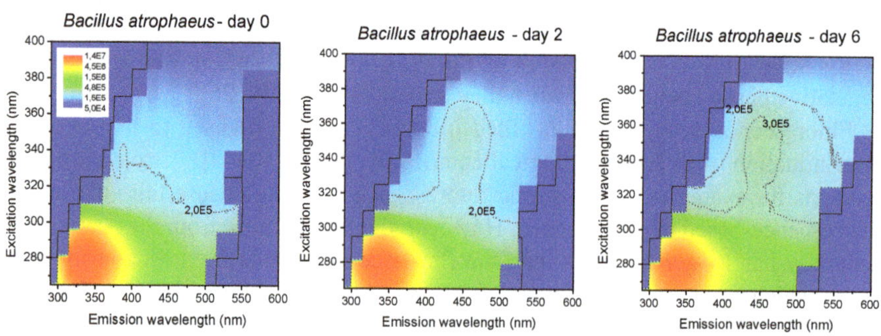

Fig. 7.5 Effect of starvation in excitation-emission spectra of *Bacillus atrophaeus* in water solution. The bacteria were smeared out on Blut Agar Basis and cultured for about 24 h. The bacteria colonies were harvested and suspended in NaCl(aq) and stored in darkness at about 26 °C for several days. The first graph shows the spectrum directly after harvesting. The *second* and the *third* graph show the spectra after 2 and 6 days, respectively. When stored in 4 °C the spectra were nearly constant

spectral and temporal detection of the emission although a significant complication is that fluorescence is highly dependent of the preparation and the environment. Fluorescence spectroscopy alone will never be adequate for accurate species determinations except in special cases. Nevertheless, so far no other method has been found better suited for early warning systems.

Overview of Fluorescence-Based Bioaerosol Detection Systems

In this section, the focus is on point detection systems. Fluorescence-based standoff detection system is covered in another chapter in this book. The following paragraphs describing the different point detection systems have been organized after the name of the systems or organizations responsible for the development rather than in a thematic or strict chronological division of the material.

There are a large number of research groups and companies that have developed fluorescence-based bioaerosol detection. Jim Ho has presented the early history of bioaerosol detection that started the development of fluorescence-based detection in the beginning of this book. A relatively new review paper of DeFreez covers most of the developed fluorescence-based systems in North America where most of the development started in the 1990s [31]. MIT Lincoln Laboratory in USA has published two overview articles that cover the technology and strategy of bioaerosol detection from triggering to confirmation giving a good introduction to the whole field including fluorescence-based detection systems [45, 46]. The collaborating groups of Yale University and US Army Research Laboratory (ARL) have recently written a book chapter where they describe their work on fluorescence based detection over more than a decade [20]. Their work is also covered in an earlier book chapter from 2007, where the development at US Naval Research Laboratory (NRL) and in the UK are also described [15].

Fluorescence Aerosol Particle Sensor

As we have seen in the history chapter in the beginning of this book, the first fluorescence aerosol particle sensor (FLAPS) was developed in the early 1990s by TSI with help of Jim Ho at Defence Research Establishment Suffield (DRES), later Defence Research and Development Canada (DRDC). The FLAPS systems have been used since more than 10 years and its user base is widespread over the world. FLAPS has gained interest both from military end-users and from test and evaluation facilities. It has almost become a standard reference equipment for biological field trials. Some of the early history on fluorescence-based biological triggers and FLAPS has also been published elsewhere [47].

The FLAPS I is a modification of the TSI product aerosol particle sizer (APS) [48]. The fluorescence channel uses a HeCd UV laser with excitation at 325 nm and detection of the emission in one band from 420 to 580 nm [49]. FLAPS II was developed by TSI in 1995 and 1996 in collaboration with DRES. For excitation a pulsed tripled Nd:YLF at 349 nm was used and emission was detected in one band from about 410 to 580 nm [50].

TSI has two commercial systems today, UV-APS [51] and FLAPS III [52]. Both have the same air sampling rate as the APS, that is, 1 l air per minute, with a sheath flow of 4 l/min. UV-APS is basically a newer version of FLAPS II with a pulsed tripled Nd:YAG at a wavelength of 355 nm and a pulse repetition frequency of up to 5 kHz as excitation. The particle size range is 0.8–15 μm.

FLAPS III uses a 405 nm excitation laser diode and the system analyses individual particles in the size range of 0.8–10 μm, with respect to light scatter and fluorescence in two wavelength bands, approximately 450–500 and 500–600 nm. Information for individual particles is not possible to resolve from the stored data, since the system measures the particle stream during a period of time, typically 3 s and summarizes the three different channels as histograms of number of particles at certain intensities.

The FLAPS III system has been integrated by Dycor in the stand-alone system C-FLAPS [53], on vehicles and on ships. This version includes an initial particle concentrator (XMX2 by Dycor), which is a high-volume system that can pull approx. 350 l/min and concentrates the particle down in an air stream of about 1 l/min that enters the particle analyzer.

Biological Agent Warning Sensor

Another important system in the early days was the biological agent warning sensor (BAWS) produced by General Dynamics in USA [54]. The system was originally developed by MIT Lincoln Laboratory from 1996 to 2000 [45, 46, 55, 56]. BAWS can be used as a stand-alone system or as a trigger for more complex systems, for example, in the joint biological point detection system (JBPDS), which includes BAWS as the trigger and immunoassays for confirmation.

The system is based on LIF using a 266-nm laser with high repetition rate 7–10 kHz. The elastic scattering (266 nm), the UV (300–400 nm) and the visible (400–600 nm) fluorescence are detected with three photomultiplier tubes (PMTs). The high repetition rate makes it possible to capture single particle fluorescence without triggering. The alarm algorithm is based on correlation from the three different detection channels.

It is difficult to find data on the performance of the operational system, but from the literature [55, 56] it seems that BAWS can probe about 20 l/min and has a sensitivity of about 10–25 agent containing particle per liter of air (ACPLA) of particles in the 1–10 μm size range. The response time is about 10 s to 1 min.

Biological Agent Real Time Sensor

The 4WARN system is manufactured by General Dynamics Canada and is a system similar to the JBPDS mentioned above. It includes a trigger based on fluorescence and a confirmation stage based on immunoassays or polymerase chain reaction (PCR). Details about the trigger in the commercial system 4WARN have not been found but are based on the system biological agent real time sensor (BARTS) developed by Computing Devices Canada, later General Dynamics Canada [57, 58]. The 4WARN systems exist in versions that can be vehicle mounted and man portable.

BARTS has resemblances with both FLAPS (mainly FLAPS II) and BAWS. The system is based on LIF using a 355-nm laser with a pulse repetition rate of 10 kHz. The system has only two detection channels: one for elastic scattering (355 nm) and another for fluorescence at longer wavelengths. Size information is obtained from the elastic scattering. Principal component analysis (PCA) of the combined fluorescence and scattering information is used in the detection algorithm [59, 60]. The system is able to operate at a flow of 30 l/min by using a particle concentrator with an output flow to the optical system of 3 l/min.

Defense Advanced Research Projects Agency Program SUVOS

An important driving force of system development during the 2000s was the Defense Advanced Research Projects Agency (DARPA) program semiconductor ultraviolet optical source (SUVOS). The goal of the program was to developed semiconductor lasers in the UV region, especially around 280 and 340 nm. Unfortunately, it appears that the technical challenges were too hard to overcome. No stable laser diodes with an emission wavelength of less than approximately 370 nm were strange syllabification. However, several companies are now producing light emitting diodes (LEDs) in deeper UV. A trend following this development was to make compact and more affordable systems utilizing the lasers and LEDs that came out of this program. Examples of these systems are BioLert [61] (laser diode at 375 or 405 nm), BAST developed by MIT Lincoln Laboratory [45, 46, 62] and TAC-BIO developed by Edgewood Chemical Biological Center [63], where the two last system were using LEDs at 365 nm. Some more details of the achievements originating from SUVOS can be found in DeFreez' review article [31].

Yale University and Army Research Laboratory

Groups from Yale University and Army Research Laboratory (ARL) have had a long collaboration on biological detection [20]. Already in 1995 they presented spectral fluorescence detection of bio-aerosol [64]. Their development system used an argon-ion laser operating at 488 nm. The year after they presented spectral fluorescence detection from individual bioaerosol particles excited with 266 nm [65].

In 1999 their system was further improved [26, 66]. Excitation at 266 or 351 nm from two different pulsed lasers was used. In order to excite the particle on-the-fly in an aerosol beam, the UV lasers had to be fired at the right time. The UV pulses were triggered by scattering from two red continuous-wave (cw) lasers upstream in the aerosol beam. The system could measure fluorescence spectra with an intensified CCD from individual dry particles with a diameter of 2–5 μm. Fluorescence spectra were measured on clean biological molecules and on BWA simulants. Several fundamental studies on fluorescence from single particles were made with this system and the dependence of, for example, particle size, concentration and excitation intensity was examined [8, 67–70]. This system was also used in environmental field trials targeting naturally occurring aerosol particles at several locations in USA [27, 28]. The instrument was able to classify about 90% of the particles into 8–10 groups with similar spectra.

Recently these groups have developed a system with dual excitation wavelength, 263 and 351 nm [18, 71]. The fluorescence spectra are measured with a 32-channel PMT array. The system is claimed to be able to measure 90,000 spectra/s. It was early pointed out that PMT arrays could improve the measuring rate [72]. The intensified CCD in their earlier setup could only measure 25 spectra/s. The present system could measure particles between 1 and 10 μm. They have also shown that the dual excitation wavelengths improve the classification capability [73]. The system has successfully been used to classify different airborne pollen and fungal particles [74]. The system operating with a single excitation wavelength at 263 nm has also been used for background measurements with results similar to their previous data on environmental aerosols [75].

The Yale group together with a group from Brown University were first to show strong fluorescence signals from single particles using light emitting diodes [76–78]. They have also developed an aerodynamic puffer able to deflect aerosol particles with pre-selected fluorescence characteristics [20, 79]. Additionally they have been involved with aerosol detection based on extraction of detailed elastic scattering features, with the so-called two-dimensional angular scattering (TAOS) [80–85].

Naval Research Laboratory

Naval Research Laboratory (NRL) has developed a number of fluorescence-based detection systems. They called their first system single particle fluorescence analyzer (SPFA). This system measured fluorescence from single particles excited with 266 nm [86, 87]. The emission was measured in two bands: 300–400 and 400–600 nm. The system was used in several field trials [88–90] among one in a subway station [91].

In 2004 a new system using two excitation wavelengths (266 and 355 nm) and three emission detection bands (300–400, 400–500 and 500–600 nm) was presented. This system was used to measure the fluorescence cross section for a number of bio-aerosol particles [10, 92].

A system called the biological aerosol-capture-enrichment (BioACE) system has recently been developed [93, 94]. The same dual excitation wavelengths as in their previous system are used but the number of detected emission bands has been increased to four by dividing the first band into two bands (300–340 and 360–400 nm). Additionally, the elastic scatting detection utilizes polarization information and an electrostatic sorting module is used to selectively collect particles triggering an alarm on the fly. The detection capability has lately been evaluated thoroughly [95]. The group has also investigated two-photon excited fluorescence of bioaerosol using a fiber laser at 524 nm [96].

MIT Lincoln Laboratory

MIT Lincoln Laboratory has been a key player in the development of fluorescence-based systems. They have developed BAWS and BAST as described previously. They have also made a very advanced biological trigger detector that combines fluorescence and laser-induced breakdown spectroscopy (LIBS) in the same instrument [97]. The instrument has three laser wavelengths 266, 355 and 1064 nm, where the two first are for the fluorescence and the third is for LIBS. The fluorescence emission is measured in four bands and the LIBS emission is measured for Mg (285 nm), Ca (422 nm), Na (589 nm), and K (766 nm). An advanced elastic scattering detection scheme is also a part of the instrument. The LIBS part of the system can measure particles down to 3 μm size. The system has a much lower false-alarm rate than the BAWS system. Two overview articles cover these achievements and give a good overview of the strategy of biological detection from triggering to confirmation [45, 46].

UK Systems

Groups from the UK have been the most active in Europe to publish work on fluorescence-based bioaerosol detectors. Much of the research has been performed at the University of Hertfordshire in collaboration with the Defence Science and Technology Laboratory (Dstl), previously Defence Evaluation and Research Agency (DERA). The first system developed combined spatial light-scattering analysis and intrinsic fluorescence measurement. A continuous-wave (cw) 266-nm laser was used for fluorescence excitation and all the emission above approximately 295 nm was detected with a single PMT. The elastic scattered light was analyzed with a 31-channel multi-pixel high-gain hybrid photodiode detector (HPD) [98, 99]. A further development was a system where the fluorescence spectra of individual particles could be acquired using a pulsed fiber laser at 266 nm with a pulse repetition rate of 80 MHz [15, 100].

Dstl presented a similar system called Fluorescence Aerosol Shape Analysis System (FLASAS). The system used a red laser diode at 635 nm for triggering and a pulsed laser with emission in the green (532 nm) and in the UV (266 nm). Elastic scattering from the green pulse, detected with a HPD, was used for size and shape analysis. The UV pulse was used for fluorescence and the emission was measured in a band between 300 and 500 nm [15, 101].

These groups have also been working on another approach for fluorescence detection using xenon lamps instead of lasers or LEDs for the excitation aiming at

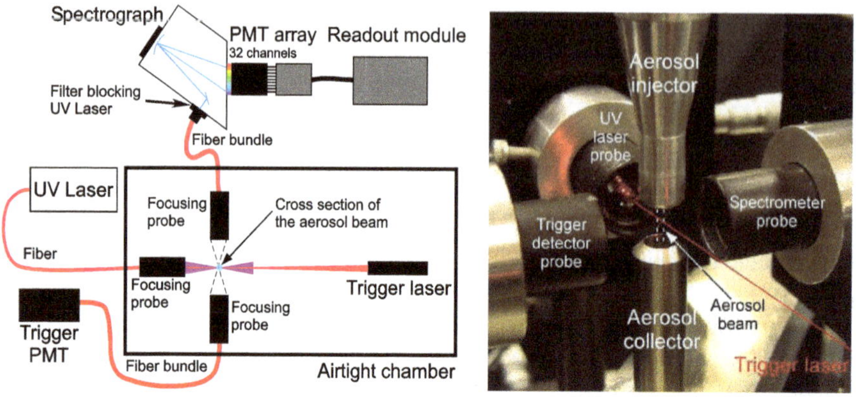

Fig. 7.6 Schematics of the setup seen from top to the *left* and a picture of the setup to the *right*

low-cost sensors. The first prototype of this kind was described in 2004 [102]. Different versions with this approach called wide issue bioaerosol sensor (WIBS) were then developed [15, 17, 103–105]. The first two versions were using fluorescence detection in two bands (320–600 and 400–600 nm) induced by two filtered xenon (Xe) lamps emitting at 280 and 370 nm, respectively. Later versions also included size and shape analysis of the forward scattered light, using a quadrant PMT. The latest version uses a 635-nm diode laser for the triggering and the scattering measurements and fluorescence detection in the 300–400 and the 420–650 nm bands. This instrument is capable of processing up to 125 particles/s (limited by the recharge time of the Xe lamps) with a corresponding flow of 0.25 l of air per minute. Several field measurements on background aerosol have been carried out with these instruments [106, 107].

The approach with Xe lamps has also been used in a commercial available system called VeroTect market by Biral in the UK (see subsequently).

Swedish Defence Research Agency (FOI)

The Swedish Defence Research Agency (FOI) initiated their activity on fluorescence-based detection in the end of the 1990s. The first measurements were made on bulk samples in liquids [108]. It was early recognized that measurements on single particle would lead to better detection and a design was presented in 2004 [42]. The FOI system is described in the literature [21, 109] and resembles the systems of Yale and ARL with spectral detection of single bioaerosol particles using a 32-channel PMT array based spectrometer. Figure 7.6 shows the schematics of the system. The fluorescence excitation wavelengths that have been examined are 266, 292, and 337 nm. Several aspects of single particle fluorescence were investigated [29, 110, 111]. The Royal Institute of Technology (KTH) designed a solid-state pulsed laser to reach tailored wavelengths between 280 and 340 nm [112–115] and different versions were used in variants of the experimental setup.

European Defence Agency Project FABIOLA

The European Defence Agency (EDA) project FABIOLA (Fluorescence Applied to BIOLogical Agents detection) investigated the fluorescence decay time, a feature little explored in fluorescence detection systems. The project is shortly described by Kopczynski et al. [13]. The final system and the results from a field trial were presented in 2008 [19]. The system had dual excitation wavelengths (293 and 337 nm) and the time decay was measured in four different wavelength bands (350–404, 426–450, 446–500, and 513–570 nm). The result from the field trial showed that the information in time-decay measurement could be used for reducing the false alarm rate. The laser [116], the aerosol concentrator and optical chamber [117] and the optical detection system [118] are presented in separate publications.

The Finish group involved in FABIOLA from Tampere University of Technology also developed a research system with a tunable laser as excitation source [119]. This system was later used for measuring cross sections of bioaerosol particles [12] and fundamental biochemicals [120]. The group has also been involved in the development of ENVI BioScout system, see subsequently.

Deep UV Excitation for Fluorescence and Raman Detection

Recent work on biological detection using deep ultraviolet excitation addresses simultaneous acquisition of fluorescence and Raman scattering spectra [121–124]. The aromatic amino acids have a very strong absorption peak of less than 240 nm compared to the strength at around 280 nm, which can be useful for fluorescence detection. More significant may be that a relatively strong Raman signal can be obtained with deep UV laser excitation wavelength [125]. The Raman signal can be strongly enhanced both due its λ^{-4} dependence and due to resonance scattering when the excitation wavelength is closely matched to electronic transition energies. A Raman cross section which is approaching the fluorescence cross section to within a factor of 1000 is claimed to be possible in the deep UV for biological materials. Since the Raman scattering and the fluorescence emission reside in different wavelength regions both kind of spectra can be simultaneously observed, thereby providing a sensor with unique sensitivity and specificity.

A sensor based on detection of a combination of fluorescence emission and Raman scattering called a targeted ultraviolet chemical, biological, and explosives (TUCBE) sensor was recently reported [126]. The TUCBE sensor is claimed to detect and identify less than 10^4 bacterial spores and 1 $\mu g/cm^2$ of explosives at a standoff range of 10 m. The sensor detects and identifies materials on surfaces in less than 1 ms and has a sample rate up to 20 Hz. Varieties of this sensor have been deployed for tests on expeditions to Antarctica, the Arctic, and the deep Ocean. The sensor uses lasers emitting at 224.3 or 248.6 nm.

Grun et al. have developed an instrument measuring two dimensional Raman maps (excitation and emission dimensions). They use pulsed UV laser excitation from 210 to 270 nm and a two stage spectrometer in their instrument denoted

SWOrRD for combined detection of both chemicals and microorganisms [127]. The instrument has been used to study the effect of growth condition on bacteria [128]. Their methods have also been applied to explosives detection [129].

It should be noted that the work of these groups have not actually addressed bioaerosol detection where the technical challenges are different. However, DeFreez has pointed out that a combination of fluorescence and Raman in the deep UV might be the future of bioaerosol detection [31].

Examples of Commercially Available Systems

There are some commercial bioaerosol detection systems that use fluorescence technology as the basic physical principle. The early systems developed in the end of 1990s have already been described previously such as the FLAPS, BAWS and BARTS systems. The two latter have mainly been used as subunits in larger systems that also include a confirmation stage based on immunoassays or polymerase chain reaction (PCR). Some examples of recent fluorescence based systems that have been launched to the market include VeroTect, ENVI BioScout, IMD and IBAC.

It is noticeable that the commercial systems only have one excitation wavelength with fluorescence detection in one or two wavelength bands. In order to improve the classification, to be able to distinguish potentially harmful aerosol from the natural background, complex signal processing algorithms are used. Diode lasers at 405 nm are mostly used, while many earlier versions used solid-state lasers (266 or 355 nm). 405 nm is not an optimal excitation wavelength but diode laser are smaller and more efficient than the corresponding solid-state lasers. Common to all modern systems is that elastic scattering, in varying degree of complexity, is used to gain complementary information in addition to the fluorescence. The following paragraphs describe the basic characteristics of some of the known systems.

VeroTect

VeroTect is manufactured by Biral in the UK [130]. The system is developed in collaboration with scientist at University of Hertfordshire and Dstl (see above). VeroTect is a bioaerosol system with two parallel detection principles; scattering and fluorescence [131, 132]. The first channel is basically an ASAS system (aerosol size and shape analyzer) that measures the numbers, sizes and symmetry factor of the aerosol particles in the size range 0.5–15 μm. This technology has been used in the British PBDS/IBDS system as a trigger of a change in the natural background of aerosols. Verotect uses an initial virtual impactor with an air flow of 33 l/min that is reduced down to 2 l/min for the fluorescence channel. A parallel channel measures the fluorescence intensity of the aerosol particles. The optical chamber is designed so that the fluorescence is measured in a larger volume over a population of particles and not on single particles.

Instead of a laser, a high intensity lamp with a filter is used that gives an excitation wavelength at 280 nm. The fluorescence is measured in two emission bands

330–650 and 420–650 nm. Subtraction of the short-band value from the broadband value is used as a measure of the 330–420 nm fluorescence. The shape and fluorescence parameters cannot be correlated on an individual particle basis due to the parallel air flows.

ENVI BioScout

ENVI BioScout is a "Detect-To-Warn System" developed by Dekati in collaboration with Tampere University of Technology in Finland. The system is marketed by Environics [133]. It is based on the detection of the UV fluorescence properties of organic particles. One approach to achieve a low false positive alarm rate is to remove large particles such as pollen from the sample flow, and then dividing the sample into two size fractions for detection; coarse particles (2–10 μm) and fine particles (smaller than 2 μm). The detection algorithm uses the information from the two size channels in such a way that it assumes that the smaller particles mainly come for natural background particles or interfering particles (from combustion processes) and the larger particles mainly from artificially disseminated bioaerosol. The airflow rate is 2 l/min, the particle size range 0.5–10 μm, the sensitivity is less than 100 ACPLA and the response time is less than 30 s.

IMD

Instantaneous microbial detection (IMD) from BioVigilant is a new detector that utilizes a 405 nm diode laser for excitation of fluorescence [134]. It has one fluorescence channel for detection and measures the particle size (0.5–10 μm) using the elastic scattering. IMD comes in two versions, IMD-A 300 and IMD-A 350, with an air flow rate of 1 and 28 l/min, respectively. The latter includes a virtual impactor.

IBAC

Instantaneous biological analyzer and collector (IBAC) developed by ICx Technologies is yet another bioaerosol detector using a diode laser at 405 nm for fluorescence excitation [31]. This sensor is now in the portfolio of FLIR System under the name FIDO B2 [135]. The airflow rate is 3.8 l/min, the particle size range down to 0.7 μm and the response time is between 30 s and 1 min.

Design Considerations of Fluorescence-Based Systems

Defense Advanced Research Projects Agency (DARPA) together with Defense Threat Reduction Agency (DTRA) in the USA have publish two reports on chemical and biological sensor standards relating to design of detection systems [136,

137]. The reports propose that the sensors should be evaluated with receiver operating characteristic (ROC) curves that capture the performance trade-off between *sensitivity*, *probability of detection*, *false positive rate*, and *response time*. ROC curves and the four key sensor metrics have also been described by Eversole elsewhere [15]. Beside these key sensor metrics, the DARPA reports suggest to include additional attributes such as unit cost, operation cost, maintenance, reliability, size, weight and power consumption. A number of different scenarios useful for evaluation are also presented.

Detection of hazardous bioaerosol particles is like finding a needle in a haystack since extremely low concentrations of BWAs are sufficient to infect people. To meet this challenge most current systems are designed to measure the characteristics of single particles. The ability to separate small fractions of harmful particles in a natural background is thereby better maintained.

Pan et al. have analyzed the excitation power required to detect single BG spores [15]. They estimate that the power flux must exceed 2.4 W/cm^2 for continuous wave (cw) excitation and that the pulse energy flux must exceed 0.12 μJ/cm^2 for pulsed excitation with a laser operating around the peak absorption wavelength 280 nm. This estimate assumes that the total fluorescence collection and detection efficiency is around 4%. For 266 and 355 nm excitations, the levels are about 10 times higher and for 400 nm about 60 times higher. Although these calculations contain some crude approximations they give a hint on the requirements for the excitation sources. Spectrally resolved detection will require even higher fluxes depending on the number of spectral channels. The power or pulse energy fluxes cannot be unlimitedly increased to raise the signal level. Nonlinear effects, photo-degradation, ablation and laser induced breakdown will limit the useful range of operation.

There are a number of sources such as lasers, light emitting diodes (LEDs) and high energy lamps, which can be used as described previously. The optimal excitation wavelength is debated. However, two optimal regions for excitation wavelength are around 280 and 340 nm. Few sources are available at those precise wavelengths so other wavelengths are often used. The sources can be cw or pulsed. At sufficiently high pulse repetition rates (PRFs) the source can be regarded to be quasi cw and triggering will not be needed to synchronize the pulse with a particle traversing the detection volume. However, the detector often still needs be triggered to achieve a sufficient signal-to-noise ratio.

Lasers are the most popular sources for bio-aerosol detection mainly due to their high optical intensity in a narrow spectral band. An essential feature is also that the radiation can be efficiently directed to the detection volume due to their low beam divergence. The third and fourth harmonic of the solid-state lasers Nd:YAG (355 and 266 nm) and Nd:YLF (351 and 263 nm) have been used in many system. Such lasers with both high power and high PRF are commercially available. GaN-based semiconductor diode lasers with wavelengths around 405 nm have also been frequently used even though this wavelength is far from optimal. Despite large efforts to lower the wavelength it has not been possible to produce stable diode lasers of less than 370 nm. Other less used alternatives include the excimer lasers at 248 (KrF), 308 (XeCl), 353 nm (XeF), the HeCd laser at 325 nm; the nitrogen gas laser at 337 nm, the Argon ion laser at 363 nm and 488 nm, and the hollow cathode lasers at 224 nm (HeAg) and 248 nm (NeCu) [15, 125]. It is possible to reach the optimal wavelengths with a solid-state laser followed by an optical parametric oscillator (OPO), see for example, [115], but there is a trade-off between the increase in the fluorescence cross section and the lower conversion efficiency of the OPO.

There are LEDs available on the market today at wavelengths down to around 260 nm. LEDs have many advantages compared to lasers and lamps. They have a higher electrical to optical power conversion efficiency and are smaller, cheaper and more robust. So far the output powers of these LEDs are just on the edge of what is needed for single particle detection. Another main drawback compared to lasers is the worse beam quality which limits the achievable radiation intensity.

The beam properties are naturally even worse for traditional lamps, for example, xenon lamps, which make them optically inefficient. Furthermore, the electrical to optical power conversion efficiency is low with much heat instead of light. However, lamps are relatively cheap and can be found with sufficient power for single particle detection [17].

Detection of the emitted fluorescence signal must be carefully designed to reach down to single bioaerosol particle sensitivity. The light gathering capacity is important. Near or more than 2π steradians are collected in many systems with different optical solutions, for example, with elliptical reflecting mirrors as in the NRL [94] FLASAS [101] and the FABIOLA [117] systems, or with combinations of spherical and elliptical mirror in the MIT systems BAWS [55] and BAST [45]. The Yale and ARL systems have a lower collection efficiency since they use a reflecting Schwarzschild objective to avoid chromatic aberration [20].

The requirements on the detectors are also high. The detector must have close to single photon sensitivity and equally important low noise characteristics. A large linear dynamic range on the order of a thousand is also necessary to cover the intensity variation for bioaerosols from 1 to 10 μm. PMTs are nearly ideal detectors and have been used in most systems. They have a high quantum efficiency of photon detection of around 30%, high gains that can be 10^6 or higher, low dark current and noise. Furthermore they have very fast response time in the order of nanoseconds and have large active areas up to 1 cm^2.

Semiconductor-based detectors are potentially more robust, smaller and generally less expensive than PMTs. In order to reach single-photon sensitivity, avalanche photo-diodes (APDs) or charge-coupled devices (CCDs) are necessary. However, APDs have a high detector-noise-per-unit active area compared to PMTs. Therefore they require the use of small active areas (sub-mm^2) which make the design of the optical system more difficult. The main disadvantages with CCDs are a large readout noise and the slow readout of the entire image. The first problem can be solved by introducing an image intensifier in the front of the CCD. The resulting intensified CCD (ICCD) is suitable for sensitive single-pulse detection, but the readout is still too slow for PRFs of more than about 100 Hz. An alternative for ICCD in spectral applications is linear PMT arrays. They have fewer elements than CCDs, but since the spectral features in fluorescence are broad this can even be an advantage in terms of signal-to-noise ratio.

Some comparisons between different detector technologies in the context of bioaerosol detection have been done by MIT Lincoln Laboratory [45]. The Yale group compares the different technologies with respect to spectral detection [15].

Other aspects which are important for the system design are the aerosol sampling and the transport to the detection volume and the detection algorithms but this is out of scope of this chapter. The aerosol sampling is treated in a previous chapter. Detection algorithms are important in the system design, but are highly dependent on the system. Since the generated data generally consist of multiple fluorescence data channels and complementary information channels, for example, from elastic scattering, the analysis of the data has to be done in a multivariate approach. Some examples of principal-component analysis made on bioaerosol detection data are found in the reference list [35, 36, 108, 138]. Other possibilities can be neural network, factor analysis or similar techniques.

Conclusions and Outlook

Fluorescence spectroscopy has proven to be a powerful method for bioaerosol detection. It has become a key method for warning sensors of adequate sensitivity and response time. However, the fluorescence signal in general has a relatively low specificity and can be significantly altered by the sample preparation and the environment. Therefore, possibly with exception for special cases, fluorescence spectroscopy by itself will not be sufficient for accurate species determinations.

A large number of bioaerosol detection systems have been developed during the years and the technology is relatively mature. Several systems are commercial available. Most of these use only one excitation wavelength with fluorescence detection in one or two wavelength bands. Research systems have explored more complex excitation and detection configurations, for example, with multiple excitation wavelengths, spectrally or temporally resolved detection. It is possible to improve on the classification ability compared to the less advanced systems. Currently, mainly unit cost and maintenance issues hold these systems back from being commercialized. New lasers and detectors can be expected to allow for further development of sensors meeting customer demands.

Almost all fluorescence-based bioaerosol detection system use elastic scattering as an additional channel of information related to the particle size or even particle shape. A general trend during the last couple of years has been to combine fluorescence with additional complementary technologies to increase the classification capability. Flame spectroscopy and laser-induced breakdown spectroscopy provide information of the elementary composition of the particles. At least one system utilizes this possibility [97]. Raman spectroscopy provides information related to the vibrational states of the molecules in the particle. A combination of Raman and fluorescence has proven viable for bulk measurements [126, 127] and could possibly be implemented in a bioaerosol detection system in the future. Finally, fluorescence has been combined with mass spectroscopy [139], potentially the strongest combination for classification ability of a fluorescence-based bioaerosol sensor.

The fundamentals of bioaerosol fluorescence are generally well understood. Nonetheless there is a need for further research mainly oriented toward practical issues such as the impact from the bioaerosol preparation and the environment. In order to be able to distinguish dangerous bioaerosol particles in a complex background there is also a need for a better understanding of naturally occurring atmospheric interfering constituents under different relevant conditions and at different locations.

Defence and security issues have been the driving force for the development of the bioaerosol detection systems. However more and more of the published work have lately moved into environmental studies [74, 106, 107, 140]. More such studies can be foreseen and adjusted systems better matched to environmental application will be developed based on the current state of the art on bioaerosol fluorescence.

References

1. Schäfer FP (1973) 1. Principles of dye laser operation. In: Schäfer FP (ed) Dye Lasers. Topics in Applied Physics, vol 1. Springer, Berlin, pp 1–89. doi:10.1007/3-540-51558-5_7
2. Measures RM (1984) Laser Remote Sensing: Fundamentals and Applications. John Wiley & Sons, New York
3. Lakowicz JR (1999) Principles of fluorescence spectroscopy. 2 edn. Kluwer Academic/Plenum Publisher, New York
4. Alimova A, Katz A, Savage HE, Shah M, Minko G, Will DV, Rosen RB, McCormick SA, Alfano RR (2003) Native fluorescence and excitation spectroscopic changes in *Bacillus subtilis* and *Staphylococcus aureus* bacteria subjected to conditions of starvation. Appl Opt 42 (19):4080–4087. doi:10.1364/AO.42.004080
5. Faris GW, Copeland RA, Mortelmans K, Bronk BV (1997) Spectrally resolved absolute fluorescence cross sections for bacillus spores. Appl Opt 36 (4):958–967. doi:10.1364/AO.36.000958
6. Seaver M, Roselle DC, Pinto JF, Eversole JD (1998) Absolute emission spectra from *Bacillus subtilis* and *Escherichia coli* vegetative cells in solution. Appl Opt 37 (22):5344–5347. doi:10.1364/AO.37.005344
7. Cheng YS, Barr EB, Fan BJ, Hargis J, P. J., Rader DJ, O'Hern TJ, Torczynski JR, Tisone GC, Preppernau BL, Young SA, Radloff RJ (1999) Detection of Bioaerosols Using Multiwavelength UV Fluorescence Spectroscopy. Aerosol Sci Technol 30 (2):186–201. doi:10.1080/027868299304778
8. Hill SC, Pinnick RG, Niles S, Fell Jr. NF, Pan Y-L, Bottiger J, Bronk BV, Holler S, Chang RK (2001) Fluorescence from airborne microparticles: dependence on size, concentration of fluorophores, and illumination intensity. Appl Opt 40 (18):3005–3013. doi:10.1364/AO.40.003005
9. Weichert R, Klemm W, Legenhausen K, Pawellek C (2002) Determination of fluorescence cross-sections of biological aerosols. Part Part Syst Charact 19 (3):216-222. doi:10.1002/1521-4117(200207)19:3<216::AID-PPSC216>3.0.CO;2-S
10. Sivaprakasam V, Huston AL, Scotto C, Eversole JD (2004) Multiple UV wavelength excitation and fluorescence of bioaerosols. Opt Express 12 (19):4457–4466. doi:10.1364/OPEX.12.004457
11. Kunnil J, Sarasanandarajah S, Chacko E, Reinisch L (2005) Fluorescence quantum efficiency of dry *Bacillus globigii* spores. Opt Express 13 (22):8969–8979. doi:10.1364/OPEX.13.008969

12. Manninen A, Putkiranta M, Saarela J, Rostedt A, Sorvajarvi T, Toivonen J, Marjamaki M, Keskinen J, Hernberg R (2009) Fluorescence cross sections of bioaerosols and suspended biological agents. Appl Opt 48 (22):4320–4328. doi:10.1364/AO.48.004320
13. Kopczynski K, Kwasny M, Mierczyk Z, Zawadzki Z (2005) Laser induced fluorescence system for detection of biological agents: European project FABIOLA. Proc SPIE 5954:595405.1–12. doi:10.1117/12.623013
14. Wlodarski M, Kaliszewski M, Kwasny M, Kopczynski K, Zawadzki Z, Mierczyk Z, Mlynczak J, Trafny E, Szpakowska M (2006) Fluorescence excitation-emission matrices of selected biological materials. Proc SPIE 6398:639806.1–12. doi:10.1117/12.687872
15. Pan Y-L, Eversole J, Kaye P, Foot V, Pinnick R, Hill S, Mayo M, Bottiger J, Huston A, Sivaprakasam V, Chang R (2007) Bio-Aerosol Fluorescence. In: Hoekstra A, Maltsev V, Videen G (eds) Optics of Biological Particles. NATO Science Series, vol 238. Springer Netherlands, Dordrecht, pp 63–164. doi:10.1007/978-1-4020-5502-7_4
16. Hill SC, Mayo MW, Chang RK (2009) Fluorescence of Bacteria, Pollens, and Naturally Occurring Airborne Particles: Excitation/Emission Spectra. ARL-TR-4722. U.S. Army Research Laboratory, Adelphi, MD, USA
17. Kaye PH, Stanley WR, Hirst E, Foot EV, Baxter KL, Barrington SJ (2005) Single particle multichannel bio-aerosol fluorescence sensor. Opt Express 13 (10):3583–3593. doi:10.1364/OPEX.13.003583
18. Huang HC, Pan Y-L, Hill SC, Pinnick RG, Chang RK (2008) Real-time measurement of dual-wavelength laser-induced fluorescence spectra of individual aerosol particles. Opt Express 16 (21):16523–16528. doi:10.1364/OE.16.016523
19. Feugnet G, Lallier E, Grisard A, McIntosh L, Hellström JE, Jelger P, Laurell F, Albano C, Kaliszewski M, Wlodarski M, Mlynczak J, Kwasny M, Zawadzki Z, Mierczyk Z, Kopczynski K, Rostedt A, Putkiranta M, Marjamaki M, Keskinen J, Enroth J, Janka K, Reinivaara R, Holma L, Humppi T, Battistelli E, Iliakis E, Gerolimos G (2008) Improved laser-induced fluorescence method for bio-attack early warning detection system. Proc SPIE 7116:71160C.1–11. doi:10.1117/12.799151
20. Huang HC, Pan Y-L, Hill SC, Pinnick RG (2010) Fluorescence-Based Classification with Selective Collection and Identification of Individual Airborne Bioaerosol Particles. In: Serpengüzel A, Poon AW (eds) Optical Processes In Microparticles And Nanostructures, A Festschrift dedicated to Richard Kounai Chang on his Retirement from Yale University. Advanced Series in Applied Physics, vol 6. World Scientific, Singapore, pp 153–167. doi:10.1142/9789814295789_0009
21. Jonsson P, Kullander F, Vahlberg C, Jelger P, Tiihonen M, Wästerby P, Tjärnhage T, Lindgren M (2006) Spectral detection of ultraviolet laser induced fluorescence from individual bioaerosol particles. Proc SPIE 6398:63980F.1–12. doi:10.1117/12.689666
22. Simard J-R, Roy G, Mathieu P, Larochelle V, McFee J, Ho J (2004) Standoff sensing of bioaerosols using intensified range-gated spectral analysis of laser-induced fluorescence. IEEE Trans Geosci Remote Sens 42 (4):865–874. doi:10.1109/TGRS.2003.823285
23. Baxter K, Castle M, Barrington S, Withers P, Foot V, Pickering A, Felton N (2007) UK small scale UVLIF lidar for stand-off BW detection. Proc SPIE 6739:67390Z.1–10. doi:10.1117/12.737730
24. Jonsson P, Elmqvist M, Gustafsson O, Kullander F, Persson R, Olofsson G, Tjärnhage T, Farsund Ø, Haavardsholm TV, Rustad G (2009) Evaluation of biological aerosol stand-off detection at a field trial. Proc SPIE 7484:74840I.1–14. doi:10.1117/12.830401
25. Farsund Ø, Rustad G, Kaasen I, Haavardsholm TV (2010) Required Spectral Resolution for Bioaerosol Detection Algorithms Using Standoff Laser-Induced Fluorescence Measurements. IEEE Sens J 10 (3):655–661. doi:10.1109/JSEN.2009.2037794
26. Hill SC, Pinnick RG, Niles S, Pan Y-L, Holler S, Chang RK, Bottiger J, Chen BT, Orr C-S, Feather G (1999) Real-time measurement of fluorescence spectra from single airborne biological particles. Field Anal Chem Technol 3 (4-5):221–239. doi:10.1002/(SICI)1520-6521(1999)3:4/5<221::AID-FACT2>3.0.CO;2-7

27. Pinnick RG, Hill SC, Pan Y-L, Chang RK (2004) Fluorescence spectra of atmospheric aerosol at Adelphi, Maryland, USA: Measurement and classification of single particles containing organic carbon. Atmos Environ 38 (11):1657–1672. doi:10.1016/j.atmosenv.2003.11.017
28. Pan Y-L, Pinnick RG, Hill SC, Rosen JM, Chang RK (2007) Single-particle laser-induced-fluorescence spectra of biological and other organic-carbon aerosols in the atmosphere: Measurements at New Haven, Connecticut, and Las Cruces, New Mexico. J Geophys Res 112 (D24):D24S19.1–15. doi:10.1029/2007jd008741
29. Jonsson P, Kullander F, Vahlberg C, Wästerby P, Tjärnhage T, Olofsson G, Lindgren M, Tiihonen M, Jelger P (2007) Ultraviolet optical techniques for early-warning detection of biological threats. In: The Proceedings of 9th International Symposium on Protection against Chemical and Biological Warfare Agents, Gothenburg, Sweden, 22–25 May 2007. Umeå, p 6
30. Farsund Ø, Rustad G, Skogan G (2012) Standoff detection of biological agents using laser induced fluorescence-comparison of 294 nm and 355 nm excitation wavelengths. Biomed Opt Express 3 (11):2964–2975. doi:10.1364/BOE.3.002964
31. DeFreez R (2009) LIF bio-aerosol threat triggers: then and now. Proc SPIE 7484:74840H.1–15. doi:10.1117/12.835088
32. Buteau S, Simard J-R, Dery B, Roy G, Lahaie P, Mathieu P, Ho J, McFee J (2006) Bioaerosols laser-induced fluorescence provides specific robust signatures for standoff detection. Proc SPIE 6378:637813/1–12. doi:10.1117/12.686010
33. Bronk BV, Reinisch L (1993) Variability of Steady-State Bacterial Fluorescence with Respect to Growth Conditions. Appl Spectrosc 47 (4):436–440
34. Campbell SD, Tremblay DP, Daver F, Cousins D (2005) Wavelength comparison study for bioaerosol detection. Proc SPIE 5778:130–138. doi:10.1117/12.610998
35. Heaton HI (2005) Principal-components analysis of fluorescence cross-section spectra from pathogenic and simulant bacteria. Appl Opt 44 (30):6486–6495. doi:10.1364/AO.44.006486
36. Kunnil J, Sarasanandarajah S, Chacko E, Reinisch L (2006) Effect of washing on identification of Bacillus spores by principal-component analysis of fluorescence data. Appl Opt 45 (15):3659–3664. doi:10.1364/AO.45.003659
37. Laflamme C, Simard J-R, Buteau S, Lahaie P, Nadeau D, Déry B, Houle O, Mathieu P, Roy G, Ho J, Duchaine C (2011) Effect of growth media and washing on the spectral signatures of aerosolized biological simulants. Appl Opt 50 (6):788–796. doi:10.1364/AO.50.000788
38. Sarasanandarajah S, Kunnil J, Chacko E, Bronk BV, Reinisch L (2005) Reversible changes in fluorescence of bacterial endospores found in aerosols due to hydration/drying. J Aerosol Sci 36 (5-6):689–699. doi:10.1016/j.jaerosci.2004.11.010
39. Santarpia JL, Pan Y-L, Hill SC, Baker N, Cottrell B, McKee L, Ratnesar-Shumate S, Pinnick RG (2012) Changes in fluorescence spectra of bioaerosols exposed to ozone in a laboratory reaction chamber to simulate atmospheric aging. Opt Express 20 (28):29867–29881. doi:10.1364/OE.20.029867
40. Dalterio RA, Nelson WH, Britt D, Sperry J, Psaras D, Tanguay JF, Suib SL (1986) Steady-State and Decay Characteristics of Protein Tryptophan Fluorescence from Bacteria. Appl Spectrosc 40 (1):86–90
41. Dalterio RA, Nelson WH, Britt D, Sperry JF, Tanguay JF, Suib SL (1987) The Steady-State and Decay Characteristics of Primary Fluorescence From Live Bacteria. Appl Spectrosc 41 (2):234–241
42. Jonsson P, Kullander F, Nordstrand M, Tjärnhage T, Wästerby P, Lindgren M (2004) Development of fluorescence-based point detector for biological sensing. Proc SPIE 5617:60–74. doi:10.1117/12.578231
43. Katz A, Alimova A, Siddique M, Savage HE, Shah M, Rosen RB, Alfano RR (2004) Time-resolved and steady-state fluorescence spectroscopy from bacteria subjected to bactericidal agents. Proc SPIE 5269:217–220. doi:10.1117/12.518656
44. Vitta P, Kurilcik N, Jursenas S, Zukauskas A, Bakiene E, Zhang J, Katona T, Bilenko Y, Lunev A, Hu X, Deng J, Gaska R (2005) Fluorescence-lifetime identification of biological agents using deep ultraviolet light-emitting diodes. Proc SPIE 5990:59900X.1–12. doi:10.1117/12.630573

45. Jeys TH, Herzog WD, Hybl JD, Czerwinski RN, Sanchez A (2007) Advanced Trigger Development. Linc Lab J 17 (1):29–62
46. Greenwood DP, Jeys TH, Johnson B, Richardson JM, Shatz MP (2009) Optical techniques for detecting and identifying biological-warfare agents. Proc IEEE 97 (6):971–989. doi:10.1109/JProc2009.2013564
47. Ho J (2002) Future of biological aerosol detection. Anal Chim Acta 457 (1):125–148. doi:10.1016/S0003-2670(01)01592-6
48. APS. http://www.tsi.com/en-1033/models/2204/3321.aspx. Accessed 31 April 2013
49. Hairston PP, Ho J, Quant FR (1997) Design of an instrument for real-time detection of bioaerosols using simultaneous measurement of particle aerodynamic size and intrinsic fluorescence. J Aerosol Sci 28 (3):471–482. doi:10.1016/s0021-8502(96)00448-x
50. Ho J, Spence M, Hairston P (1999) Measurement of biological aerosol with a fluorescent aerodynamic particle sizer (FLAPS): correlation of optical data with biological data. Aerobiol 15 (4):281–291. doi:10.1023/A:1007647522397
51. UV-APS. http://www.tsi.com/en-1033/models/2200/3314.aspx. Accessed 31 April 2013
52. FLAPS III. http://www.tsi.com/en-1033/models/2234/3317.aspx. Accessed 31 April 2013
53. C-FLAPS. http://www.dycor.com/Portals/39/pdf/dycor_cflaps_feb2011.pdf. Accessed 31 April 2013
54. Lynch EJ, Bogucki MI, Gardner PJ, Hyttinen L (2005) Biological agent warning sensor (BAWS): laser-induced fluorescence as the joint biological point detection system trigger. Proc SPIE 5795:75–78. doi:10.1117/12.609918
55. Reyes FL, Jeys TH, Newbury NR, Primmerman CA, Rowe GS, Sanchez A (1999) Bio-aerosol fluorescence sensor. Field Anal Chem Technol 3 (4–5):240–248. doi:10.1002/(SICI)1520-6521(1999)3:4/5%3C240::AID-FACT3%3E3.0.CO;2-%23
56. Primmerman CA (2000) Detection of biological agents. Linc Lab J 12 (1):3–32
57. Luoma G, Cherrier P, Zheng C, Piccioni M, Wong A (2001) Development of a novel biological agent real time sensor (PS-BARTS) based on fluorescence particle sizing. In: Proceedings of the 7th International Symposium on Protection against Chemical and Biological Warfare Agents, Stockholm, Sweden, 15–19 June 2001. FOI, Umeå, p 12
58. Luoma G, Cherrier PP, Piccioni M, Tanton C, Herz S, DeFreez RK, Potter M, Girvin KL, Whitney R (2002) A fluorescence particle detector for real time quantification of viable organisms in air. Proc SPIE 4576:32–39. doi:10.1117/12.456967
59. Retfalvi LA, Newman E, Boryski M, Kacelenga R (2004) The challenges of effective biological agent detection in homeland security applications. In: The proceedings of the 8th International Symposium on Protection against Chemical and Biological Warfare Agents, Gothenburg, Sweden, 2–6 June 2004. FOI, Umeå, p 17
60. Mudigonda NR, Kacelenga R (2006) Biological agent detection based on principal component analysis. Proc SPIE 6218:62180P.1–9. doi:10.1117/12.669522
61. Wilson GA, DeFreez RK (2004) Multispectral diode laser induced fluorescence biological particle sensor. Proc SPIE 5617:46–52. doi:10.1117/12.578854
62. Campbell SD, Jeys TH, Eapen XL (2007) Bioaerosol optical sensor model development and initial validation. Proc SPIE 6538:65380P.1–9. doi:10.1117/12.717075
63. Cabalo J, DeLucia M, Goad A, Lacis J, Narayanan F, Sickenberger D (2008) Overview of the TAC-BIO detector. Proc SPIE 7116:71160D.1–11. doi:10.1117/12.799843
64. Hill SC, Pinnick RG, Nachman P, Chen G, Chang RK, Mayo MW, Fernandez GL (1995) Aerosol-fluorescence spectrum analyzer: real-time measurement of emission spectra of airborne biological particles. Appl Opt 34 (30):7149–7155. doi:10.1364/AO.34.007149
65. Chen G, Nachman P, Pinnick RG, Hill SC, Chang RK (1996) Conditional-firing aerosol-fluorescence spectrum analyzer for individual airborne particles with pulsed 266-nm laser excitation. Opt Lett 21 (16):1307–1309. doi:10.1364/OL.21.001307
66. Pan Y-L, Holler S, Chang RK, Hill SC, Pinnick RG, Niles S, Bottiger JR (1999) Single-shot fluorescence spectra of individual micrometer-sized bioaerosols illuminated by a 351- or a 266-nm ultraviolet laser. Opt Lett 24 (2):116–118. doi:10.1364/OL.24.000116

67. Pan YL, Pinnick RG, Hill SC, Niles S, Holler S, Bottiger JR, Wolf JP, Chang RK (2001) Dynamics of photon-induced degradation and fluorescence in riboflavin microparticles. Appl Phys B 72 (4):449–454. doi:10.1007/s003400100532
68. Hill SC, Pinnick RG, Niles S, Fell Jr NF, Pan Y-L, Bottiger J, Bronk BV, Holler S, Chang RK (2002) Fluorescence from airborne microparticles: Dependence on size, concentration of fluorophores, and illumination intensity - Erratum. Appl Opt 41 (21):4432. doi:10.1364/AO.41.004432
69. Pan Y-L, Hill SC, Wolf JP, Holler S, Chang RK, Bottiger JR (2002) Backward-enhanced fluorescence from clusters of microspheres and particles of tryptophan. Appl Opt 41 (15):2994–2999. doi:10.1364/AO.41.002994
70. Pan Y-L, Hartings J, Pinnick RG, Hill SC, Halverson J, Chang RK (2003) Single-particle fluorescence spectrometer for ambient aerosols. Aerosol Sci Technol 37 (8):628–639. doi:10.1080/02786820390195433
71. Pan Y-L, Pinnick RG, Hill SC, Huang H, Chang RK (2008) Dual-wavelength-excitation single-particle fluorescence spectrometer/particle sorter for real-time measurement of organic carbon and biological aerosols. Proc SPIE 7116:71160J.1–8. doi:10.1117/12.801774
72. Pan YL, Cobler P, Rhodes S, Potter A, Chou T, Holler S, Chang RK, Pinnick RG, Wolf JP (2001) High-speed, high-sensitivity aerosol fluorescence spectrum detection using a 32-anode photomultiplier tube detector. Rev Sci Instrum 72 (3):1831–1836. doi:10.1063/1.1344179
73. Pan Y-L, Hill SC, Pinnick RG, Huang H, Bottiger JR, Chang RK (2010) Fluorescence spectra of atmospheric aerosol particles measured using one or two excitation wavelengths: Comparison of classification schemes employing different emission and scattering results. Opt Express 18 (12):12436–12457. doi:10.1364/OE.18.012436
74. Pan Y-L, Hill SC, Pinnick RG, House JM, Flagan RC, Chang RK (2011) Dual-excitation-wavelength fluorescence spectra and elastic scattering for differentiation of single airborne pollen and fungal particles. Atmos Environ 45 (8):1555–1563. doi:10.1016/j.atmosenv.2010.12.042
75. Pan Y-L, Pinnick RG, Hill SC, Chang RK (2009) Particle-Fluorescence Spectrometer for Real-Time Single-Particle Measurements of Atmospheric Organic Carbon and Biological Aerosol. Environ Sci Technol 43 (2):429–434. doi:10.1021/es801544y
76. Pan Y-L, Boutou V, Chang RK, Ozden I, Davitt K, Nurmikko AV (2003) Application of light-emitting diodes for aerosol fluorescence detection. Opt Lett 28 (18):1707–1709. doi:10.1364/OL.28.001707
77. Davitt K, Song YK, Nurmikko AV, Jeon SR, Gherasimova M, Han J, Pan YL, Chang RK (2005) UV LED arrays for spectroscopic fingerprinting of airborne biological particles. Phys Status Solidi C 2 (7):2878–2881. doi:10.1002/pssc.200461591
78. Davitt K, Yoon-Kyu S, Patterson WR, III, Nurmikko AV, Gherasimova M, Jung H, Pan Y-L, Chang RK (2005) 290 and 340 nm UV LED arrays for fluorescence detection from single airborne particles. Opt Express 13 (23):9548–9555. doi:10.1364/OPEX.13.009548
79. Pan Y-L, Boutou V, Bottiger JR, Zhang SS, Wolf J-P, Chang RK (2004) A puff of air sorts bioaerosols for pathogen identification. Aerosol Sci Technol 38 (6):598–602. doi:10.1080/02786820490465450
80. Holler S (1999) Real-time Airborne Microparticle Characterization: Two-dimensional Angular Optical Scattering (TAOS) and UV Fluorescence Spectroscopy. Ph.D., Yale University, New Haven, CT, USA
81. Holler S, Auger JC, Stout B, Pan Y, Bottiger JR, Chang RK, Videen G (2000) Observations and calculations of light scattering from clusters of spheres. Appl Opt 39 (36):6873–6887. doi:10.1364/AO.39.006873
82. Pan Y-L, Aptowicz KB, Chang RK, Hart M, Eversole JD (2003) Characterizing and monitoring respiratory aerosols by light scattering. Opt Lett 28 (8):589–591. doi:10.1364/OL.28.000589
83. Holler S, Zomer S, Crosta GF, Pan Y-L, Chang RK, Bottiger JR (2004) Multivariate analysis and classification of two-dimensional angular optical scattering patterns from aggregates. Appl Opt 43 (33):6198–6206. doi:10.1364/AO.43.006198

84. Fernandes GE, Pan YL, Chang RK, Aptowicz K, Pinnick RG (2006) Simultaneous forward- and backward-hemisphere elastic-light-scattering patterns of respirable-size aerosols. Opt Lett 31 (20):3034–3036. doi:10.1364/OL.31.003034
85. Sindoni OI, Saija R, Iati MA, Borghese F, Denti P, Fernandes GE, Pan Y-L, Chang RK (2006) Optical scattering by biological aerosols: experimental and computational results on spore simulants. Opt Express 14 (15):6942–6950. doi:10.1364/OE.14.006942
86. Seaver M, Eversole JD, Hardgrove JJ, Cary WK, Roselle DC (1999) Size and Fluorescence Measurements for Field Detection of Biological Aerosols. Aerosol Sci Technol 30 (2):174–185. doi:10.1080/027868299304769
87. Eversole JD, Roselle D, Seaver ME (1999) Monitoring biological aerosols using UV fluorescence. Proc SPIE 3533:34–42. doi:10.1117/12.336868
88. Eversole JD, Hardgrove JJ, Cary WK, Choulas DP, Seaver M (1999) Continuous, rapid biological aerosol detection with the use of UV fluorescence: Outdoor test results. Field Anal Chem Technol 3 (4–5):249–259. doi:10.1002/(SICI)1520-6521(1999)3:4/5<249::AID-FACT4>3.0.CO;2-O
89. Eversole JD, Cary Jr. WK, Scotto CS, Pierson R, Spence M, Campillo AJ (2001) Continuous bioaerosol monitoring using UV excitation fluorescence: Outdoor test results. Field Anal Chem Technol 5 (4):205–212. doi:10.1002/fact.1022
90. Eversole JD, Scotto CS, Spence M, Campillo AJ (2003) Continuous bioaerosol monitoring using UV excitation fluorescence. Proc SPIE 4829:532–533. doi:10.1117/12.525515
91. Birenzvige A, Eversole J, Seaver M, Francesconi S, Valdes E, Kulaga H (2003) Aerosol characteristics in a subway environment. Aerosol Sci Technol 37 (3):210–220. doi:10.1080/02786820300941
92. Sivaprakasam V, Huston AL, Scotto C, Eversole JD (2004) Multiple UV wavelength excitation and fluorescence of bioaerosols. Proc SPIE 5585:71–78. doi:10.1117/12.571296
93. Eversole JD, Sivaprakasam V, Pletcher TA, Keller D (2008) Single aerosol particle selection and capture using laser scattering and fluorescence. Proc SPIE 7116:71160F.1–11. doi:10.1117/12.799890
94. Sivaprakasam V, Pletcher T, Tucker JE, Huston AL, McGinn J, Keller D, Eversole JD (2009) Classification and selective collection of individual aerosol particles using laser-induced fluorescence. Appl Opt 48 (4):B126–B136. doi:10.1364/AO.48.00B126
95. Sivaprakasam V, Lin H-B, Huston AL, Eversole JD (2011) Spectral characterization of biological aerosol particles using two-wavelength excited laser-induced fluorescence and elastic scattering measurements. Opt Express 19 (7):6191–6208. doi:10.1364/OE.19.006191
96. Sivaprakasam V, Lou JW, Currie M, Eversole JD (2011) Two-photon excited fluorescence from biological aerosol particles. J Quant Spectrosc Radiat Transf 112 (10):1511–1517. doi:10.1016/j.jqsrt.2011.02.010
97. Hybl JD, Tysk SM, Berry SR, Jordan MP (2006) Laser-induced fluorescence-cued, laser-induced breakdown spectroscopy biological-agent detection. Appl Opt 45 (34):8806–8814. doi:10.1364/AO.45.008806
98. Barton JE, Hirst E, Kaye PH, Clark JM (2000) Simultaneous light scattering and intrinsic fluorescence measurement for bioaerosol detection. J Aerosol Sci 31 (SUPPL 1):S967–S968. doi:10.1016/S0021-8502(00)90977-7
99. Kaye PH, Barton JE, Hirst E, Clark JM (2000) Simultaneous light scattering and intrinsic fluorescence measurement for the classification of airborne particles. Appl Opt 39 (21):3738–3745. doi:10.1364/AO.39.003738
100. Hirst E, Kaye PH, Foot V, Clark JM, Withers PB (2004) An instrument for the simultaneous acquisition of size, shape, and spectral fluorescence data from single aerosol particles. Proc SPIE 5617:416–423. doi:10.1117/12.578269
101. Foot VJ, Clark JM, Baxter KL, Close N (2004) Characterising single airborne particles by fluorescence emission and spatial analysis of elastic scattered light. Proc SPIE 5617:292–299. doi:10.1117/12.578198

102. Kaye PH, Hirst E, Foot V, Clark JM, Baxter K (2004) A low-cost multi-channel aerosol fluorescence sensor for networked deployment. Proc SPIE 5617:388–398. doi:10.1117/12.578283
103. Kaye PH, Stanley WR, Foot V, Baxter K, Barrington SJ (2005) A dual-wavelength single particle aerosol fluorescence monitor. Proc SPIE 5990:59900N.1–12. doi:10.1117/12.629868
104. Foot VE, Kaye PH, Stanley WR, Barrington SJ, Gallagher M, Gabey A (2008) Low-cost real-time multi-parameter bio-aerosol sensors. Proc SPIE 7116:711601.1–12. doi:10.1117/12.800226
105. Stanley WR, Kaye PH, Foot VE, Barrington SJ, Gallagher M, Gabey A (2011) Continuous bioaerosol monitoring in a tropical environment using a UV fluorescence particle spectrometer. Atmos Sci Lett 12 (2):195–199. doi:10.1002/asl.310
106. Gabey AM, Gallagher MW, Whitehead J, Dorsey JR, Kaye PH, Stanley WR (2010) Measurements and comparison of primary biological aerosol above and below a tropical forest canopy using a dual channel fluorescence spectrometer. Atmos Chem Phys 10 (10):4453–4466. doi:10.5194/acp-10-4453-2010
107. Gabey AM, Stanley WR, Gallagher MW, Kaye PH (2011) The fluorescence properties of aerosol larger than 0.8 μm in urban and tropical rainforest locations. Atmos Chem Phys 11 (11):5491–5504. doi:10.5194/acp-11-5491-2011
108. Tjärnhage T, Strömqvist M, Olofsson G, Squirrell D, Burke J, Ho J, Spence M (2001) Multivariate data analysis of fluorescence signals from biological aerosols. Field Anal Chem Technol 5 (4):171–176. doi:10.1002/fact.1018
109. Jonsson P, Kullander F, Wästerby P, Tiihonen M, Lindgren M (2005) Detection of fluorescence spectra of individual bioaerosol particles. Proc SPIE 5990:59900M.1–15. doi:10.1117/12.630141
110. Jonsson P, Kullander F, Tiihonen M, Nordstrand M, Tjærnhage T, Wæsterby P, Olofsson G, Lindgren M (2005) Development of fluorescence-based LIDAR technology for biological sensing. Mater Res Soc Symp Proc 883:51–62. doi:10.1557/PROC-883-FF1.6
111. Jonsson P, Kullander F, Vahlberg C, Gustavsson O, Tiihonen M, Jelger P, Wästerby P, Tjärnhage T, Lindgren M (2006) Spectral Detection of Ultraviolet Laser Induced Fluorescence from Dry Biological Particles. In: Proceedings of the 7th Joint Conference on Standoff Detection for Chemical and Biological Defense, Williamsburg, VA, USA, 23–27 October 2006. Williamsburg. p 10
112. Tiihonen M, Pasiskevicius V, Laurell F, Jonsson P, Lindgren M (2004) A compact OPO/SFG laser for ultraviolet biological sensing. Proc SPIE 5332:134–142. doi:10.1117/12.530292
113. Tiihonen M, Pasiskevicius V, Laurell F, Hammarström P, Lindgren M (2004) A UV laser source for biological and chemical sensing. Proc SPIE 5240:127–136. doi:10.1117/12.509641
114. Tiihonen M, Pasiskevicius V, Laurell F, Lindgren M (2004) A novel UV-laser source for fluorescence excitation of proteins. Proc SPIE 5617:261–268. doi:10.1117/12.568436
115. Tiihonen M, Pasiskevicius V, Laurell F (2007) Tailored UV-laser source for fluorescence spectroscopy of biomolecules. Opt Laser Eng 45 (4):444–449. doi:10.1016/j.optlaseng.2005.03.016
116. Feugnet G, Grisard A, Lallier E, McIntosh L, Hellström J (2008) Advanced double-pulse UV source for laser-induced fluorescence of bioaerosols. Proc SPIE 7116:71160O.1–7. doi:10.1117/12.799143
117. Rostedt A, Putkiranta M, Marjamaki M, Keskinen J, Janka K, Reinivaara R, Holma L (2006) Optical chamber design for aerosol particle fluorescent measurement. Proc SPIE 6398:63980G.1–10. doi:10.1117/12.689803
118. Battistelli E, Paolinetti R, Pompei C, Puccini S (2008) The optical detection system of FABIOLA. Proc SPIE 7116:71160G.1–9. doi:10.1117/12.800131
119. Manninen A, Putkiranta M, Rostedt A, Saarela J, Laurila T, Marjamaki M, Keskinen J, Hernberg R (2008) Instrumentation for measuring fluorescence cross sections from airborne microsized particles. Appl Opt 47 (2):110–115. doi:10.1364/AO.47.000110

120. Putkiranta M, Manninen A, Rostedt A, Saarela J, Sorvajärvi T, Marjamäki M, Hernberg R, Keskinen J (2010) Fluorescence properties of biochemicals in dry NaCl composite aerosol particles and in solutions. Appl Phys B 99 (4):841–851. doi:10.1007/s00340-010-4073-z
121. Grun J, Manka CK, Nikitin S, Zabetakis D, Comanescu G, Gillis D, Bowles J (2007) Identification of bacteria from two-dimensional resonant-Raman spectra. Anal Chem 79 (14):5489–5493. doi:10.1021/ac070681h
122. Bhartia R, Hug WF, Salas BC, Sijapati K, Lane AL, Reid RD, Conrad PG (2006) Biochemical detection and identification false alarm rate dependence on wavelength using laser induced native fluorescence. Proc SPIE 6218:62180J.1–9. doi:10.1117/12.674404
123. Bhartia R, Hug WF, Salas EC, Reid RD, Sijapati KK, Tsapin A, Abbey W, Nealson KH, Lane AL, Conrad PG (2008) Classification of Organic and Biological Materials with Deep Ultraviolet Excitation. Appl Spectrosc 62 (10):1070–1077. doi:10.1366/000370208786049123
124. Bhartia R, Salas EC, Hug WF, Reid RD, Lane AL, Edwards KJ, Nealson KH (2010) Label-Free Bacterial Imaging with Deep-UV-Laser-Induced Native Fluorescence. Appl Environ Microbiol 76 (21):7231–7237. doi:10.1128/aem.00943-10
125. Hug WF, Bhartia R, Taspin A, Lane A, Conrad P, Sijapati K, Reid RD (2005) Status of miniature integrated UV resonance fluorescence and Raman sensors for detection and identification of biochemical warfare agents. Proc SPIE 5994:59940J.1–12. doi:10.1117/12.628923
126. Hug WF, Reid RD, Bhartia R, Lane AL (2009) Performance status of a small robot-mounted or hand-held, solar-blind, standoff chemical, biological, and explosives (CBE) sensor. Proc SPIE 7304:73040Z.1–8. doi:10.1117/12.817881
127. Grun J, Bowles J, Gillis D, Kunapareddy P, Lunsford R, Manka CK, Nikitin S, Wang Z (2010) Tunable multi-wavelength resonance-Raman detection of bacteria and chemicals in complex environments. Proc SPIE 7687:768706.1–12. doi:10.1117/12.863209
128. Kunapareddy N, Grun J, Lunsford R, Gillis D, Nikitin S, Wang Z (2012) Multi-wavelength resonance Raman spectroscopy of bacteria to study the effects of growth condition. Proc SPIE 8358:83580B.1–7. doi:10.1117/12.918652
129. Comanescu G, Manka CK, Grun J, Nikitin S, Zabetakis D (2008) Identification of explosives with two-dimensional ultraviolet resonance Raman spectroscopy. Appl Spectrosc 62 (8):833–839
130. VeroTect. http://www.biral.com/content/Biral_and_biodetection. Accessed 28 June 2013
131. Shelton MJ, Evans SP, Smith PD, Simpson IA, Kaye PH, Clarke JM (2004) Real-time biological agent detection using particle size, shape and fluorescence characterisation. Proc SPIE 5617:284–291. doi:10.1117/12.573636
132. Clark JM, Shelton MJ, Evans SP, Smith PD, Simpson IA, Kaye PH (2005) A new real-time biological agent characterisation system. Proc SPIE 5990:59900Z.1–8. doi:10.1117/12.634065
133. ENVI BioScout. http://www.environics.fi/index.php/biological-detection/envi-bioscout. Accessed 28 June 2013
134. IMD. http://www.biovigilant.com/products/. Accessed 28 June 2013
135. IBAC. http://gs.flir.com/products/icx-detection/biological/ibac/. Accessed 28 June 2013
136. Carrano JC, Jeys T, Cousins D, Eversole J, Gillespie J, Healy D, Licata N, Loerop W, O'Keefe M, Samuels A, Schultz J, Walter M, Wong N, Billotte W, Munley M, Reich E, Roos J (2004) Chemical and biological sensor standards study. Defense Advanced Research Projects Agency, Arlington VA. Available from: http://www.dtic.mil/cgi-bin/GetTRDoc?Location=U2&doc=GetTRDoc.pdf&AD=ADA458370
137. Carrano J, Jeys T, Eversole J, Gillespie J, Licata N, Loerop W, Munley M, O'Keefe M, Roos J, Samuels A, Schultz J, Shatz M, Wong N, D'Amico F, Casale AM, Holster SE, McGrath JF, Metrovich A, Murphy C, Nelson-Patel K, Reich E, Riisager T (2010) Chemical and Biological Sensor Standards Study II. Advanced Research Projects Agency and Defense Threat Reduction Agency, Arlington VA. Available from: http://www.dtra.mil/docs/system-documents/Chem_Bio_Sensor_Standards_Study_Vol_2_Oct_2010.pdf

138. Kunnil J, Sarasanandarajah S, Chacko E, Swartz B, Reinisch L (2005) Identification of Bacillus spores using clustering of principal components of fluorescence data. Aerosol Sci Technol 39 (9):842–848
139. Van Wuijckhuijse AL, Stowers MA, Kleefsman WA, Van Baar BLM, Kientz CE, Marijnissen JCM (2005) Matrix-assisted laser desorption/ionisation aerosol time-of-flight mass spectrometry for the analysis of bioaerosols: Development of a fast detector for airborne biological pathogens. J Aerosol Sci 36 (5–6):677–687
140. Huffman JA, Treutlein B, Pöschl U (2010) Fluorescent biological aerosol particle concentrations and size distributions measured with an Ultraviolet Aerodynamic Particle Sizer (UV-APS) in Central Europe. Atmos Chem Phys 10 (7):3215–3233. doi:10.5194/acp-10-3215-2010

Chapter 8
Bioaerosol Detection with Atomic Emission Spectroscopy

Nicolas Leone, Damien Descroix and Salam Mohammed

Introduction to Atomic Emission Spectroscopy-Based Methods

Atomic emission spectroscopy (AES) methods refer to elementary analysis of a sample by thermal excitation and analysis of the emitted radiation. According to modern atomic theory, atoms and molecules may exist in certain discrete levels of internal energy. To excite an atom or a molecule from the ground state to a higher energy level, some energy must be absorbed, in this case from a thermal source, for example, a flame or plasma. The atoms or the molecules are then in an unstable state from which the atoms or molecules may return to their original state by delivering some energy under radiation. The energy of the radiation is directly related to the electronic transitions which are also intrinsic of the excited compounds. These characteristic radiations (spectral lines, molecular bands) can then be analyzed by means of spectroscopic devices. The wavelength of the emission gives the elementary identity, whereas the intensity is proportional to the quantity of the elements.

AES is a broad area that includes several analytical chemistry techniques for elementary analysis used in both academic and industrial instrumentations within biological and chemical sciences. The different techniques are often named according to the way of producing the thermal excitation. Some examples are flame emission spectroscopy (FES) , spark and arcs AES, and different plasma techniques named after the way the plasma is generated, for example, inductively coupled plasma (ICP), direct current plasma (DCP) , microwave induced plasma (MIP) and laser-induced breakdown spectroscopy (LIBS) . This chapter is focused on FES and LIBS since mainly these two techniques have been used in bioaerosol detection.

N. Leone (✉) · D. Descroix
Physical Detection Department, DGA CBRN Defence, 5 rue Lavoisier, BP n°3,
91710, Vert le Petit, France
e-mail: nicolas.leone@intradef.gouv.fr

S. Mohammed
Division of CBRN Defence and Security, FOI—Swedish Defence Research Agency,
Cementvägen 20, SE-901 82, Umeå, Sweden

P. Jonsson et al. (eds.), *Bioaerosol Detection Technologies,* Integrated Analytical Systems, 143
DOI 10.1007/978-1-4419-5582-1_8, © Springer-Verlag New York 2014

AES techniques have several advantages as a biological trigger, where the main asset is fast analysis, essentially because no sample preparation is necessary. Secondly, the excitation–relaxation phenomena are rapid (for example a few μs for thermal plasma). Real-time analysis is then possible by means of the current optoelectronic devices and related data processing. The elementary analysis of the matter makes it possible to use these technologies in both chemical and biological detection. In bioaerosol detection the AES techniques can give an efficient and adequate warning function. However, they will likely never be a very specific biological identifier, since microorganisms have complex structures that cannot be determined by its elementary composition only.

A bacterial cell is a living structure composed of a genome, a cytoplasm, and various membranes. This biological structure is made of chemical elements in complex molecules (DNA, proteins, lipids, and so on) with mainly organic elements (e.g., C, H, O, and N). The cytoplasm contains salts for the metabolism (mineral elements like Mg, Na, Fe, K, or Ca). All these elements may be used as markers of certain living cells by AES means. The purpose of AES in biological detection is to excite the matter and to analyze the characteristic photon emissions via obtained spectra. The next step is to try to look after the signatures by using different statistical methods such as chemometrics. The expected observables could be a specific compound (like P in DNA), atomic ratios (content of Na versus the one of K inside the cytoplasm). However, certain observables are likely to be modified or affected by changing parameters, for example, growth conditions, dispersion ways and particle background.

To get some information about aerosols and more especially about bacteria (particles of interest), it is necessary to pass particle by particle through the energetic source in order to avoid chemical coincidences. By analyzing a spectrum from a single particle (commonly in the 250–950 nm wavelength window, depending on the experimental configuration), a chemical fingerprint can be qualitatively, and sometimes even quantitatively resolved. These different signatures could be used to separate for example, bacteria from other interfering natural or artificial atmospheric particles such as dust, soot, oil, smoke, salt particles, pollen, molds, fungi, etc.

Many features are common for the AES techniques described in this chapter, that is, FES and LIBS. First, the matter is excited by a thermal source (flame or plasma) where the sample is vaporized, atomized, and even ionized by the highly energetic source. The resulting emission is then analyzed in a similar way. However there are some important differences.

One important issue is that FES is a continuous method where all the particles passing through the burner are analyzed. On the contrary, LIBS is a discrete method. In order to induce the breakdown, a high energy pulsed laser is required. This type of lasers is limited by the pulse repetition frequency, typically up to 50 Hz. Furthermore the duration of the plasma emissions is a few microseconds. To detect a particle with LIBS, this one must be within the plasma volume. It is therefore necessary to trigger the laser pulse when a particle is present within this volume in order to get a sufficient hit rate.

Another important issue is due to the difference of energetic sources. For FES, a flame is generated in a burner with a combustive and a fuel. The obtained temperatures can be in the range 1500–5000 K. The temperature depends on the nature of the fuel, the combustive, and their proportions. Concerning the thermal plasma induced by a pulsed focused laser, plasma temperatures greater than 10000 K can be reached. Since the elementary excitation is a function of temperature and dissociation energy level, this difference in temperature allows the LIBS to excite and to detect more elements. Furthermore, LIBS is a purely optical process. It is then possible to use it both in point detection as well as stand-off detection, while FES could only be used in point detection.

Hence, LIBS theoretically presents higher potential for bioaerosol detection in some aspects. However in order to use the full LIBS potential, a better spectral resolution is needed compared to FES, which implies high resolution spectrographs and intensified CCD (charge couple device) cameras. Also the triggering makes a LIBS system more complex. Comparatively, FES is a simpler technology. Therefore the cost of a FES system is lower. As a consequence, LIBS is still an emerging technique with only a few operational demonstrators developed so far, while operational FES systems have been used for a long time for bioaerosol detection. Existing FES systems are transportable and even man-portable.

An AES detector discussed here requires: sampling system, reactor (flame or plasma), optical sensors, and reliable data processing. Each device is used for monitoring ambient air in order to potentially detect the presence of hazardous biological agents. In the biological detection field, the challenge is to develop sensitive methods or systems to detect low concentrations of bio agents—ideally down to a single biological particle per liter of air—dispersed within a natural, variable and complex atmospheric background. To improve this detection, it is often necessary to concentrate the particles of interest. Moreover, an adequate, robust, and reliable algorithm should be designed and produced to differentiate artificially sprayed bio agents versus atmospheric (background) particles.

This chapter is organized as follows. Firstly, the FES is described in general terms. The advantages and drawbacks of this technique are discussed. The emphases are put on some important concepts about flame transformation processes. A commercially available FES system called MAB (*Moniteur d'Alerte Biologique*) is described. Experimental results with this system in bioaerosol detection application are presented.

LIBS technique is then presented briefly. The basic LIBS apparatus are described and emphases are put on the elemental sensitivity and on the sampling concerns linked to the random interaction between the laser-induced plasmas and the particles.

Both FES and LIBS use similar signal processing strategies to analyze the detected emissions. Here, we list various kinds of multi- and megavariate algorithms used to sort and classify the different analyzed particles: a nonpredictive method (principle component analysis—PCA), and a predictive one (partial least squares—PLS).

Finally, the general conclusions of FES- and LIBS-based systems as bioaerosol detectors are discussed. Potential technical improvements are suggested for both techniques to further enhancement of the bioaerosol detection.

Flame Emission Spectroscopy

General Characteristics

FES, often called flame photometry, has been a very powerful analytical method for many years with applications in several scientific and technological areas [1] such as industry [2], environment [3–5], energy, biology, medicine, defense and so on. The instruments based on this technique are also called spectrophotometers or flame photometric detectors (FPD or FES).

FESs are often developed and used for laboratory analysis, where the main purpose is to characterize a suspension or a solution. The FES devices usually consist of four main parts: a nebulizer, a burner, some optical devices, and a signal processing unit [6]. A compressed air sprayer is used to nebulize the solution or liquid suspension to be analyzed. Air containing the spray and fuel gas are mixed inside the burner. They produce a flame characteristically colored according to the nature of the excited vapors. The global flame spectrum and the superimposed lines from the sample are collected by means of a dedicated optical device. The spectral line intensity is measured and the signal is processed by algorithms with a computer. The obtained results give qualitative and quantitative (if proper calibration has been initially performed) information about the compounds present in the sample.

In the defense domain, several systems have been developed and used to detect chemical warfare agents for example, the portable chemical contamination control apparatus (AP2C) made by the French company Proengin [7]. Earlier systems were fielded in Israel (CHASE) and France (DETALAC). A flame photometer appears quite simple in terms of design and operation compared to other physical method-based analytical devices such as for example, mass spectrometers [8]. The possibilities to simultaneously detect two or more elements in a particle (or vapor) and to measure minute quantities of sample material are one of the remarkable advantages of FES. The minimal detectable mass for sodium is about 10^{-17} g [9].

Compared to alternative analytical methods (chromatography, mass spectrometry, etc.), flame photometry is sensitive, accurate and rapid:

- The method allows analyzing small quantities and the precision is generally very good for minute concentrations of emissive elements in a spray or aerosol. FES is often used for trace element or impurity analysis [10]. Its sensitivity, that is the best detection limits of some 70 elements, is in the range from 10^{-5} to 10^{2} ppm in mass,
- The flame photometry is very well appropriate to analyze airborne particles [3, 4] as these are already in the aerosol form. Thus, it is not necessary to spray a

solution or to prepare a sample. Furthermore, an internal reference standard is not necessary. FES-based techniques are very rapid and dozens of spectra can be measured and recorded per second [3].

- Although flame photometry has been applied to more than 70 elements, its usual application is often restricted to the metallic elements such as alkali or alkaline-earth metals, because they are easily detectable and require less energy to be excited [10]. In bacteria, some of these elements are present and detectable [11].

Even if a flame is a complex chemical and physical system, it presents numerous advantages compared to other excitation sources (spark, ICP, LIBS, etc.). It can for example remain constant and stable during hours.

FES has other advantages in biological detection. It enables continuous monitoring [3], and simultaneous real time detection [12]. The flame photometers can be portable as for example, the AP2C, reliable, with low maintenance level, and insensitive to harsh conditions. Another advantage is the relatively low cost of the technology in comparison with other alternatives, for example, mass spectroscopy, laser-based, and fluorescence-based systems (LIBS, LIF, etc.).

A limitation with FES is that in order to obtain information about the elementary composition of a single particle, it is necessary to burn particle by particle. An aerosol particle will be detected only if it contains emissive compounds with concentrations larger than the limits of detection for each emissive element [3].

Also, the flame temperature plays an essential role in the possibility to detect a compound in a particle. The higher the flame temperature is, the more compounds are possible to detect. Some flames can only detect alkali and earth alkali (Na, K, Ca…) compounds present in various cells [13]. Since many particles in the atmosphere contain sodium [14] and potassium [5] compounds, these two elements are not specifically present only in the bacteria. The knowledge of the sodium–potassium ratios and the presence in high quantity of certain particles can allow to discriminate between the atmospheric aerosol and man-made particle dispersions [14, 15]. Some particles present in the air may have the same FES signature as the bacterial particles and can therefore also be interfering. However, the bacterium flame spectra present several well separated lines. The resolution and the identification of the individual elements are then easy, even in the presence of other elements.

With other techniques such as spark flame, oxygen-cyanogen flames (4600 °C) [1] or laser-induced plasmas (15000 °C) [16–18], higher temperatures can be obtained. The maximum temperature is an important parameter in the fuel choice, but not the only criterion. The choice can be made with other criteria like simplicity of using, advantages of certain fuels, low cost, safety and so on.

We describe how the FES signal from an aerosol sample is obtained in the following sections. The associated processes in analytical flame emission are also described. Emphasis is laid on experimental or qualitative description rather than the theoretical aspects of the subject. FES is not a biologically selective method in the sense that the technique can perform "identification", but it can be used for classification of bioaerosols and as a triggering device, which will be shown with experimental examples.

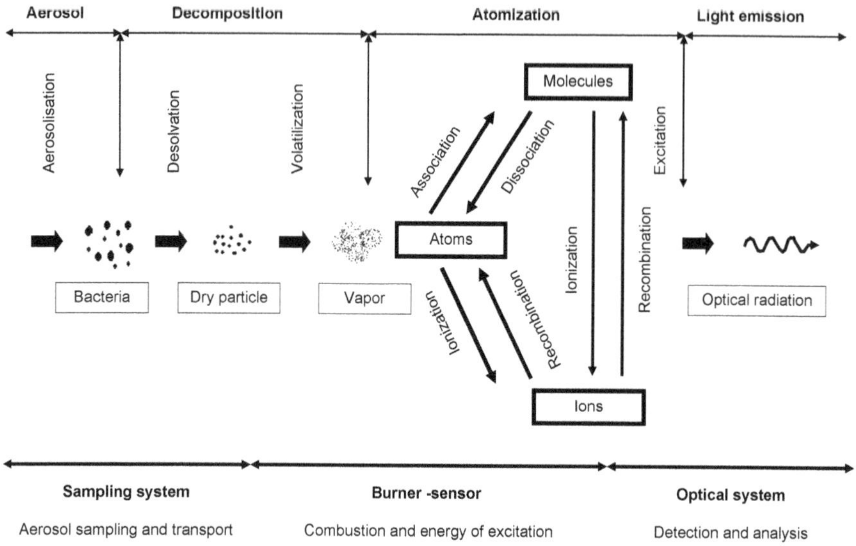

Fig. 8.1 Aerosol transformation processes in the different parts of a FES aerosol monitor

Flame Emission Spectroscopy Transformation Processes

The transformation process in FES is described in Fig. 8.1. Once the particle is sampled, it is transported toward the burner. The particle then enters the flame and it is converted into excited atomic or molecular vapor able to radiate. The flame is produced within a burner by combustion of a fuel gas (e.g., C_2H_2 or H_2) with an oxidant gas (e.g., air or O_2) at constant pressure in order to produce a steady flame [1]. The heat of the flame makes any liquid in the particles to start evaporating before passing the flame. Because of the high flame temperature, the remaining dry aerosol particles vaporize from solid or molten state into the gas state (vapor). In this state the elements to be analyzed (e.g., Ca) may be present as free neutral atoms (Ca) or molecules (CaO) or as positive ions (Ca^{2+}). Vaporization is an irreversible process and in the gaseous state, reversible reactions and processes usually set up a certain balance between the atomic, molecular and ionic species [19]. The following successive stages are: decomposition, atomization and light emission. The atomization is the formation of free atoms by molecular dissociation and/or ion recombination, and *vice versa* (i.e., ionization or association). In the vapor state, the elements/compounds are excited by thermal energy and may emit atomic or ionic spectral lines. From the excited energy levels, spontaneous emission of characteristic light may occur during deexcitation. The flame temperature plays an essential role in the excitation process and usually also in the atomization of the dry aerosol sample. The suitable line or band from the element/molecule is selected from the flame spectrum by a spectral optical system and detected by means of a photocell [1, 19], or an optoelectronic system [20].

The time necessary for a complete evaporation of a particle strongly depends on the initial particle diameter. This time is also a function of the flame temperature, the chemical properties of gases, physical characteristics of the flame, the boiling point of the liquid, and the thermal conductivity of the vapor. For example, it takes 3 μs to completely evaporate a 1 μm-diameter water droplet with a flame temperature of about 2000 K. For sodium chloride (NaCl) in water, a NaCl residue may remain after the water has totally evaporated. This solid particle is then further heated until its melting point ($T_{melting}$(NaCl) = 1100 K), and to its boiling point ($T_{boiling}$(NaCl) = 1750 K). When the flame temperatures are hotter than the boiling point, the molten particles are not stable and salt residues evaporate quickly and completely.

Neutral atoms or molecules may be ionized after evaporation, giving a positive ion and a free electron. The proportion of ions produced depends on the flame temperature and the ionization energy. The ionization energy is the minimum amount of energy required to remove an electron from an atom or a molecule. The alkali atoms have a strong tendency to form ions and the alkaline-earth elements are less easily ionized compared to the alkali atoms because the ionization energy is higher.

The intensity of characteristic emission lines from the atoms, ions or molecules excited in the flame is related to the amount of that particular element/molecule. It is very important that the flame is stable and highly reproducible. If the characteristics change, the emission process is affected. The main requirement of a burner is therefore the production of a steady and robust flame when supplied with fuel and oxidant gases.

The flame theory is well developed, despite the fact that a flame is a highly complex system [21]. Several fast reactions occur both simultaneously and consecutively in the flame. The separation of these reactions in space and time is more or less impossible. In general the flame can be simply defined as a system, composed of high temperature gases in which chemical and physical processes are taking place. The optimum experimental conditions can, however, be defined and empirically found.

In general, the flames can be classified in two categories: stationary and travelling flames [1]. The stationary flames use a burner. The position of flame in space is well defined and independent of time. These flames can be further divided into two groups: flame diffusion (e.g., AP2C) [9] and pre-mixed flame (e.g., aerosol sodium detector—ASD) [3]. The two components of the diffusion flame are mixed in the burner by diffusion and convection. In a pre-mixed flame, the combustible and combustive gases are mixed in the desired ratio prior to the formation of the flame. In a travelling flame, the position of the flame in space changes with time. This type of flame is used for instance in internal combustion engines.

The most important parameter of a flame in FES is its temperature [1, 13, 21]. In general the flame temperature is a measure of the energy of the molecules of the gaseous system constituting the flame. The temperature value is primordial because it largely determines the accessibility of elements (spectral lines) in emission analysis. A hotter flame also reduces potential problems with incomplete evaporation and dissociation [3].

The flame temperature depends on the gas mixture type and the ratio of the components (gas and fuel) combination. For a given mixture, there is a relation between the temperature and the mixture ratio. First of all, the temperature increases according to the mixture ratio to reach a maximum, which is got for the stoichiometric ratio. The most common fuels in FES are hydrogen and acetylene. Often, the maximum temperature is somewhat lower compared to the theoretical maximum as a result of heat losses, incomplete combustion, etc.

There are several reasons to choose a hydrogen (H_2) flame for FES. Even if the hydrogen flame has not the highest temperature, it gives the best combination of emission line signal to background [1, 9], since it gives less radiation of its own to interfere with spectrum lines of the elements. The flame has very little intrinsic radiation in the visible region of the spectrum if clean and dust-free hydrogen (analysis purity) is used. Only the OH bands give interfering radiation [1]. In contrast to a hydrocarbon-based flame, hydrogen cannot produce carbon deposits around the burner orifices. Other H_2 flame advantages are: compactness, that is to say, portability with a cartridge (e.g., AP2C) or a small bottle (e.g., MAB). H_2 procurement or local production is easy and the use is simple (just a spark to ignite). This implies, however, that the fuel must be handled extremely carefully. Finally, the pure fuel is not expensive.

The combustion is an exothermic and irreversible reaction which involves gases. During this reaction, numerous chemical species interact. The combustion can be characterized as one (or multiple) strongly exothermic irreversible reaction(s) between a fuel (reducer) and a combustive (oxidizing) [22], according to the pattern: fuel and combustive give combustion products with thermal energy.

The proportion between the fuel and the combustive plays a considerable role where the stoichiometric mixture corresponds to complete combustion. The phenomenon of combustion can be schematized by chain reaction process where the molecules of the gas are initially activated by the ignition [13, 21, 23, 24].

Flame Spectrophotometer for Bioaerosol Detection

The French company Proengin has developed in collaboration with the French Ministry of Defense (DGA) a portable biological alarm detector. A device is named as MAB (*Moniteur d'Alarme Biologique* or biological alarm monitor), and uses hydrogen flame spectroscopy techniques. It samples and concentrates the particle-laden air containing potentially threat particles. Samples are finally analyzed by an optoelectronic system. Spectral data are then processed with real time software. Figure 8.2 gives a simplified description of the MAB and its main characteristics [25].

The main building blocks are the sampling system, the burner, and the spectrometer; see also in section Flame Emission Spectroscopy.

The sampling system consists of an inlet and two kinds of impactors (real and virtual). The largest particles are separated in the inlet and the real impactor [26–28].

Specifications	Characteristics
Size (LxWxH)	360 x 160 x 450 mm
Total height	800 mm (with the sampling system)
Weight	14 kg
Volume	22 l
Storage temperature	-39 - + 71°C
Operating temperature	-10 - + 40°C
Autonomy H_2	10 days (bottle module) 2 days (cartridge module)
Sampling volume rate	15 l / min
Power supply	12-32 V DC, 30 W
Remote data	RS-485
Response time	< 1 min
Software	MABCOM

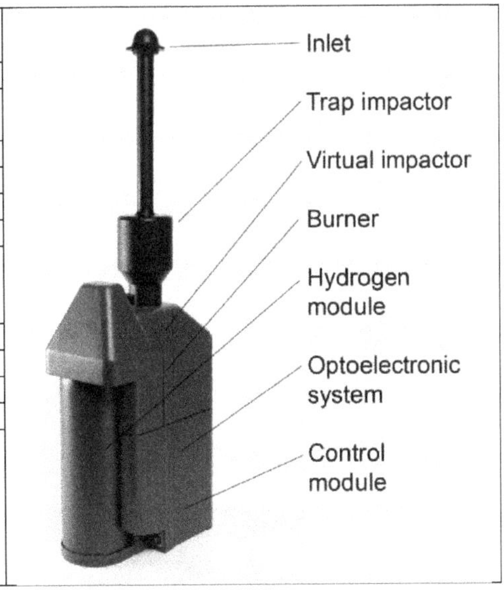

Fig. 8.2 MAB description and characteristics (with permission from Proengin company)

The smaller particles, the fraction that is able to reach the lungs and possibly cause disease continues in the flow. The next stage is a virtual impactor that concentrates the particles of interest and increases the sensitivity [29, 30]. The concentration factor is equal to the ratio of the total flow rate, 15 liters per minute (lpm) versus the minor flow rate 1.5 lpm, resulting in a theoretical concentration factor of 10 [29].

The MAB burner consists of several components, sleeves and nozzles, etc. in order to get a steady flow of the sampled air mixed with the fuel (H_2), and to get a steady flame in the chamber. Within this chamber, the mixture of air (containing the particles) and pure hydrogen is combusted. The hydrogen flow rate is 2 l/h, while the analyzed air flow rate in the burner is about 5 l/h.

The optical system collects the light radiating from the steadiest part of the flame. The intensity of the spectral line(s) of interest is then measured by a photosensitive detector. In the MAB, particle emission lines are detected by a Peltier-cooled CCD (charge coupled device), 512 channels array. The spectral window is between 340 and 800 nm.

Bioaerosol Detection

Living cells contain ions and metabolites from exchanges with the environment. Analytical studies on cells, and more especially with bacteria, succeeded in characterizing qualitatively and quantitatively these ions [11, 17]. Among these elements, certain compounds can be detected by FES.

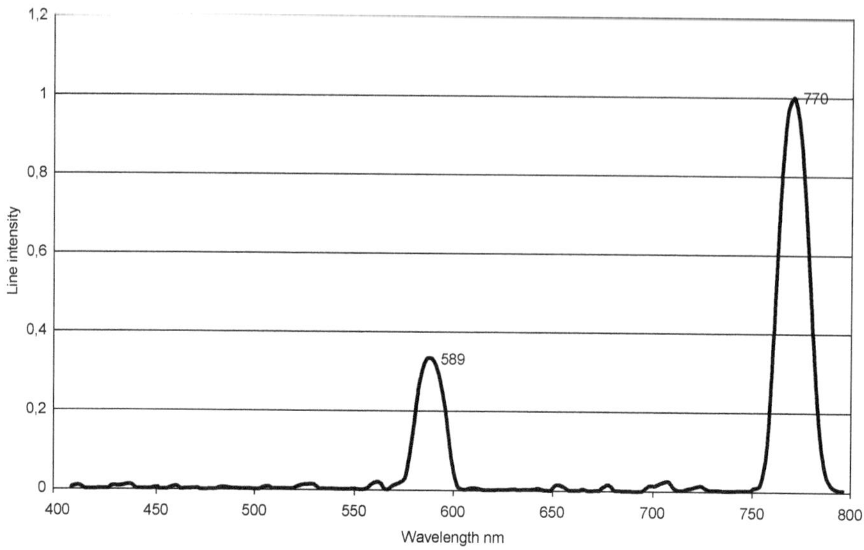

Fig. 8.3 Typical spectrum of *Bacillus atrophaeus* from FES

Bacterium Spectral Analysis

When the spectral signature of bacteria detected by the MAB is observed qualitatively, there are two significant peaks corresponding to the wavelengths of potassium (K) at 770 nm and of sodium (Na) at 589 nm. Figure 8.3 shows a typical spectrum of *Bacillus atrophaeus,* noted Bg [31]. The potassium peak is stronger than the sodium one for all the studied bacteria. Lines corresponding to other emissive elements (like Ca, P and S) are not always observed [32].

The results show that there is also a relation between the detected particle number and the line intensity for every emissive compound (i.e., Na and K mainly). The emission intensity is proportional to the quantity of burnt matter. This linear relation is true for every observed emissive element [32].

For a given bacterium and for different particle concentrations, the Na to K intensities ratio is constant (Fig. 8.4). The correlation coefficient is close to 1. The slope, however, is different from one preparation of agent to another. Bg1, Bg2 and Bg3 correspond here to 3 different cultures. This linear relation is also observed with other metallic compounds when the bacterium concentrations are high. At higher concentrations, when lines from more elements are significant, there are more such ratios improving the specific "fingerprint" for the respective preparation.

Various identically prepared bacteria contain about the same proportions of emissive elements. When looking at the Na–K ratio, there seems to be very small differences between species. The ratio can differ between preparations (growth media etc.), but for each preparation, the ratio is constant.

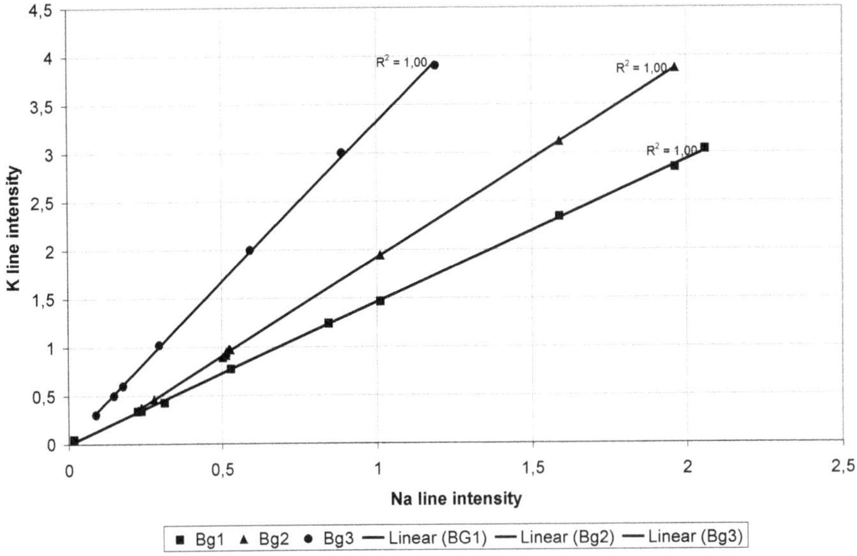

Fig. 8.4 Na and K line intensity versus different preparations of one type of bacteria

Biological Atmospheric Background

In order to see if the MAB can be used as bioaerosol detector, it is necessary to know the environment in which it will run and to characterize its behavior with atmospheric particles. For this purpose, the MAB was used on different outside locations to detect and to register particle spectra coming from different aerosol environments (as seaside, countryside, and so on). There are a lot of detectable particles in the atmosphere containing one or several emissive compound(s). The spectra from atmospheric aerosol mainly contains some sodium and/or some potassium and/or some calcium with various ratios between elements.

Figure 8.5a shows a spectrum of a particle type that is commonly observed in the atmosphere. These particle spectra were found in a great number of different kinds of regions: in highly urbanized area, as well as industrialized zones, rural and maritime zones. Sometimes, sodium- and potassium- containing particles can also be strongly present in the air. Their spectra can then bear very close resemblance to spectra from bacteria. Other types of particles (containing calcium) are shown on the Fig. 8.5b.

The analysis using PCA shows, however, that it is possible to differentiate most of the dispersed biologic particles from the particles of the atmospheric background.

Biological Aerosol Detection in the Field

Some field trial experiments with bacteria and other BWA simulants have been conducted in France and also in Sweden. The agents were in most cases dispersed using

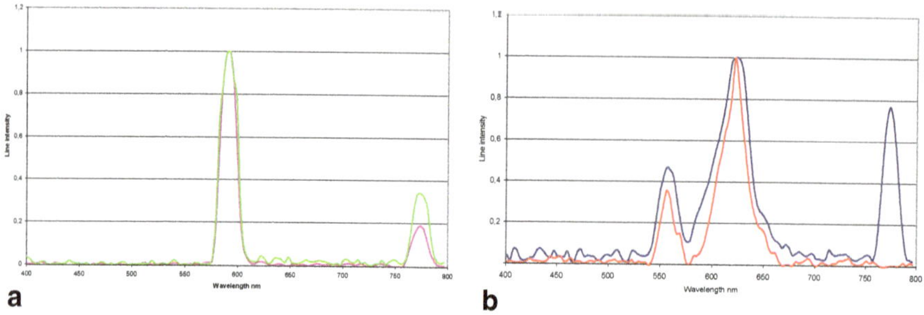

Fig. 8.5 Spectra of atmospheric particles; **a** sodium particles; **b** other (Ca-containing) particles

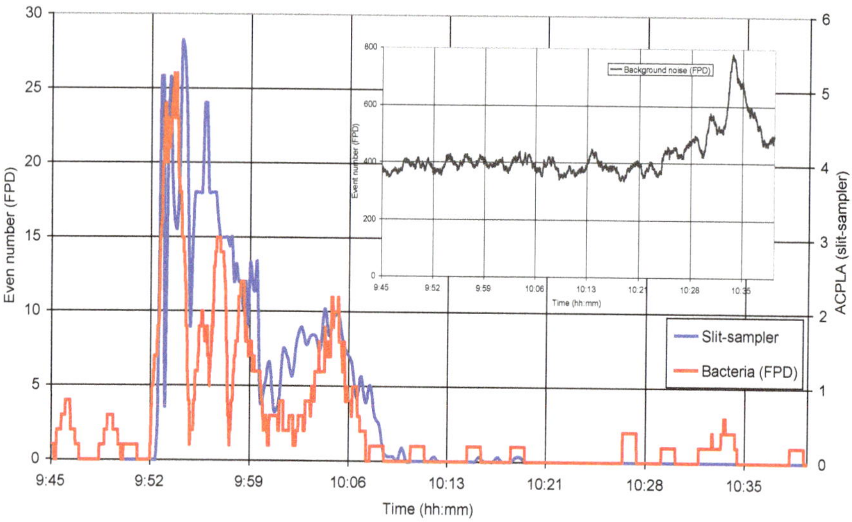

Fig. 8.6 Comparison between MAB results (*red*) and slit sampler reference (*blue*) in a field trial. The inset graph shows the nonbiological particle count (*black*)

agricultural spray nozzles. As a reference for biological particle counting, a set of slit samplers was used to monitor the time evolution of the concentration of bacteria. Figure 8.6 shows the comparison between MAB results and reference from the slit samplers. All these data are processed and the final results are given with a time averaging over one minute [12].

The plot on the upper right inset corresponds to the signal coming from the atmospheric background. These events are interpreted as particles without any biological signature, for example, a nonbiological emissive particle count. In the same plot, the peak just before 10:35 is due to a background variation without any bio signature and therefore of no interest for alarming purposes.

On the main plot there is the final comparison of the MAB results (red line) and those coming from the slit sampler (blue line—delayed information). The correlation is quite good and clearly shows what kind of important information we can get with a FES spectral device. Nothing is visible with a basic particle counter (e.g., corresponding to the addition of the black and red line), but a clear detection of a specific event can be observed when the spectral signature is considered.

Conclusions and Prospects

FES techniques can be used for point detection to warn. A portable alarm detector has been developed, the MAB. Results from experiments carried out on various bacteria and simulants are quite promising. They clearly demonstrate that the system is able to discriminate a bio agent embedded in the ambient background. The correlation with slit samplers also shows that the detectable concentration is compatible with a 5–25 ACPLA challenge mentioned in various NATO documents as the minimum detectable concentration for a bio detection device to achieve [12].

A better knowledge of the atmospheric background particles should allow a more reliable detection. More exhaustive studies combining FES and systematic biologic analysis in atmospheric sampling campaigns would allow the understanding of the false alarm causes, and possibly to remedy it by further developments of algorithms.

The described results show the numerous possibilities of the FES. The detection and the analysis of only the sodium and the potassium compounds in the particle are limiting. To access other elements, it is necessary to determine and operate the burner under optimal conditions, or to obtain a more energetic flame to excite more elements. That would be a way to obtain an increased sensitivity as well as selectivity. Another AES analysis method, LIBS, which is more energetic and gives the possibility to detect more elements, will be discussed in the next section.

Laser-Induced Breakdown Spectroscopy (LIBS)

General Characteristics

LIBS relies on a multielementary emission technique where the individual atoms and ions are dissociated from their parent molecules in highly energetic plasma [33]. The induced plasma is the result of the interaction between a high power focused laser beam (typically of a few ns duration with pulsed Nd:YAG laser) and the sample (gaseous, solid or liquid), in ambient conditions of pressure and temperature. All the molecules constituting the matter are dissociated inside a complex system then formed with atoms, ions and electrons, that is to say thermal plasma with a composition representative of the target of interest (gas, liquid, solid, or aerosols,

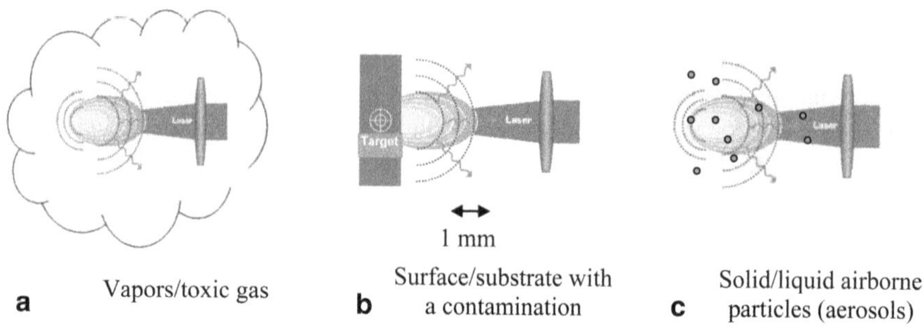

Fig. 8.7 Examples of possible interactions between focalized laser pulse and various target

etc.). At the end of the laser pulse, free atoms and ions decay back to lower states by emitting characteristic light resulting in discrete spectral lines.

If induced in air with vapors or aerosols like in Fig. 8.7a, c, the plasma stands for a destructive sampling volume of a few mm^3. For surface analyzes like in Fig. 8.7b, the size of an ablated surface is on the order of around tens of microns depending the laser pulse energy. The relaxation of the plasma takes a few microseconds (10^{-6} s), which offers the potentiality to do real time analyses.

Compared to the FES technique described formerly with temperature continuously around 1500–2000 K, the plasma temperature could reach up to 10000 K—and even more—during its life time (a few ns if ns laser pulse is used). The breakdown of the matter creates plasma with very high temperatures, which lets the atomization and the ionization of molecules. It includes even the more electronegative atoms such as chlorine and fluorine. The thermal characteristics give the access to the detection of almost every element of the periodic classification, with mass limit of detection from the part per billion (ppb) to the part per million (ppm). However the short lifetime of thermal plasma and some related effects (local shockwave due to plasma expansion) make the interaction with particulate matter such as bioaerosols complex (see the following parts).

LIBS is a powerful and versatile analytical tool that can be used to analyze every kind of phase like solid and bulk matter, aerosols, gas, and vapors. LIBS is used worldwide to probe a variety of materials in fields such as environmental analysis [34, 35], aerosols analysis [36, 37], forensics [38, 39], biomedical [40], and pharmaceutical analysis [41, 42]. In the industrial application area, LIBS makes it possible to perform quality control of materials during manufacturing, process monitoring, and sorting of scrap materials for recycling [43]. LIBS is also expected to have potential applications for remote sensing and stand-off detection of explosive substances [44], and for samples in hostile or hot environments, for example, nuclear reactor buildings [45, 46]. For instance, LIBS on bulk samples is considered to be a nondestructive or rather a minimally destructive technique, because it only consumes small amounts of materials (up to a few ng) during the analysis.

Additional advantages of the LIBS technique are that sample preparation is generally not required, which minimizes any possibility of chemical or biological contamination during the preparation steps. Moreover LIBS has the ability to depth profile a surface sample by repeatedly fire the laser at the same location, thus effectively dig deeper into the sample with each shot. LIBS provides also rapid, real time measurements with results in less than one second, making it particularly useful for, for example, on-line industrial monitoring.

On the other hand, like all other analytical techniques, LIBS has several limitations. One limitation is the laser shot-to-shot fluctuation which causes differences in the plasma properties and hence degrades the LIBS accuracy. By averaging spectra from several laser shots, this effect may be suppressed. Another limitation is that LIBS sampling surface or volume is limited with the analyses of small amounts of a bulk sample. Then the result might be nonrepresentative of the underlying bulk composition. Additional limitation is the signal interferences from different elements that can cause poor sensitivity in LIBS signal.

However, LIBS has been developed rapidly since the last decade. Emphasis has been put on components and system miniaturization in order to reach maturity for commercialization. Increased research activities and interest from several groups and laboratories such as the National Aeronautics and Space Administration (NASA), European Space Agency (ESA) as well as some military labs worldwide have also contributed in LIBS development. The Mars Science Laboratory (MSL) is a good example, with their hybrid LIBS-Raman system [47, 48] used by the 2012 NASA Curiosity rover to determine the composition of rocks on Mars.

LIBS-Raman, LIBS-LIBS and LIBS-LIF hybrid systems have also been developed and used to provide complementary spectroscopic information in different fields, for example, standoff detection of explosives [49]. Another significant development of the technique is the use of double- or multi-pulse, in order to further improve the sensitivity. Advances in new variations of the technique, for example, molecular LIBS to observe emission from excited molecules resulting from sample ablation/desorption, by selecting appropriate laser energy and measuring delay, are another examples of LIBS recent development [50].

Basic Experimental Set-Up

Basic LIBS apparatus is considered to be one of the most experimentally simple and quite low expensive spectroscopic analytical tools to determine the elementary composition of a sample. A typical LIBS set-up is shown in Fig. 8.8a:

- High energetic laser as an excitation source, for example, a pulsed Nd:YAG laser often used in the fundamental near infrared wavelength (1064 nm). Pulse duration is a few ns (10^{-9} s) with repetition rate up to 50 Hz. Other pulsed laser wave-

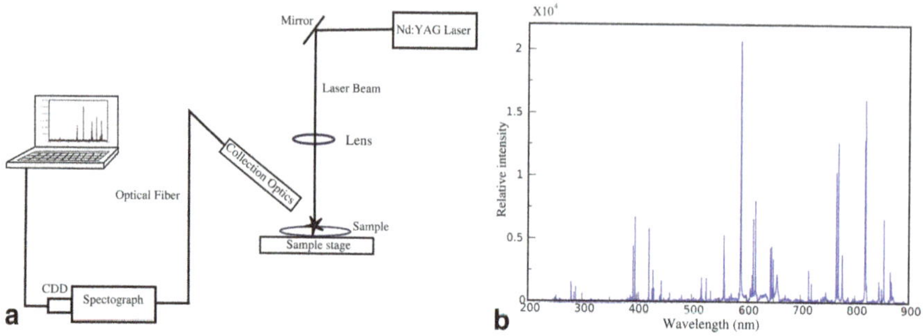

Fig. 8.8 **a** Schematic set-up for a LIBS system for surface/bulk material analysis and **b** an averaged emission spectrum from a dry *Bacillus atrophaeus* spore preparation

lengths and durations (like fs 10^{-15} s pulse duration) are also used for different applications,

- Optics used to focus the laser beam in order to form the plasma locally or up to several meters from samples; and to capture the light emitted from the plasma,
- A spectrometer to temporally and spectrally resolve the emissions lines coming from the different elements composing the sample of interest.

Laser Properties

Laser properties such as pulse energy, pulse duration and wavelength can influence the analytical qualities of the LIBS measurement. However, independent of the type of laser that is used, a power density in the order of 10^9 W · cm^{-2} is usually required in order to generate the plasma. If the laser energy is very close to the breakdown threshold, the pulse-to-pulse fluctuation can degrade plasma reproducibility, which may result in poor measurement precision. On the other hand, increasing the laser energy generates very dense and hot plasma. This results in an increase in the continuum emission and may decrease the elementary signal intensities due to self-absorption processes. Additionally, the laser wavelength can also affect the plasma-formation throughout the collision-induced ionization process. For instance, when the laser wavelength decreases (i.e., 266 nm), multiphotonic processes increase and breakdown thresholds decrease. Furthermore, depending on the laser pulse duration (ns/ps/fs), the threshold intensity can also be modified [51].

Detection Parameters (Delay and Integration Time)

At the early time of plasma relaxation, the emissive resulting light is dominated by a strong continuum emission where the intensity changes slightly as a function of wavelength. If the plasma emission light is not temporally resolved, the continuum light can conceal the elementary spectral lines of interest [43]. It means that the analytical time scales has to be controlled, that is to say, by using detector with delays and gates generators with typical range between 1 and 50 μs following the breakdown [52].

LIBS on Bio-Aerosol Particles

In spite of recent advances in the use of LIBS technique to detect and characterize biological materials (bacteria and spores), this technique has still some fundamental challenges and limitations when it comes to the detection and classification of μm-sized bio-aerosol particles [18, 43, 50, 53]. In this case, the hit ratio between discrete plasma and particles must be maximized. For this aim, a LIBS set-up is designed so that particles are sampled in an air stream in order to be directed towards the optical focus and plasma location. A dedicated chamber is then built via an aerosol collection and concentrating system. The apparatus is technically quite challenging because the positioning and timing of the LIBS laser pulse respected to the location of the aerosol particle has to be strictly coordinated. Limits of detection, single-shot analytic precision and low hit rates (<<1% for un-triggered system) are just few examples of these challenges [52–57]. The detection limit is mainly determined with the very small amount of material representing the particle and also with the surrounding air that will contribute to a great extent with emission signals. By considering particle sizes less than 2–3 μm, the signal to noise ratio is less favorable [55]. If the LIBS system has no synchronization regarding particle position and laser firing, the statistical chance for a successful excitation of a single particle is very low according to a Poisson's law. Typical Nd:YAG lasers can only operate with a pulse repetition rate of a few 10's of Hz. This leads to that only a few particles are analyzed in a given time frame. One way to upgrade the situation is to add an extra upstream continuous wave (cw) trigger laser beam which can be used to predict when the particle is in the correct position (see Fig. 8.9). This has also been demonstrated in a recent work [56] where a significant improvement was achieved in the numbers of successful hits. Aerosol particles are collected and concentrated via a virtual impaction stage and then injected into the optical chamber with a sheath flow nozzle. A cw laser is used to detect scattered light from the particles passing by. A timing system predicts the instant so that the LIBS laser is firing when the particle is in the correct position. This system improves the hit rate significantly compared to a nontriggered solution.

In LIBS, the plasma-particle process means an interaction between the particle compounds and the laser-induced plasma. Such interaction is time and space dependent and has been investigated in a series of review articles by Hahn and Omenetto [50, 52, 57]. The plasma-particle interaction process is a very important aspect in the LIBS quantitative analysis. The former depends on the laser characteristics and sample properties, while the latter is space and time dependent. They might cause together so called matrix effect. The matrix effect is a chemical background noise generated by the elemental and molecular composition of a sample other than the element of interest. This effect can have a considerable effect on the

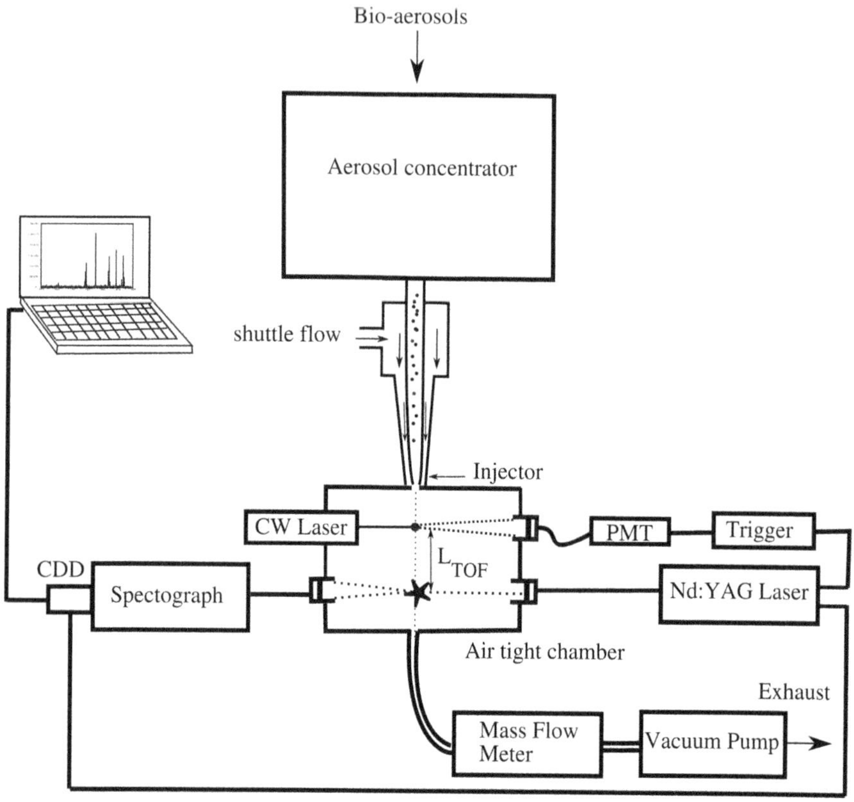

Fig. 8.9 Schematic illustration of a triggered aerosol LIBS system

LIBS signal and the quality of the results. The matrix effects can be divided into two types:

- **Physical Matrix Effects:** this effect depends on the physical properties of the sample and relates to ablation processes. The differences between thermal conductivity, absorption and latent heat of vaporization of different matrices can affect the emitted signal. For instance due to the matrix effect, the signal strengths from silicon can be different in water and in soil, though they have same concentrations [43],
- **Chemical Matrix Effects:** when the emission intensity of an ionized element is affected by the presence of additional element emission, this phenomenon is called the chemical matrix effect. For instance if nitrogen N from the ambient air recombines with the carbon atoms in the analyte sample, it will disturb the carbon signal signatures [18].

Understanding the physical processes of plasma–particle interaction will not only benefit LIBS technique to be an outstanding analytical tools, but also will contribute to other techniques such as laser ablation inductively coupled optical emission spectrometry (LA-ICP-OES) and mass spectrometry (LA-ICP-MS) which are based on similar physical processes. Data analysis of LIBS spectra is similar to the techniques used for FES and is described in the following section.

According to the quantity of bacteria analyzed by LIBS (bacterial growths on Petri dishes or single airborne bioaerosols), some elements composing the bacteria can be detected by LIBS: Ca, K, Na, Mg, C, P, Fe, etc. Of course, the data processing data must be done carefully because the nutritive media can also be present as interfering lines.

Signal Processing of Emission Spectroscopy Data

Atomic emission spectroscopy (AES) techniques are not targeting specific biological molecules because the elementary composition of the sample is primarily measured. However, the relative ratio between certain relevant elements such as Na, K, Ca and Mg, etc. can be used as indicators that it is a particle with biological origin.

The particle analysis by AES methods can often generate a great number of spectral information data points depending on the wished spectrographic resolution. Spectra are usually constituted by several hundreds, even up to millions of pixels (i.e., in the case of the use of a broadband Echelle spectrometer and a whole matrix in an ICCD camera chip). To analyze spectra and to process the data, several tools and statistical methods can be used:

- Analysis of various characteristic emission lines ratios, by using linear regression as statistical tool,
- Nonpredictive multi and/or megavariate data analysis method for pattern recognition. Example of such method is principal components analysis (PCA),
- Predictive multi and megavariate data analysis method such as partial least squares (PLS) ,
- More complex method like neural network for self-calibration of classification, etc.

The data processing is based on the analysis of spectral emission lines. The study of spectra is typically done in several steps: examination of the specific emission lines to select discriminating observables, study of the global spectra of particles and bacteria, study of all the detected particles, and the use of multi and megavariate data analysis methods [58, 59].

In general, a spectrum is composed of several hundreds, even up to millions of pixels representing intensity as function of wavelength. However, in the case of the developed, commercial FES, it has been found that a spectrum of 512 pixels can be reduced to 128 pixels without losing the information needed for classification. The spectrum data is often filtered and normalized according to some mechanisms, that is, versus the most intense lines before analysis.

The FES particle detection system can generate a data set with several dozens of spectra per second and each spectrum is characterized by several hundreds of pixels. Nowadays, with the use of powerful computers and mathematical statistics tools (e.g., Matlab [12], SIMCA P [31], Bionumerics [60]), it is possible to process large data sets nearly in real time [59].

Every particle spectrum can usually be characterized by one or several specific emission lines. When a spectrum has several lines, their relative ratios are also compared, considering that the intensity of each line is a quantitative variable. A particle can then be characterized by ratios of different elements.

PCA is a multivariate projection and factorial method that is designed to extract and display the variation in a complex data matrix [31], by losing as little information as possible. PCA can be used to find statistical similarities between observations and to learn which variables are the most important to explain the statistical variation in a data set. This makes PCA a perfect tool for variable selection data reduction [61].

PLS is also a multivariate method like PCA, but it is also a predictive method using linear regression. The first step is to create a model based on measurements of known observations as a calibration set. These reference spectra are obtained from measurements of similar samples (salts, bacteria, etc.), which then form clusters of points in the discriminating multidimensional space (projection in the PCA space) [12]. The second stage is to compare the unknown spectrum with the defined groups in the databank by linear regression and by cross validation. The objective is to classify the unknown sampled particles into a known group or particle family already established and included in the library.

General Conclusion and Outlooks

Through this chapter, the described atomic emission spectroscopy (AES)-based technologies, like FES and LIBS, have been demonstrated to be used for bioaerosol point detection. AES-based systems exhibit potentialities to directly monitor the particles and to trigger an alarm. The detection is only based on spectroscopic and elementary analysis. Therefore AES-based apparatus will never be biologically specific. However, two improvement directions can be lead in order to enhance the sensitivity, the selectivity, and to reduce false alarm rate: hardware and software. These technologies present numerous assets to be used on the field and to answer to the biohazard threats.

The AES technology can be improved in terms of temperature and combustion efficiency. Concerning FES, it means either to work on the properties of the burner, or to change fuel and combustive nature. This would aim at increasing the flame temperature because it is the most important parameter for sensitivity and selectivity. As for the LIBS, one of the most important key issues is to improve the plasma–particle hit ratio by triggering synchronically the plasma generation with the particle presence for an efficient detection. Likewise both the temperature and the electronic density inside the plasma could be increased by means of other laser pulse regimes (i.e., double pulses configuration to increase the initial vaporization of the particle) or by means of another plasma gas more energetic than air (argon for instance). But this will also make the LIBS-based sensor more complex.

The AES technologies are of interest to diagnose some elements (overall metallic) contained in aerosolized bacteria. To reach this information, the particles are destroyed by means of high energetic sources (flame or laser-induced plasma) . The metallic information is not exactly specific of biological compounds. In order to enhance the specificity, it will be possible to couple AES with another technology. In the case of coupling them with another analytical method (like LIF, Raman, etc.), it is necessary to put a nondestructive technology before AES systems. Combining molecular and atomic information could permit to increase specificity and selectivity, and improve the detection quality (decreasing of false negative and positive alarm rates).

At this moment, important work must be done on the spectral exploitation. A spectrum contains a lot of observables. The goal is therefore to find more discriminating data extraction. Statistical tools based on multivariate data analysis, like PCA or PLS for instance, are used to sort and classify certain observables. Further statistical analytical methods should be investigated and even coupled in order to improve bioaerosol detection ability. For example, if several algorithms give an alarm at the same instant and if the weighting coefficients are confident, it is possible to warn with greater confidence. To obtain a good and reliable detection algorithm it is also of great importance to establish a large databank with known bioaerosols and interfering aerosols to compare with unknown sampled particles.

The AES technologies both present assets for an operational deployment and for field detection: real time and fast response, high sensitivity and selectivity, few sampling steps, transportability, affordable costs, and low maintenance. Respected to the LIBS, the FES presents a higher technological readiness level (TRL). Operational FES-based sensors have been designed for operational users like military forces or civilian security members. The FES-based detector MAB from Proengin is for instance devoted to monitor the particulate atmospheric background. If an abnormal atmospheric composition is detected by MAB, the resulting alarm could trigger other more specific analytical techniques to identify the unknown agent, and simultaneously initiate protective acts (wearing of respiratory mask, gloves, etc.) and medical countermeasures. On the contrary, LIBS has not reached the maximal TRL and only a few demonstrators have been developed. Moreover LIBS presents a few drawbacks in term of sensitivity. Nevertheless because LIBS is a versatile, an *in situ*, and a fast technique, it exhibits high potentialities to complement other bioaerosol detection techniques.

Both fundamental differences and complementary uses between FES and LIBS have been discussed. AES has shown a good ability to detect particle and to sort bacteria versus aerosol background. In operational conditions, however AES must face up to the main trade-off: atmospheric background changes and the biodiversity. Both technologies seem to be good candidates for operational bioaerosol point detection.

References

1. Dean JA (1960) Flame Photometry. McGraw-Hill series in advanced chemistry. McGraw-Hill, New York
2. Binek B, Dohnalova B, Przyborowski S, Ullmann W (1967) Using the scintillation spectrometer for aerosols in research and industry. Staub 27:379–383
3. Clark CD, Campuzano-Jost P, Covert DS, Richter RC, Maring H, Hynes AJ, Saltzman ES (2001) Real-time measurement of sodium in single aerosol particles by flame emission: laboratory characterization. J Aerosol Sci 32 (6):765–778. doi:10.1016/S0021-8502(00)00120-8
4. Huntzicker JJ, Hoffman RS, Ling C-S (1978) Continuous measurement and speciation of sulfur-containing aerosols by flame photometry. Atmos Environ 12 (1–3):83–88. doi:10.1016/0004-6981(78)90190-7
5. Guichard JC, Lamauve M (1979) La mesure du diamètre chimique des particules au compteur à scintillation—application a l'étude de l'aérosol atmosphérique. Atmos Environ 13 (4):511–517. doi:10.1016/0004-6981(79)90144-6
6. Mavrodineanu R (ed) (1970) Analytical flame spectroscopy: selected topics. Macmillan & Co. Ltd., London
7. Levet R (2002) Faire face au risque chimique et biologique. L'Armement 77:78–84
8. Wind F (2002) Contribution de la spectrométrie de masse à la détection et à l'identification de bactéries: Intérêt du couplage avec la pyrolyse. Ph. D., Conservatoire National des Arts et Métiers, Paris
9. Suzanne P (1987) Application de la photométrie de flamme au comptage de particules d'un aéosol. In: 4èmes Journées d'études sur les aérosols: Paris, 1 et 2 décembre 1987, Paris and Vert le Petit, 1–2 December 1987. GAMS, Paris, pp 33–42
10. Herrmann R, Alkemade CTJ (1963) Chemical analysis by flame photometry (trans: Gilbert Jr. PT), vol 14. Chemical analysis, 2 edn. John Wiley & Sons, New York
11. Madigan MT, Martinko JM, Paker J (1997) Biology of microorganisms. Brock, 8 edn. Prentice Hall, Upper Saddle River
12. Descroix D, Lancelin H, Adam P (2004) Detection of biological aerosols with the biological alarm monitor (MAB). Paper presented at the 4th Singapore International Symposium On Protection Against Toxic Substances (SISPAT), Singapore, 4–10 December 2004
13. Gaydon AG, Wolfhard HG (1960) Flames: their structure, radiation and temperature. 2 edn. Chapman and Hall Ltd, London
14. Baxter K, Castle MJ, Withers PB, Clark JM (2004) The UV LIDAR for stand-off airborne biological weapons detection. Paper presented at the 8th International Symposium on Protection Against Chemical and Biological Warfare Agents, Gothenburg, 2–6 June 2004
15. Adam P, Descroix D, Chiaroni JP (1998) Flame photometry for biological detection. Paper presented at the 6th International Symposium on Protection Against Chemical and Biological Warfare Agents, Stockholm, 10–15 May 1998
16. Hahn DW, Lunden MM (2000) Detection and Analysis of Aerosol Particles by Laser-Induced Breakdown Spectroscopy. Aerosol Sci Technol 33 (1–2):30–48. doi:10.1080/027868200410831
17. Hybl JD, Lithgow GA, Buckley SG (2003) Laser-Induced Breakdown Spectroscopy Detection and Classification of Biological Aerosols. Appl Spectrosc 57 (10):1207–1215
18. Morel S, Leone N, Adam P, Amouroux J (2003) Detection of Bacteria by Time-Resolved Laser-Induced Breakdown Spectroscopy. Appl Opt 42 (30):6184–6191. doi:10.1364/AO.42.006184
19. Alkemade CTJ (1970) From sample to signal in emission flame photometry; an experimental discussion. In: Mavrodineanu R (ed) Analytical flame spectroscopy: selcted topics. Macmillan & Co. Ltd., London,
20. Adam P (2005) French technological trends for B and C detection. Paper presented at the 5th CBRN seminar, Avignon, France, 10–12 May 2005

21. Pungor E (1967) Flame photometry theory. The Van Nostrand series in analytical chemistry. Van Nostrand Company Limited, London
22. Veynante D (1999) Flamme de diffusion laminaire -Sciences de base. Technique de l'ingénieur BE 2 (Article BE8320)
23. Pruvot P (1972) Spectrophotométrie de flammes. Gauthier—Villars, Paris
24. Alkemade CTJ, Herrmann R (1979) The Flame. In: Alkemade C (ed) Fundamentals of analytical flame spectroscopy. Wiley, New York,
25. Proengin www.proengin.com. Accessed 16 January 2014
26. Descroix D, Lancelin H, Scurrah K, Attoui MB (2003) Design and calibration of a standalone impactor bacteria-pollen device for flame spectroscopy. Paper presented at the European Aerosol Conference, Madrid, 31 August–5 September 2003
27. Marple VA, Willeke K (1976) Inertial impactor: theory, design and use. In: Liu BYH (ed) Fine particles: aerosol generation, measurement, sampling, and analysis. Academic Press, New York, pp 412–446
28. Marple VA, Rubow KL, Olson BA (1993) Inertial, gravitational, centrifugal and thermal collection techniques. In: Willeke K, Baron PA (eds) Aerosol measuremen: principles, techniques and applications. Van Nostrand Reinhold, New York, pp 206–259
29. Descroix D, Lancelin H, Scurrah K, Attoui MB (2004) Design and calibration of a virtual impactor that concentrates the particle-laden stream for a flame spectroscopy device. Paper presented at the European Aerosol Conference, Budapest, 6–10 September 2004
30. Kim MC, Lee KW (2000) Design Modification of Virtual Impactor for Enhancing Particle Concentration Performance. Aerosol Sci Technol 32 (3):233–242. doi:10.1080/027868200303768
31. Descroix D, Gustafson I, Lancelin H, Olofsson G, Rännar S, Tjärnhage T (2004) Biological Aerosol Classification with Spectroscopic Flame Photometry and Principal Component Analysis. Paper presented at the 8th International Symposium on Protection Against Chemical and Biological Warfare Agents, Gothenburg, 2–6 June 2004
32. Descroix D (2005) Application de la spectrophotométrie de flamme à la détection des aérosols biologiques dans l'air ambiant par analyse multivariée. Ph. D., Paris 12, Créteil
33. Cremer DA, Radziemski LJ (2006) Handbook of laser-induced breakdown spectroscopy John Wiley & Sons, Chichester. doi:10.1002/0470093013
34. Bublitz J, Dölle C, Schade W, Hartmann A, Horn R (2001) Laser-induced breakdown spectroscopy for soil diagnostics. Eur J Soil Sci 52 (2):305–312. doi:10.1046/j.1365-2389.2001.00375.x
35. Burakov VS, Raikov SN, Tarasenko NV, Belkov MV, Kiris VV (2010) Development of a laser-induced breakdown spectroscopy method for soil and ecological analysis (review). J Appl Spectrosc 77 (5):595–608. doi:10.1007/s10812-010-9374-9
36. Hahn DW, Flower WL, Hencken KR (1997) Discrete Particle Detection and Metal Emissions Monitoring Using Laser-Induced Breakdown Spectroscopy. Appl Spectrosc 51 (12):1836–1844
37. Boyain-Goitia AR, Beddows DCS, Griffiths BC, Telle HH (2003) Single-Pollen Analysis by Laser-Induced Breakdown Spectroscopy and Raman Microscopy. Appl Opt 42 (30):6119–6132. doi:10.1364/AO.42.006119
38. Dockery CR, Goode SR (2003) Laser-Induced Breakdown Spectroscopy for the Detection of Gunshot Residues on the Hands of a Shooter. Appl Opt 42 (30):6153–6158. doi:10.1364/AO.42.006153
39. Barnett C, Cahoon E, Almirall JR (2008) Wavelength dependence on the elemental analysis of glass by Laser Induced Breakdown Spectroscopy. Spectrochim Acta B 63 (10):1016–1023. doi:10.1016/j.sab.2008.07.002
40. Liu X-Y, Zhang W-J (2008) Recent developments in biomedicine fields for laser induced breakdown spectroscopy. J Biomed Sci Eng 1:147–151. doi:10.4236/jbise.2008.13024
41. St-Onge L, Kwong E, Sabsabi M, Vadas EB (2002) Quantitative analysis of pharmaceutical products by laser-induced breakdown spectroscopy. Spectrochim Acta B 57 (7):1131–1140. doi:10.1016/S0584-8547(02)00062-9

42. Green RL, Mowery MD, Good JA, Higgins JP, Arrivo SM, McColough K, Mateos A, Reed RA (2005) Comparison of Near-Infrared and Laser-Induced Breakdown Spectroscopy for Determination of Magnesium Stearate in Pharmaceutical Powders and Solid Dosage Forms. Appl Spectrosc 59 (3):340–347
43. Miziolek AW, Palleschi V, Schechter I (eds) (2006) Laser Induced Breakdown Spectroscopy (LIBS): Fundamentals and Applications. Cambridge University Press, Cambridge
44. Gottfried J, Lucia F, Jr., Munson C, Miziolek A (2009) Laser-induced breakdown spectroscopy for detection of explosives residues: a review of recent advances, challenges, and future prospects. Anal Bioanal Chem 395 (2):283–300. doi:10.1007/s00216-009-2802-0
45. Whitehouse AI, Young J, Botheroyd IM, Lawson S, Evans CP, Wright J (2001) Remote material analysis of nuclear power station steam generator tubes by laser-induced breakdown spectroscopy. Spectrochim Acta B 56 (6):821–830. doi:10.1016/S0584-8547(01)00232-4
46. Hernandez C, Roche H, Pocheau C, Grisolia C, Gargiulo L, Semerok A, Vatry A, Delaporte P, Mercadier L (2009) Development of a Laser Ablation System Kit (LASK) for Tokamak in vessel tritium and dust inventory control. Fusion Eng Des 84 (2–6):939–942. doi:10.1016/j.fusengdes.2008.12.033
47. Sallé B, Cremers DA, Maurice S, Wiens RC (2005) Laser-induced breakdown spectroscopy for space exploration applications: Influence of the ambient pressure on the calibration curves prepared from soil and clay samples. Spectrochim Acta B 60 (4):479–490. doi:10.1016/j.sab.2005.02.009
48. Sallé B, Cremers DA, Maurice S, Wiens RC, Fichet P (2005) Evaluation of a compact spectrograph for in-situ and stand-off Laser-Induced Breakdown Spectroscopy analyses of geological samples on Mars missions. Spectrochim Acta B 60 (6):805–815. doi:10.1016/j.sab.2005.05.007
49. Moros J, Lorenzo JA, Laserna JJ (2011) Standoff detection of explosives: critical comparison for ensuing options on Raman spectroscopy-LIBS sensor fusion. Anal Bioanal Chem 400 (10):3353–3365. doi:10.1007/s00216-011-4999-y
50. Hahn DW, Omenetto N (2012) Laser-Induced Breakdown Spectroscopy (LIBS), Part II: Review of Instrumental and Methodological Approaches to Material Analysis and Applications to Different Fields. Appl Spectrosc 66 (4):347–419. doi:10.1366/11-06574
51. Cabalín LM, Laserna JJ (1998) Experimental determination of laser induced breakdown thresholds of metals under nanosecond Q-switched laser operation. Spectrochim Acta B 53 (5):723–730. doi:10.1016/S0584-8547(98)00107-4
52. Hahn DW (2009) Laser-Induced Breakdown Spectroscopy for Analysis of Aerosol Particles: The Path Toward Quantitative Analysis. Spectroscopy 24 (9):27–33
53. Dixon PB, Hahn DW (2004) Feasibility of Detection and Identification of Individual Bioaerosols Using Laser-Induced Breakdown Spectroscopy. Anal Chem 77 (2):631–638. doi:10.1021/ac048838i
54. Leone N, Fath G, Adam P (2007) Advances in the detection of chemical and biological aerosolized pollutants by means of a field-transportable laser-induced breakdown spectroscopy-based detector. High Temp Mater Process 11 (1):125–147. doi:10.1615/HighTempMatProc.v11.i1.110
55. Hybl JD, Tysk SM, Berry SR, Jordan MP (2006) Laser-induced fluorescence-cued, laser-induced breakdown spectroscopy biological-agent detection. Appl Opt 45 (34):8806–8814. doi:10.1364/AO.45.008806
56. Tjärnhage T, Gradmark P-Å, Larsson A, Mohammed A, Landström L, Sagerfors E, Jonsson P, Kullander F, Andersson M (2013) Development of a laser-induced breakdown spectroscopy instrument for detection and classification of single-particle aerosols in real-time. Opt Commun 296 (0):106–108. doi:10.1016/j.optcom.2013.01.044
57. Hahn DW, Omenetto N (2010) Laser-Induced Breakdown Spectroscopy (LIBS), Part I: Review of Basic Diagnostics and Plasma-Particle Interactions: Still-Challenging Issues Within the Analytical Plasma Community. Appl Spectrosc 64 (12):335A–366A
58. Eriksson L, Johansson E, Kettaneh-Wold N, Wold S (2001) Multi- and megavariate data analysis: principles and applications. Umetrics Academy, Umeå

59. Lebart L, Morineau A, Warwick KM (1984) Multivariate descriptive statistical analysis: correspondence analysis and related techniques for large matrices (trans: Berry EM). Wiley series in probability and mathematical statistics. John Wiley & sons, New York
60. Descroix D, Attoui MB (2007) Application of the flame spectrophotometry in the biological Aerosol detection in the airborne particle, using statistical Multi- and megavariate analysis. Paper presented at the European Aerosol Conference, Salzburg, 9–14 September 2007
61. Georgin J-P (2002) Analyse interactive des données (ACP, AFC) avec Excel 2000: théorie et pratique. Collection Didact Statistique. Presses Universitaires de Rennes, Rennes

Chapter 9
Mass Spectrometry Techniques in the Analysis of Bioaerosols: Development and Advancement

Rabih E. Jabbour, Samir V. Deshpande, A. Peter Snyder and Mary M. Wade

Introduction

Bioaerosols are airborne particles that may contain pathogenic species, that can cause serious risks to various government and public sectors. The major health concern due to bioaerosols is that certain communicable diseases are transmitted through airborne particles, including viruses, bacteria, and fungi. Moreover, biological warfare agents can be disseminated as bioaerosol particles and could pose severe safety issues for military operations as well as serious economic and health concerns to the public. It had been estimated that when 100,000 persons are exposed to bioaerosol, it would cost approximately $ 26 billion in decontamination and medical treatment [10, 32]. Thus, it is imperative to develop and implement real-time detection and accurate identification technologies for the monitoring of bioaerosols. Initially, bioaerosol analysis was mainly based on traditional microbiological methods involving collection and culturing followed by agar plate counting and often susceptibility tests to antibiotics and microbial phages [57]. Moreover, newly developed molecular biology techniques are being utilized for bioaerosol analyses such as immunoassay, polymerase chain reaction (PCR) amplification, and other gene-based techniques. Although microbiological methods showimpressive sensitivity, they are time- and labor-intensive for sample preparation and analysis, sensitive to shelf life time of the reagents, and beholden to the availability of specific reagents.

Alternatively, several mass spectrometry (MS) techniques have been developed and improved in their sensitivity, fieldability, and compatibility to bioaerosol analysis and characterization in real-time settings, and these have shown promising capabilities. Pyrolysis MS techniques were initially used for analyses in which the bioaerosol particles were heated and their volatile molecules exposed to electron

R. E. Jabbour (✉) · A. P. Snyder · M. M. Wade
U.S. Army Edgewood Chemical Biological Center,
Aberdeen Proving Ground, MD, 21010-5424, USA
e-mail: rabih.e.jabbour.civ@mail.mil

S. V. Deshpande
Science and Technology Corporation, Edgewood, MD, 21040, USA

P. Jonsson et al. (eds.), *Bioaerosol Detection Technologies,* Integrated Analytical Systems,
DOI 10.1007/978-1-4419-5582-1_9, © Springer-Verlag New York 2014

ionization or chemical ionization, followed by mass spectral detection to establish biomarkers for bacterial differentiation [24, 51–53]. Concurrently, the development of laser desorption–ionization, matrix-assisted laser desorption–ionization (MALDI) [14, 69], and electrospray ionization (ESI) coupled with MS [11, 14, 36, 69, 70] made it possible to analyze large biomolecules such as proteins and oligonucleotides in bioaerosol particles that resulted in enhanced selectivity and identification accuracy. Moreover, several fielded studies showed that MS-based techniques are capable of detecting single bioaerosol particles [25, 27].

This chapter will address the characteristics of various MS techniques used in bioaerosol analysis. An overview of bioareosol MS is presented to provide general perspectives on the application of MS in bioaerosol detection and identification. Also, the capabilities of MS techniques and the nature of their output and impact on the detection and identification of bioaerosols will be discussed. Exploration of the advantages and drawbacks of the applications for different MS techniques in the analysis of bioaerosols is addressed.

Overview of Bioaerosol MS

Spurny [57], provides a review for the detection of bioaerosols. The analysis of bioaerosol sampling includes the stresses that a particle experiences during the collection process. Major stresses and changes that can occur to a bioaerosol particle during the collection includes desiccation; loss of viability due to radiation, oxygen, ozone, prehumidification, or high water vapor concentration; and temperature. These stresses affect the phospholipid, protein, and nucleic acid components of bacteria. Bulk and individual particle interrogation may be performed depending on the analysis technology.

The information content from data reduction of a bioaerosol MS analysis depends on whether the bioaerosol is collected in bulk, as individual particles, or captured in liquid with subsequent sample analysis. Bioaerosols must be converted into a vapor state prior to entrance into a mass spectrometer. The primary method used to accomplish this is pyrolysis or heating of the solids [55]. The solid sample can be either in bulk [25, 27] or individual particle form [51–53]. Both require heating to break up the solid into individual molecules in order to be analyzed by MS. A hot filament, Curie point wire, heated tube, ribbon filament, and laser are the major methods used to heat a solid substance [9, 56]. The tube and wire methods can be performed either without or with derivatization reagents applied directly to the sample prior to heating [1, 2, 12, 13, 25, 27]. A derivatization reagent is necessary if gas chromatography (GC) is part of the MS analysis. The derivatization reagent reacts with acid and alcohol functional groups on compounds to produce esters and ethers, respectively, so as to provide better volatility to very low vapor pressure biochemicals resident in or extracted from bacteria [1, 12, 13, 28].

Bulk bioaerosols usually are deposited in microgram amounts [27]. MS can analyze bulk amounts [25, 27] as well as single particles [24, 51–53, 26] of bioaerosols depending on the interface configuration. However, both sample forms need to be vaporized by a pyrolyzer device (*vide supra*). Another separation technique prior to a mass spectrometer, such as GC or liquid chromatography (LC), requires bulk or microgram amounts in order to provide for enough substance to pass through the column. Single bioaerosol particles need to be vaporized inside the MS vacuum chamber by impacting onto a heated surface prior to analysis. Alternatively, single particles can be vaporized by a laser beam situated perpendicular to the aerosol beam with appropriate triggering mechanisms. Laser pyrolysis features very high heating rates, and single particles can be detected at approximately 10^{-12} g. An early version of this heating technology was called laser microprobe mass analyzer (LAMMA). Single particles were captured in a vacuum ion trap and visually observed with a microscope. A laser beam was positioned to pyrolyze the particle and was analyzed by time-of-flight (TOF)-MS. A limitation in single particle analysis by MS (PAMS) systems is the fact that heating or pyrolysis vaporization techniques generally produce masses of only a few hundred Daltons. Differences in the spectra from different bacteria were not significant, and multivariate data analysis methods are virtually mandatory for pyrolysis of complex substances such as bacteria [7, 8, 39, 66, 68].

Particle Analysis by MS (PAMS)

Biological compounds and substances by definition are associated with very large molecular weight material outside of the detection range of a mass spectrometer, especially when the biomaterial is an intact bacterial cell. Therefore, the key to an analysis is the conversion of the intact cell to small parts, pieces, and components that lie within the range of an MS system. In addition, the initial sample is desired as a stream of aerosol particles.

In the early 1980s, Sinha et al [51–53] adapted the technique of pyrolysis, where a sample was placed on a heated surface such as a filament, boat, or Curie point wire and either slow or rapid heating to 400–500°C produces decomposition of the biological substance into significantly smaller pieces. These pieces usually comprise the fundamental monomer building blocks and fragments thereof. The method, PAMS, produces an aerosol particle beam by simple expansion through multiple skimmers at differentially pumped lower pressure regions to remove the gas load (Fig. 9.1). Isokinetic sampling of the particles allowed for a significant increase in particle transmission to approximately 70% reaching the mass spectrometer with respect to that entering the particle beam tube. The transmitted particles hit a hot vee-shaped rhenium filament in a 10^{-6} Torr ionization source. An electron beam ionizes the hot neutrals, and the ions are extracted into the quadrupole mass spectrometer (QMS). *Pseudomonas putida*, *Bacillus subtilis* (BG), and *Bacillus cereus* were investigated (Fig. 9.2). All three provided very similar mass spectra including

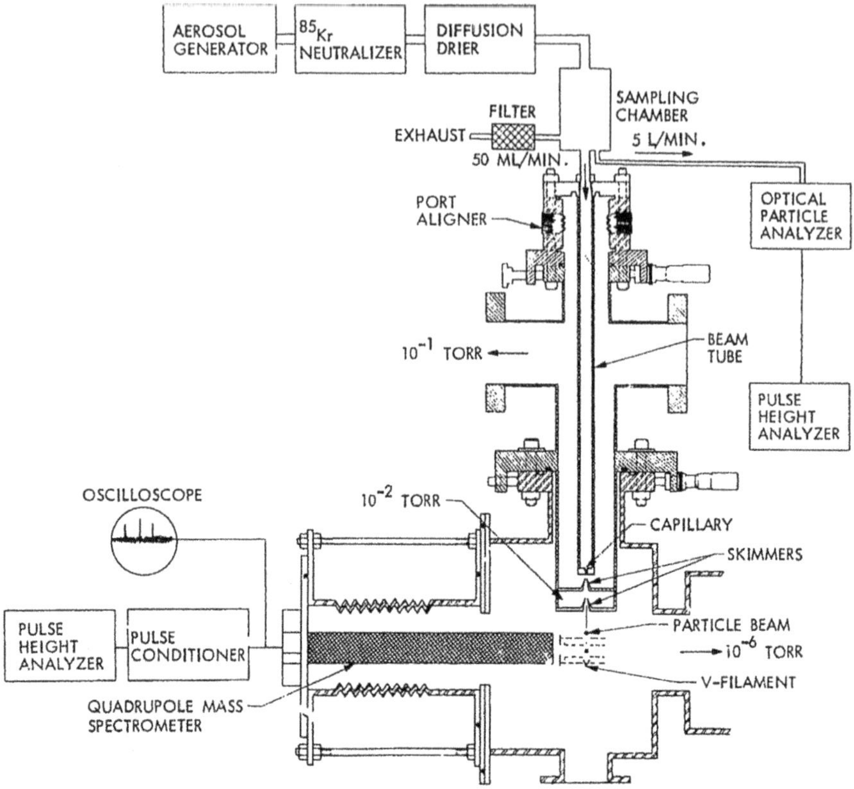

Fig. 9.1 Particle analysis by mass spectrometry (PAMS) system. (Reprinted with permission from Sinha et al. [52]. Copyright 1984 Elsevier Science Publishers B.V., Amsterdam)

a small peak at m/z 282 and masses of less than m/z 169. This work provided the first evidence for MS analysis of beams of bacterial particles.

Early Bioaerosol Ion Trap MS

Gieray et al. [24] extended the work by Sinha et al. [51–53] in that the QMS was replaced with an ion trap tandem mass spectrometer. Five different bacteria were used for bioaerosols. Mass spectra were characterized as a peak at every mass, and masses were observed up to m/z 220 in the positive ion mode. Figure 9.3 shows mass spectra for *Enterobacter aerogenes*, BG, and *Micrococcus lysodeikticus*. Individual bacterial particles produced considerable variations from cell to cell, that is, spectrum to spectrum. The variations included total number of peaks and peak height ratios. Space charge saturation of the ion trap usually occurred, and this can represent a cause of poor spectral resolution and reproducibility. Negative ion spectra resulted in masses up to m/z 300.

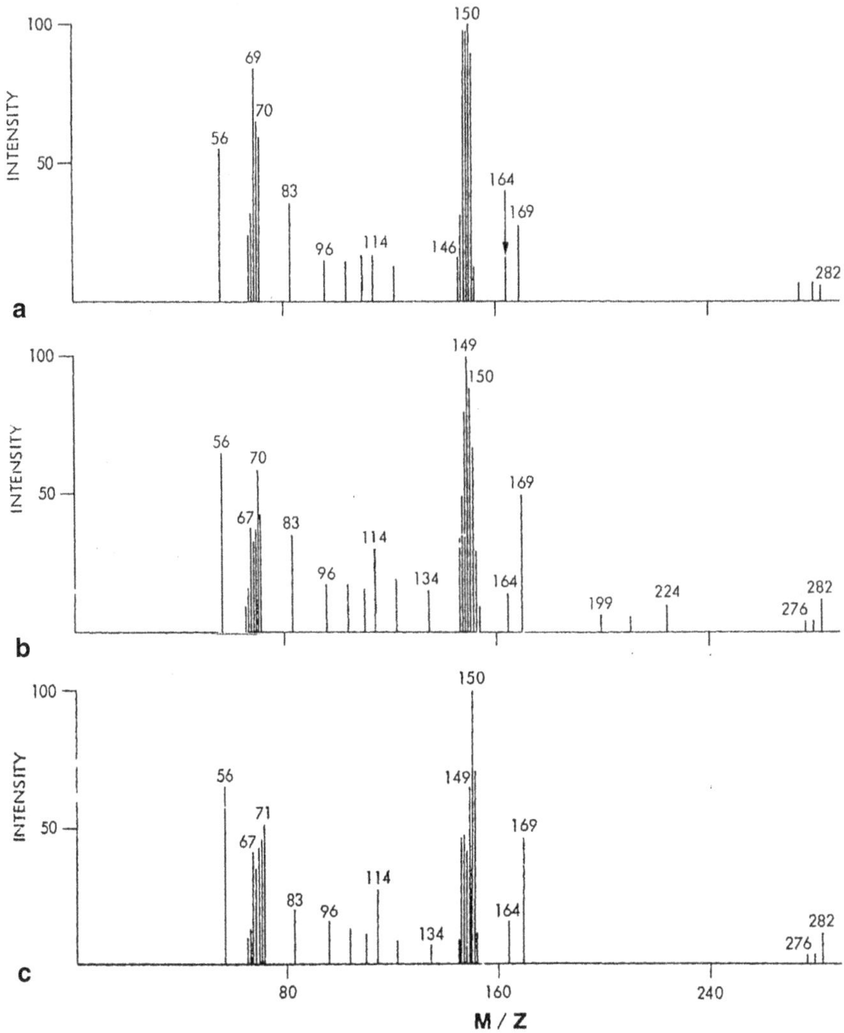

Fig. 9.2 Mass spectra of particle analysis by mass spectrometry (PAMS) system of particles of (**a**) *Pseudomonas putida*, (**b**) *Bacillus cereus*, and (**c**). *Bacillus subtilis*. (Reprinted with permission from Sinha et al. [52]. Copyright 1984 Elsevier Science Publishers B.V., Amsterdam)

Bioaerosol Matrix-Assisted Laser Desorption–Ionization Time of Flight Mass Spectroscopy

Initial Studies

In the mid-1990s, Murray and Russell provided a series of papers showing that the MALDI-TOF-MS technique could analyze a stream of aerosols. The core principle of MALDI is that a solid or liquid sample analyte is mixed with one of a series of

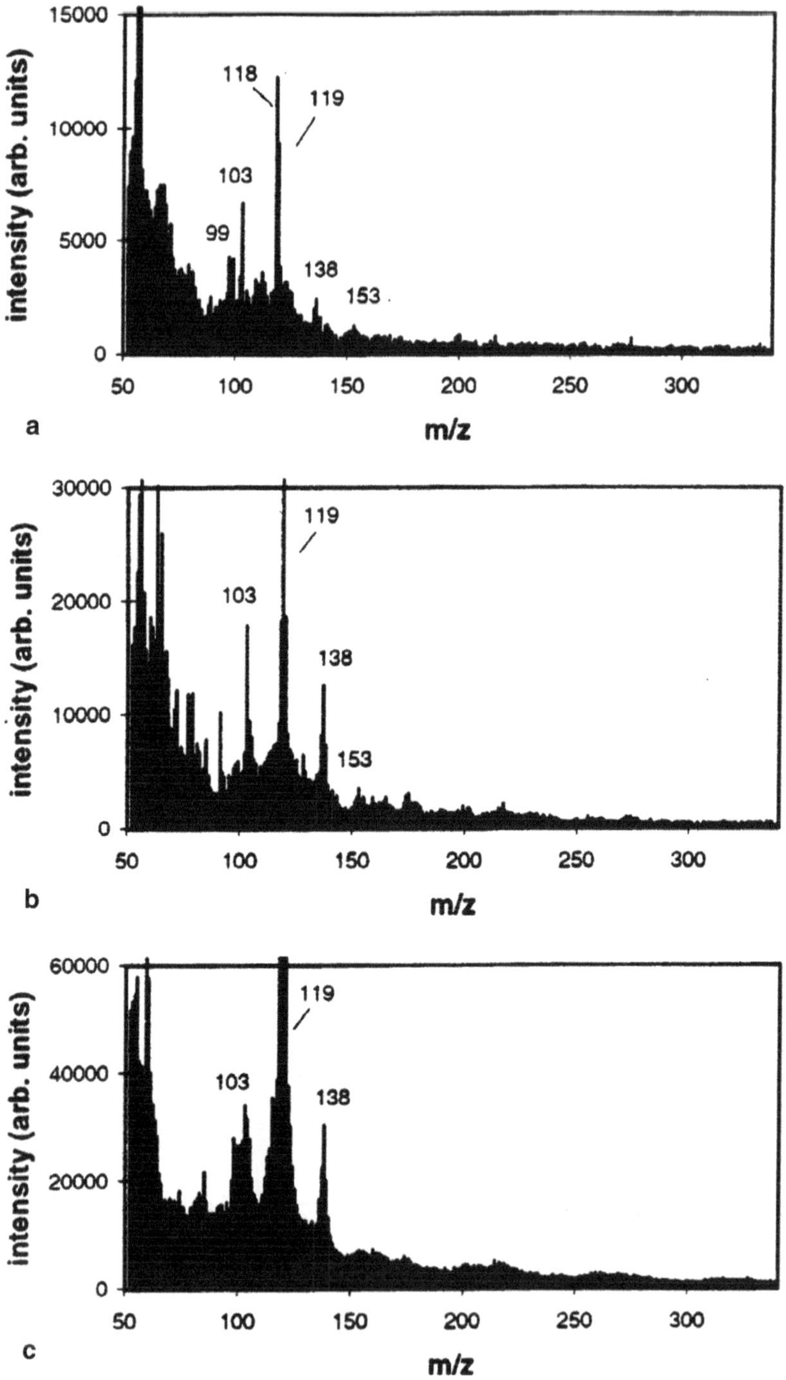

Fig. 9.3 Laser ablation ion trap positive ion mass spectra of (**a**) *Enterobacter aerogenes*, (**b**) *Micrococcus lysodeikticus*, and (**c**) *Bacillus subtilis* var. *globigii*. (Reprinted with permission from Gieray et al. [24]. Copyright 1997 Elsevier Science Publishers B.V., Amsterdam)

aromatic functional group-containing chemicals or matrices. A portion is placed on a support and inserted into the vacuum chamber of an MS system, and a laser impinges on the surface to "lift" or vaporize the sample as well as ionize the sample.

Significantly, Murray and Russell [41] also showed that a mixture of matrix and sample can be aerosolized. The liquid enters the nebulizer at up to 1 mL/min, and the aerosol particles are initially wet. Therefore, the particles had their solvent adduct molecules removed as the particles passed through a drying tube. A desolvation gas stream also helped dry the particles. The released solvent and carrier gas were removed with high-speed pumps prior to particle introduction into the TOF-MS vacuum. The particle beam was collimated, and a perpendicular ultraviolet (UV) laser beam ablated and ionized the particles. The ions were directed into and analyzed by a TOF-MS system [41]. Fundamental parameters were investigated and compared to the surface, bulk MALDI procedure. Investigations included matrix to analyte ratio, performance of different organic matrix ion yield at different laser energies, and intensity with respect to the temperature of the drying tube. The size of the matrix-covered particle also was analyzed.

The particle size did not depend on the amounts of matrix or analyte mixed together or the matrix substance. The length of the drying tube did have an effect where a 6 or 25 cm tube length produced 50 and 100 μm-sized aerosol particles, respectively. With the 25 cm tube, the particle size did not vary from 25 to 500 °C drying tube temperature. Thus, it appeared that the drying gas was a more important parameter compared to the tube temperature for final particle size characterization.

Aerosols of a mixture of analyte–MALDI matrix started with bovine insulin as a test compound [41]. The concentration of matrix and protein analyte was investigated in order to obtain a good signal to noise ratio (S/N). Two considerations were addressed, namely the prevention of nozzle clogging and matrix–analyte ratio. It was found that ~2 mg/mL each of 4-nitroaniline matrix and bovine insulin protein were the highest amounts that could be used without clogging the nebulizer orifice and for satisfactory ion intensity. A minimum ratio of matrix and analyte was about 50 : 1 to prevent clogging the aerosol orifice. Note that for regular surface MALDI, matrix and analyte amounts span the 100–50,000 ratio range.

Several matrices were also investigated and the α-cyano-4-hydroxycinnamic acid (CHCA) matrix provided an M^+intensity of bovine insulin 12× higher than that of the 4-nitroaniline. However, the M^+signal was very broad compared to 4-nitroaniline. Therefore, 4-nitroaniline was deemed as the best compromise matrix.

General comparisons were made to surface MALDI. Compared to the surface version, aerosol MALDI has a (a) lower mass resolution, (b) higher limit of detection, (c) mass peak range <300 Da, and they mostly arose from protonated and alkylated solvent clusters and not the matrix, (d) better performance with a lower matrix to analyte ratio, and (e) better S/N with higher laser irradiance.

The following provides a summary of improvements and refinements in the bioaerosol MALDI technique. This development consisted of transforming the exit of an LC column into an aerosol spray [40]. Prior to the exit, the LC effluent of a protein mixture was mixed with a matrix solution in a mixing tee. The matrix–analyte mixture subsequently entered the nebulization chamber at a rate of 0.5–1.0 mL/min

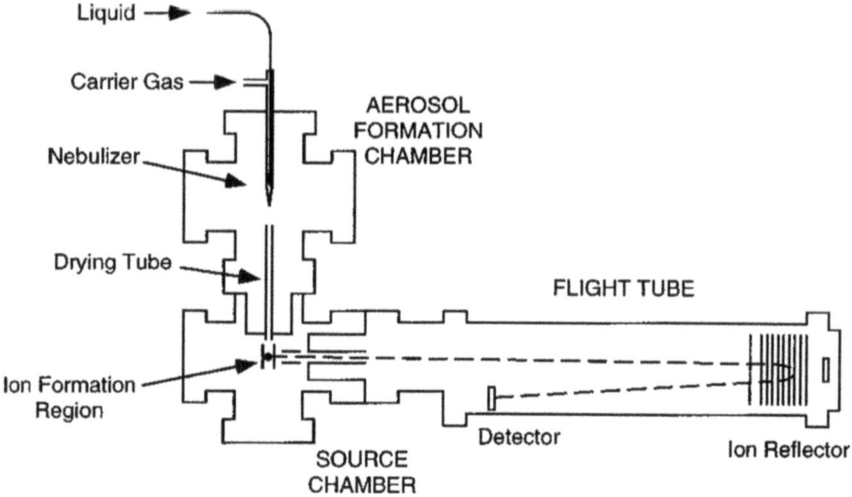

Fig. 9.4 An aerosol matrix-assisted laser desorption–ionization (MALDI) reflectron time-of-flight (TOF) mass spectrometer. Aerosol particles are formed in the vacuum chamber and passed through a heated drying tube into the ion source. The laser vaporizes the individual particles and forms ions. The ions are accelerated into the MS detection chamber. (Reprinted with permission from Fei et al. [16]. Copyright 1996 American Chemical Society)

and sprayed toward the entrance of a TOF-mass spectrometer. Analytes consisted of a mixture of nanomole concentrations of bradykinin, gramicidin S, and myoglobin. A three-dimensional (3D) retention time–flight time plot revealed the three distinct protein peaks.

Reflectron Time of Flight Mass Spectroscopy

Aerosol-MALDI still had some challenges such as the relatively large amounts of sample needed for satisfactory detection and mass resolution. A continuous aerosol beam necessarily uses a relatively large amount of sample. Even though only femtomole amounts of sample were actually ionized, hundreds of nanomoles of biological substance were expended. Using a mixture of bradykinin, angiotensin II, gramicidin D peptides, and vitamin B_{12} [16], improvements were made in aerosol-MALDI-TOF-MS. Mass resolution was a problem, and it was reasoned that thelinear TOF-MS be replaced with a reflectron TOF-MS system. A reflectron takes into consideration the spread of ion energies. The higher energy ions follow a longer drift tube flight path to reach the ion detector with respect to the lower energy ions (Fig. 9.4). A reflectron causes the ions to have the same energy. This allows them to arrive at the detector at the same time and therefore increases the signal strength that is manifested by a sharper peak in the mass spectrum. Mass resolution of the bradykinin, angiotensin II, and gramicidin D ions were 270, 330, and 310, respectively (Fig. 9.5). Earlier work [41, 42] produced a resolution on the

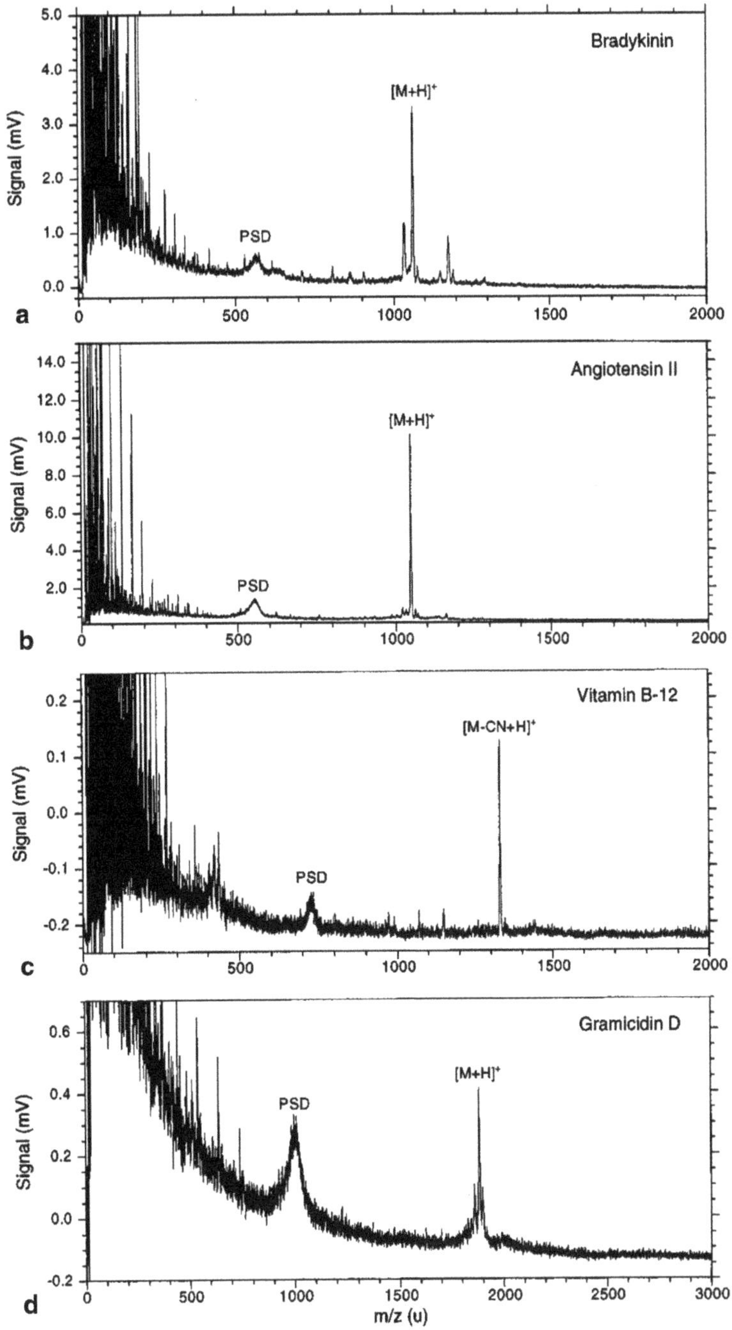

Fig. 9.5 Mass spectra from the MALDI reflectron of (**a**) bradykinin peptide, (**b**) angiotensin II, (**c**) gramicidin D, and (**d**) vitamin B_{12}. (Reprinted with permission from Fei et al. [16]. Copyright 1996 American Chemical Society)

order of only six for bovine insulin. A reflectron greatly improved mass resolution of peptides and proteins. The mass resolution was highest at 440 m/Δm when the laser energy was at a value of 10 mJ, which was the lowest energy above the threshold of ion appearance. Resolution decreased to approximately 300 for laser energies at 200 mJ for the peptides. The 300–400 resolution values are >10× that for aerosol MALDI-linear TOF.

Salient information on mass resolution includes the time duration of ion formation, initial ion spread in space, and ion source pressure. A reflectron can compensate or provide path lengths such that all the ions result in similar energies. The initial ion energies can be 10–20% in variance. The laser itself causes an approximate 6% energy spread of isobaric ions. Thus, it appears in a practical sense that the spread of ions is not the limiting cause of mass resolution. However, ion-neutral collisions in the laser ionization region can cause a distribution of energies, which is not taken into consideration in the 6% energy spread approximation from the laser itself. This ion spread causes the $[M+H]^+$peak to widen.

Ionization Efficiency

Mentioned earlier was the overwhelming amount of gas (pressure) found when nebulizing a liquid such as a mixture of analyte and MALDI matrix [3]. Excess gaseous material was pumped away prior to the ion source since this excess neutral gas species can cause considerable problems for efficient ion resolution in the mass spectrometer. Ion-neutral collisions cause nonuniform ion energies, and it introduces a spatial spread of ions. The latter deteriorates the focusing of the ions. These are functions of the ion yield, and further work was performed to refine the analyte concentration and matrix to analyte ratio with respect to their earlier work [16, 40–42]. A mixture of bradykinin and gramicidin S peptides and bovine insulin and myoglobin proteins were used [3]. A bioaerosol particle includes the biological analyte as well as salt, buffer, trifluoroacetic acid modifier, organic matrix, and residual solvent species. These compounds can affect the ionization efficiency of the biological analyte component and provide excess neutral molecule presence to destabilize the ion current, broaden peak widths, and reduce detection efficiency. It was shown that the ion signal for aerosol MALDI was nonlinear with respect to the analyte concentration, and this was attributed to residual solvent in the particles [3]. With increasing particle size, it was shown that solvent content increased causing a decrease in ion signal. Evidence also suggested that the presence of solvent in the aerosol particles did not affect the ionization efficiency in a uniform manner for different matrices and analytes. In order to overcome this problem, it was suggested that internal standards mixed in with the aerosol process may be required.

Three significant points for bioaerosol MALDI were observed: (a) matrix–analyte ratios for optimal ion yield were 10–100 : 1, (b) desolvation or removal of solvent to produce drier particles allowed for greater biological analyte ion signal, and (c) sensitivity was analyte dependent. Based on the number of residues, hydrophobicities, and surface area of each analyte, a correlation was established between particle size and ionization efficiency.

A similar effect for online bioaerosol matrix particle generation was performed by Mansoori et al. [38] where the dried droplets exiting the heated particle drying tube were ablated and ionized in flight with subsequent ion extraction into a TOF-MS system. Leucine enkephalin and gramicidin S were used for the analyses. Improvements implemented were a uniform size, shape, and composition of the monodispersed aerosol compared to the heterogeneous (or polydisperse) distribution of these parameters for aerosol particles [3, 16, 40–42]. Further, light scattering was used to detect each particle so as to trigger the laser to ablate the matrix-coated analyte for on-line MALDI. 3-nitrobenzylalcohol (3-NBA, liquid) and 2,5-dihydroxybenzoic acid (DHB, solid) were the matrices used. Intense $[M–H+2Na]^+$peaks were observed. The threshold for ionization of leucine enkephalin was approximately 0.2 mJ. Up to 2 mJ of laser energy or fluence provided only a minimal increase in the $[M+Na]^+$signal area. This was most likely due to the monodisperse, uniform nature of the particles at 4.7 μm. Gramicidin S also displayed $[M+Na]^+$ions.

Matrix Coating of Bioaerosol in the Vapor Phase

In the early 2000s, two other methods were devised by a pair of research groups to advance bioaerosol MALDI [30, 31, 33, 62]. In one of the developments [30], particles entered a chamber containing a heated, saturated vapor of organic matrix. As the aerosol penetrated the vapor in the chamber, the matrix condensed on the analyte particles. The matrix-laden particles exited the chamber and were collected on a surface for 5 min. Laser impingement created the ions for TOF-MS analysis. This process resulted in an off-line aerosol MALDI-TOF-MS method [30, 33].

In the work by Jackson and Murray [30], size distributions of particles were investigated with and without matrix vapor in the saturation container, and these two conditions caused the mean of the mass-weighted particle size distribution to decrease from 1.54 to 1.34 μm, respectively. Gramicidin S and gramicidin D peptides with 3-NBA matrix were used. As in all the above analyses [30, 31, 33, 62], the M^+, $[M+H]^+$, and alkali adduct M^+ ions were observed.

A 5-min, off-line deposition of picolinic acid (PA)-coated erythromycin produced an intense $[M+Na]^+$ peak at 757 Da and that of the on-line analysis produced no high-mass information [31]. The mass spectra mainly consisted of <200 Da masses. On-line analysis included a container saturated with MALDI matrix vapor for condensation on bioaerosol particles. On-line, single particle analyses proceeded where ions were continuously extracted from the ion source for analysis. On-line aerosol MALDI analysis of NBA-coated erythromycin showed a significantly smaller $[M+H]^+$ peak at 735 Da [31] compared to the same spectral peak for PA. On-line analysis of NBA-coated gramicidin S particles resulted in a relatively small $[M+Na]^+$ peak at 1164.5 Da, while PA matrix did not produce a molecular ion. Off-line analysis of gramicidin D coated with PA matrix resulted in a poor spectrum with significant low mass interference.

Kim et al. [33] presented off-line *Escherichia coli* aerosol MALDI-TOF-MS information. Two matrices, CHCA and sinapinic acid (SA), were used. Three different methods of combining *E. coli* and matrix were investigated with off-line MALDI-TOF-MS analysis. Each of the six aerosol permutation procedures along with the standard dried droplet method (*vide infra*) with both matrices resulted in eight experiments yielding several different distributions of ions shown in figure 9.6a–g in the 3000–10,000 Da range [33]. Mass spectra from the three techniques were compared to that from the standard dried droplet method, which merely is mixing the initially impacted bacteria with a drop of MALDI organic compound (matrix) and subsequent drying. The four methods were as follows:

1. Standard dried droplet method: An amount of *E. coli* bacteria from a suspension was pipetted onto a MALDI target plate. One microliter of saturated liquid matrix was added to the spot and dried. This was performed for both CHCA (Fig. 9.6a) and SA (Fig. 9.6b) matrices. Matrix solution was added to the dried bacteria to produce the MALDI phenomenon. The deposited bacteria required the matrix solution to be added to produce the MALDI phenomenon (Figs. 9.6a, b).
   The mass spectra in Fig. 9.6a, b were used for comparison to the subsequent bioaerosol-generated MS in Fig. 9.6c–i. Figure. 9.6a, b shows between 30 and 40 peaks. The CHCA matrix generated many more peaks $<$m/z 6000 than that for SA. A reproducibility of $>95\%$ was observed for peaks having an S/N >3.
2. Bioaerosol method 1: A suspension of lyophilized *E. coli* was aerosolized with a collision nebulizer, and an Andersen impactor was used to collect and concentrate the bioaerosol onto a MALDI target plate. One microliter of saturated liquid matrix was added on top of the spot and dried. This occurred for both CHCA and SA matrices. The impacted dry aerosol required the matrix solution to be added to produce the MALDI phenomenon (Figs. 9.6c, d).
   Figure 9.6c, d resulted from CHCA and SA as organic matrix, respectively. Compared to Fig. 9.6a, b, c, and d were obtained using one-third higher laser fluence to obtain adequate signal above m/z 6000. Fewer peaks are observed for identification utility in Fig. 9.6c, d compared to Fig. 9.6a, b.
3. Bioaerosol method 2: Four hundred microliters of matrix were applied to a MALDI target surface and dried. Afterwards, the bacterial aerosol was impacted directly onto the dried matrix, and 200 μL of a 1 : 1 mixture of acetonitrile—water not added (Fig. 9.6e with SA matrix) and added (Fig. 9.6f (CHCA) and 9.6 g (SA)) to the impacted bioaerosol with subsequent MALDI-MS mass spectral analysis.
   In Fig 9.6e, f, and g *E. coli* aerosols were employed for MALDI analysis. However, the order of deposition onto a solid metal target of matrix and aerosol was reversed from that shown in Fig. 9.6c, d. Figure 9.6e shows no signal $>$m/z 3000; however, mass spectral signals appeared with the addition of a solvent mixture (Figs. 9.6f, g). Figure 9.6b and g are very similar in appearance. Figure 9.6f is lower in resolution, lower in the number of peaks, and generally of lower quality than that in Fig. 9.6a.

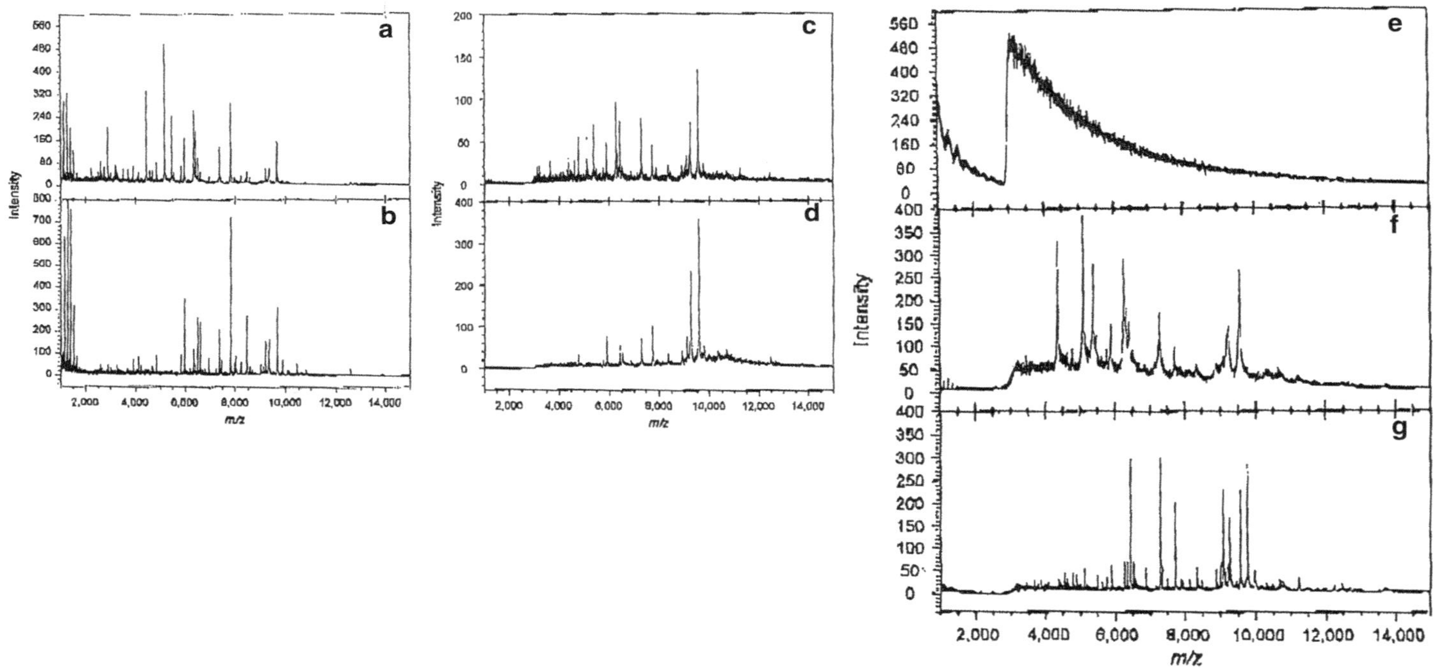

Fig. 9.6 Matrix-assisted laser desorption–ionization (MALDI) time-of-flight mass spectra of *Escherichia coli* bacteria and (**a**) α-cyano-4-hydroxycinnamic acid (CHCA) and (**b**) sinapinic acid (SA) matrices. Mass spectra of *E. coli* by placing 1 μL of (**c**) CHCA and (**d**) SA matrix on top of the collected bioaerosol particles. Mass spectra of bioaerosols deposited on a target surface already coated with (**e**) SA matrix with no additional treatments, (**f**) CHCA matrix and 200 μL of solvent, and (**g**) SA matrix and 200 μL of a 1 : 1 acetonitrile/water solvent. Bioaerosols placed onto a target surface that was precoated with CHCA and SA matrices with a sprayed solution of 1 : 1 acetonitrile/water solvent. (Reprinted with permission from Kim et al. [33]. Copyright 2005 John Wiley & Sons, Ltd.)

4. Bioaerosol method 3: Bioaerosols were impacted onto the MALDI target surface and organic matrix was spray-deposited onto the bacteria. The solvent mixture was then sprayed onto the target by a collision nebulizer and dried. Then MALDI-MS mass spectral analysis then took place (Fig. 9.6f (CHCA) and g (SA)).
   Relatively few peaks are observed in Fig. 9.6f and g compared to Fig. 9.6a–g. The broad mass peaks in Fig 9.6f and g indicate low mass resolution and relatively poor mass spectral fidelity for the bioaerosol method.

The laboratory work [33] presented above has ramifications with respect to translation into field portable and transportable instrument scenarios. Solvents and organic matrices are liquids as well as expendable components. They necessitate reservoirs built into the instrumentation. Pre-coated MALDI solid targets would not be reliable, instead, the matrix and solvent need to be applied to the deposited bioaerosol in a real-time fashion. Therefore, these expendable liquids need to be replaced periodically, and the liquids necessitate an increase in size and complexity of the system.

In another development, Stowers et al. [62] applied a similar online MALDI-TOF-MS concept except with whole bacteria analyte, and the particles were analyzed in real time. That is, triggering was performed and an excimer laser was turned on and off to ablate and ionize the particles in the airstream. This was accomplished by splitting a HeNe laser into two beams perpendicular to the particle stream. The HeNe laser beams detected a particle, and this was used to turn on the excimer laser at the appropriate time to ablate and ionize the particle. BG spores were aerosolized and entered a chamber containing heated, vaporized organic matrix. Matrix coating of the particle took place as it traversed the vapor. Upon exiting the chamber, the aerosol BG particles were ablated and ionized. The ions were focused into a TOF-MS. 3-NBA did not produce signals > 200 Da for BG. PA matrix, however, produced a single peak at 1224 Da, which was tentatively attributed to a portion of the peptidoglycan. That was the only observed peak > 1000 Da.

Harris et al. [26] extended the above on-line aerosol MALDI technique by replacing the TOF-MS with an ion trap MS system (Fig. 9.7). An aerodynamic six lens system was also incorporated prior to the laser ionization source. Matrices PA and 3-NBA were used. Solid matrices were heated and saturated in a condenser where the particles passed through. The vapor phase matrix condensed onto the exterior of each particle to form a coating. Sharp MH^+ bradykinin peaks with NBA and PA matrices were observed for single particles in Fig. 9.8a and b, respectively. Bradykinin was deposited on a probe, and Fig. 9.8c presents an average mass spectrum from 256 separate laser-derived mass spectra. A significant increase in mass spectral peaks was observed with 3-NBA and leucine enkephalin, and the base peaks were MH^+ and $[M+Na]^+$. With the aerodynamic lens and ion trap MS systems, attomole sensitivity was observed for 1 μm particle diameters, which was significantly better than the performance of an aerosol MALDI-TOF-MS [30, 31, 33].

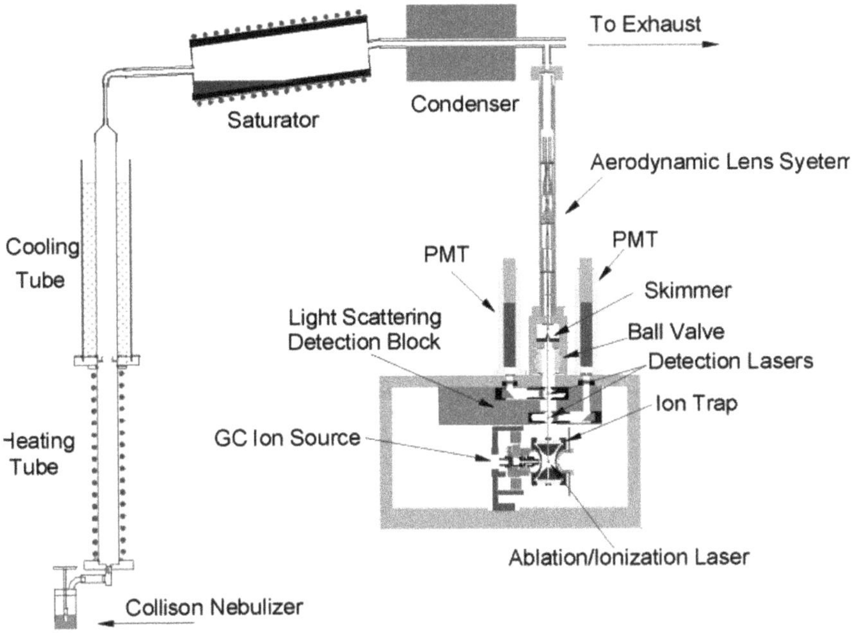

Fig. 9.7 A bioaerosol generator, online organic matrix condensation module onto the aerosol particles, and aerosol ion trap mass spectrometer. (Reprinted with permission from Harris et al. [26]. Copyright 2005 American Chemical Society)

Nanoparticle Bioaerosol

Wilson et al. [67] published information where a system similar to the aforementioned [26] was built, except that a TOF-MS system replaced the ion trap-MS. Test biochemicals included tryptophan and Phe-Gly-Gly biochemicals and no MALDI matrix was used. Instead the pure biochemicals were transformed into nanoparticles and passed through a five-aperture aerodynamic lens system. The particles entered a 10^{-5} Torr ion source. The aerosol impacted onto a heated copper tip and the plume was ionized by a vacuum UV photon beam from a synchrotron (photoionization). Sharp intense M^+ ions (204 Da) were observed at the lowest bioaerosol vaporization temperature of 373 K and 8 eV photoionization energy. Increasing the temperature to 573 K with an 8 eV photoionization energy resulted in essentially the methylene indole fragment, $C_9H_8N^+$, at m/z 130. Vaporization of Phe-Gly-Gly showed an intense M^+ at m/z 279 with essentially no fragmentation at 323 K with an S/N = 800. An observation concerning the use of nanoparticle generated biological compounds is that decomposition of the compounds into small fragments is minimized. A logistic convenience is that a complex system to jet cool the desorbed molecules is not necessary because the desorbed molecules are relatively cool so as to prevent significant thermal decomposition. This internal energy deposition can be controlled so as to vary the amount of decomposition if desired.

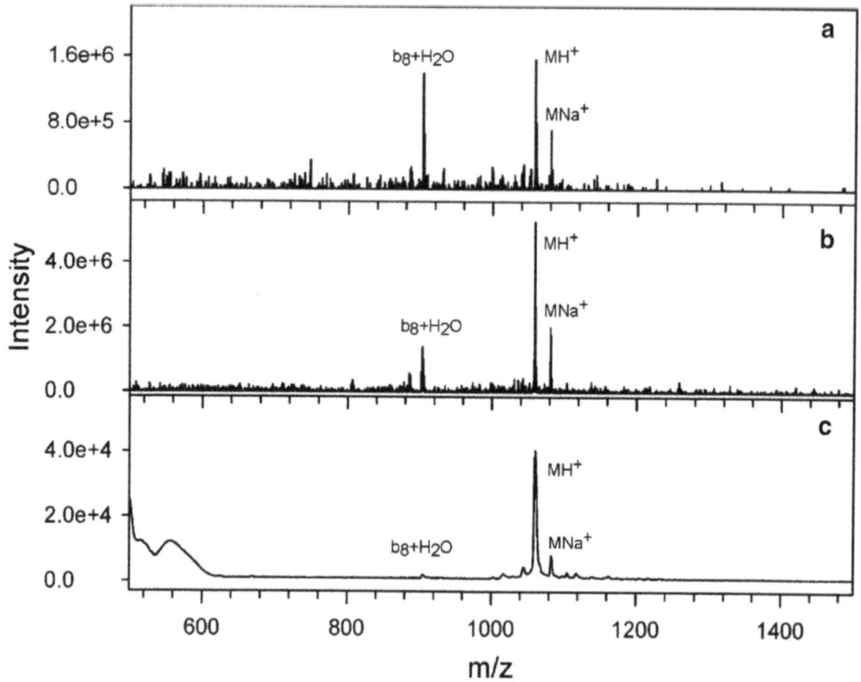

Fig. 9.8 Single-particle MALDI mass spectrum of (**a**) bradykinin coated with 3-nitrobenzyl alcohol, and (**b**) bradykinin coated with picolinic acid; (**c**) MALDI mass spectrum of an average of 256 laser shots performed on separate regions of bradykinin placed onto a probe. (Reprinted with permission from Harris et al. [26]. Copyright 2005 American Chemical Society)

Addition of Ultraviolet Fluorescence Triggering

An improvement was provided with respect to the aforementioned bioaerosol on-line and off-line TOF-MS literature [24, 26, 67]. The improvement consisted of selective MALDI of biological particles as guided by UV fluorescence triggering of those particles [65]. Ferrulic acid was the MALDI matrix and insulin, myoglobin, cytochrome c, and BG organism were the bioaerosols. $[M+H]^+$ ions were observed for the protein analytes in an off-line collection of particles with premixed ferrulic acid and biological substance. This method provides masses of proteins of up to 20,000 Da, which can increase the efficiency of data analysis methods for better identification figures of merit.

Single Particle vs. Bulk Mass Spectra

An interesting situation occurred where the mass spectrum of an aerosol particle of myoglobin and an averaged spectrum of 40 particles were compared [34]. Both spectra had a mass resolution <50 and both $[M+H]^+$ and $[M+2H]^{2+}$ species were observed at 16,952 and 8477 Da, respectively. Intensities were higher in the averaged spectrum. However, the averaged spectrum had a lower mass resolution. It was expected that an averaged mass spectrum should have better mass resolution than a single-shot spectrum. This can be explained by the different ionization regions for different particles as they travel through the width of the laser beam waist. Different ionization regions mean different ion flight path lengths and hence different ion flight times. Ions generated from surface MALDI do not experience this anomaly.

Single aerosol particles of *Erwinia herbicola* (EH) vegetative cells were investigated by online aerosol TOF-MS. A HeNe fluorescence emission system was used to trigger the 308 nm excimer laser to ablate and ionize the particles. SA was pre-mixed with the bacteria to a 1 mg/mL concentration and dried. The powder was nebulized into the vacuum system, and sharp peaks were observed in the 5000–20,000 Da range.

Bioaerosol Mass Spectrometry

Initial System

A project developed at Lawrence Livermore National Laboratories (LLNL) documented success in tackling the challenge of detection and identification of bacterial aerosols by TOF-MS. The impetus for developing a bioaerosol-TOF-MS system originated in the modification of a system based on the analysis of ambient inorganic and organic aerosols [35, 63] naturally found, generated, and released into the environment. Temporal characteristics of ambient firework pyrotechnic [35] releases were monitored. Combinations of inorganic ions as well as select ambient hydrocarbons were noted. The instrument was called an aerosol TOF-MS system (ATOFMS), and it was used to investigate ambient atmospheric fine (100–300 nm) and ultrafine (< 100 nm) particles consisting of simple organic compounds. Pyrotechnic material that was sampled from the ambient atmosphere included alkali and alkaline ions.

Other ambient analytical challenges [4, 23] included wood smoke, particles containing masses attributed to oxygenated linear hydrocarbons [50] such as aldehydes, and acetonitrile car emissions in an on-line fashion. In the present discussion, microliters of BC, *Bacillus thuringiensis* (BT), and *B. subtilis* variant niger (BG) suspensions were pipetted onto a sample probe and dried. Either an infrared (IR) or UV laser was used to effect ionization.

The bioaerosol-TOF-MS system developed at LLNL does not use matrix, liquid, or solid consumables [46, 59]. In the development of this system (Fig. 9.9), a collision nebulizer provided BG aerosols, and they were dried by passing through a drier desiccant tube. The spores passed through a sizing region consisting of two lasers. A frequency-quadrupled, Q-switched Nd:YAG ablated, desorbed, and ionized each particle. Both positive and negative ions are scanned because two separate TOF-MS tubes emanate 180° from the ion source. The system was called the bioaerosol MS (BAMS) system. Typical mass spectra for a BG spore shows positive ions < 150 Da

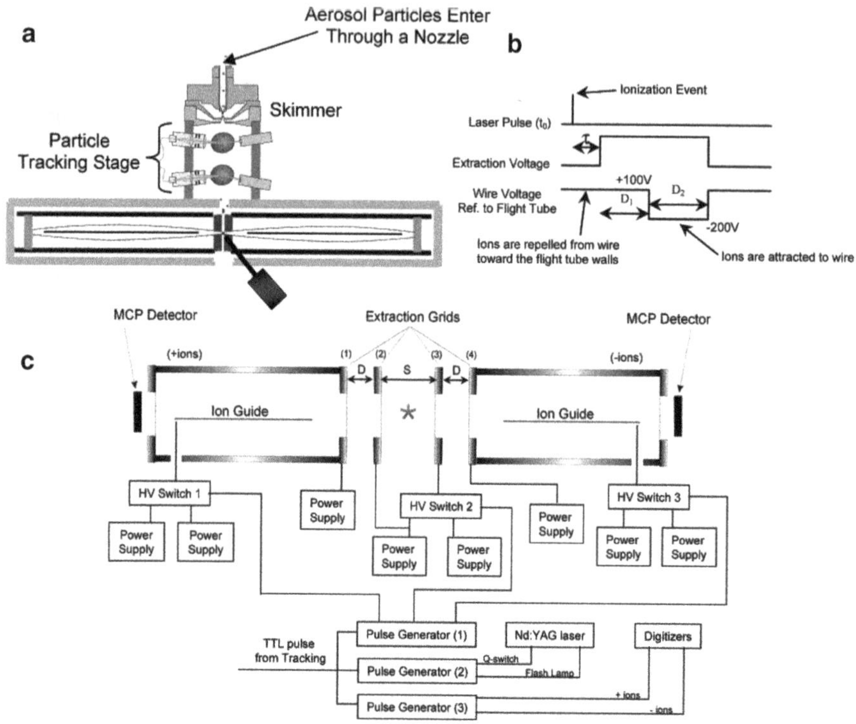

Fig. 9.9 (**a**) Schematic of the bioaerosol mass spectrometry system (BAMS), (**b**) timing diagram of the BAMS for delayed ion extraction and pulsed ion guide wires, and (**c**) schematic of the pulsing and wiring that controls the triggering of the delayed ion extraction and ion guide wires. (Reprinted with permission from Russell et al. [46]. Copyright 2005 American Chemical Society)

and negative ions <200 Da (Fig. 9.10). This was the first recording of positive and negative ion mass spectra from the same biological particle. In the negative mode, as the laser energy increased, lower numbers of masses appeared, and this trend was reversed in the positive mode. The high mass peak intensities (>160 Da) increased when the laser energy was increased. Figure 9.10 shows that at relatively low laser energies, very few peaks were observed in the positive mode. At higher laser energies, an increased number of peaks are observed predominately in the low mass range (<66 Da). It was shown that the laser fluence or energy was a very important variable in the generation of quality TOF mass spectra.

The development of the BAMS [17] saw scrutiny on the concept of analyzing every particle in succession. Spores of BG and BT were used with no liquid reagent. Laser desorption/ionization TOF-MS was reported for aerosols of spores, and rapid data analysis was highlighted. The data analysis starts with a positive and negative mass spectrum of 350 masses (elements) or measurable mass bins for each spectrum. The spectrum was treated as a vector, thus similar spectra are equivalent to similar vectors. In that case, the angle between the two vectors is small. The experimental spectra are then compared to a database of the organism. The positive

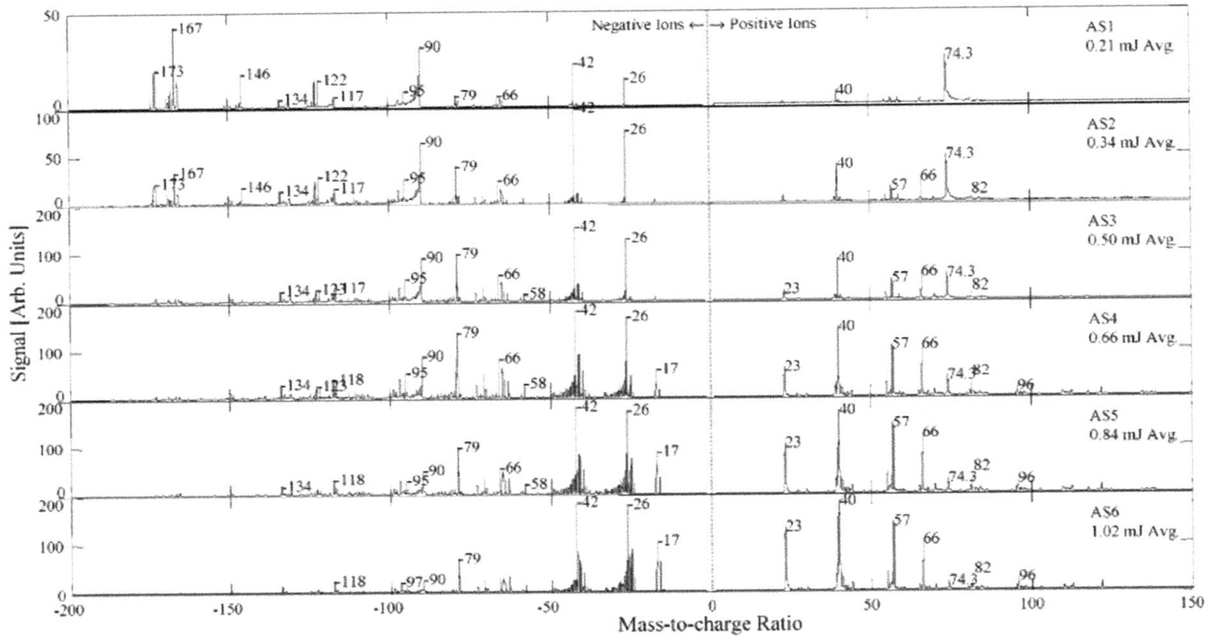

Fig. 9.10 Averages of 1000 spectra (AS) collected at each average laser energy given to the right of each spectrum. The highest intensity mass peaks in the negative ion mode are in AS1 and steadily decrease from AS2 to AS6. The highest intensity mass peaks in the positive ion mode, however, steadily increase and are greatest in AS6. The spectral differences result from changes in the average fluence produced by each laser setting. Therefore, the laser profile must introduce spectral variability from shot to shot since successive spores interact with different regions (and different fluences) in the nonuniform profile and absorb different amounts of energy. (Reprinted with permission from Steele et al. [59]. Copyright 2003 American Chemical Society)

and negative vectors (mass spectra) of an experimental analyte are compared with that of database spectra. A match occurs if both positive and negative vectors are within a 46° angle. If multiple standards match, then the closest negative ion spectrum match was considered as the experimental analyte. Positive spectra showed peaks <100 Da and negative spectra highlighted peaks <200 Da. Reproducibility depended strongly on laser wavelength and fluence. The wavelength of 266 nm was chosen because it was absorbed by dipicolinic acid (DPA) in the spores. The fluence or power was optimized to yield minimum variability from spectrum to spectrum. Spectral displays were averaged from 600 to 1000 particles. Different bacterial growth media for both spores saw minimal mass spectral differences within each species.

Determination of individual particles was made in real time in two steps. First, a prescreening stage eliminated nonbacterial particles by data analysis of the acquired spectra. The microbial-related spectra then went through mass-related criteria to refine and provide a database match to the experimental spectra. BG spores were identified 93.2% when compared to BT. Various commonly found and commercially available white powders were separately mixed with BG and BT to test the data analysis algorithm for interference properties. The spores recognized were as follows: 91% of the time with Gold Bond powder; 86% with growth media; 78% with Equal sweetener; 56% with fungal spores; and 46% with Knox gelatin [44].

Work then focused on efficient ionization characteristics of bacteria based on model biochemical compounds known to be found in bacteria [45]. Amino acids, DPA, and peptides were aerosolized and analyzed. Arginine and glycine amino acids are found in spore coats, and hence they were chosen as model compounds. A concern was the relative smallness of the particle compared to the laser beam cross-section. The cross-section of the beam fluence was not uniform, and the particles may have traveled through different areas of the cross-section, experiencing differential laser fluences. This was caused by the divergence of the particles as they travel through space. Thus, particles of the same size will have a similar divergence from the orifice.

Not all particles in an experiment produced a mass spectrum. Some percentage of the particles experience too wide a divergence from the laser beam, because some particles pass through a low-fluence region of the laser beam cross section. The highest effective ionization probability (EIP), or ratio of particles yielding spectra to the total number of particles, was approximately 40% for the biochemicals. DPA provided 35%, phenylalanine 31%, tryptophan had 30%, tyrosine had 23%, histidine had 20%, and arginine had 10%. The higher EIP substances have the UV absorbing aromatic side chains. Peptides were investigated and the one that had and did not have aromatic side chains yielded EIP values of 26% and <6%. As observed, an UV absorbing functional group in a compound provides a necessary avenue for ionization of the particle. This may be equivalent to the performance of a conventional MALDI organic matrix. Mixing a nonaromatic and an aromatic-containing compound together for bioaerosol analyses provided an increase in the EIP of the nonaromatic-containing compound. A MALDI-like effect

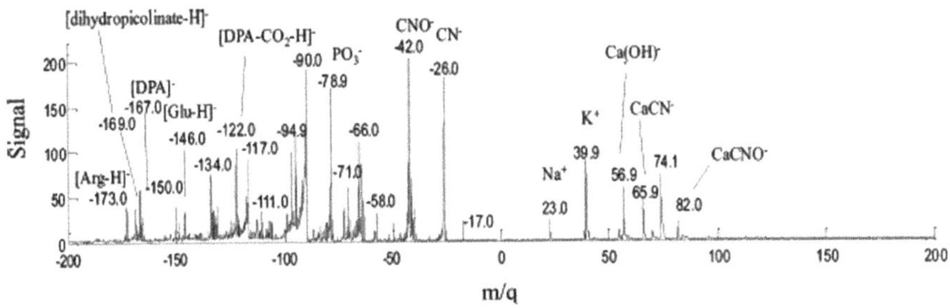

Fig. 9.11 Average of 1000 spectra of *Bacillus atrophaeus* spores using a dual-polarity time-of-flight mass spectrometer. (Reprinted with permission from Czerwieniec et al. [6]. Copyright 2005 American Chemical Society)

was suggested for these observations. Thus, acidities and basicities of the functional groups of particle analytes were shown to be important in the production of useable MS.

Isotope-Labeled Growth Media

Identification of substances is a very difficult task that produce peaks in a mass spectrum from the laser desorption/ionization of bacteria such as in BAMS, because any one particular peak usually arises from many compounds or fragments of larger compounds. An interesting analysis ensued where BG was grown in isotopically enriched media for targeting peaks in mass spectra of individual aerosol particles [6]. BG was grown until mid-log phase in regular media and then was transferred into either ^{15}N- or ^{13}C-labeled growth media. Growth proceeded until the stationary post-log phase was reached. Spectra were acquired in the positive and negative modes by TOF-MS. Figure 9.11 shows the nonlabeled media spectra of BG spores, comprising an average 1000 particle spectra. The dominant information was in the negative ion spectrum.

Major peaks were observed at m/z 167 and 166, which were tentatively identified as the molecular ion (DPA) $^-$ and $[M-H]^-$, respectively. A loss of CO_2 shows peaks at 123 and 122 for the pair of compounds. It is well known that DPA can comprise between 5 and 15% of the dry weight of a spore [5, 15, 29, 48]. A few other compounds are labeled, and Fig. 9.12a, b shows ^{15}N- and ^{13}C-labeled spectra, respectively. A number of peaks did not shift and are most notably the alkali and alkaline metal ions and their compounds as well as PO_3^{3-} when compared to Fig. 9.11. The DPA and DPA-CO_2 peaks shifted one mass unit higher. The arginine peak shifted four mass units higher because it contains four nitrogen atoms. The ^{15}N incorporation appeared complete in that there were was no mixture of labeled and unlabeled nitrogen-containing compounds. For ^{13}C-labeled bacterial aerosols, an envelope about each identified mass was observed and indicated only partial ^{13}C uptake in a percentage of the molecules in the spores. This phenomenon is observed

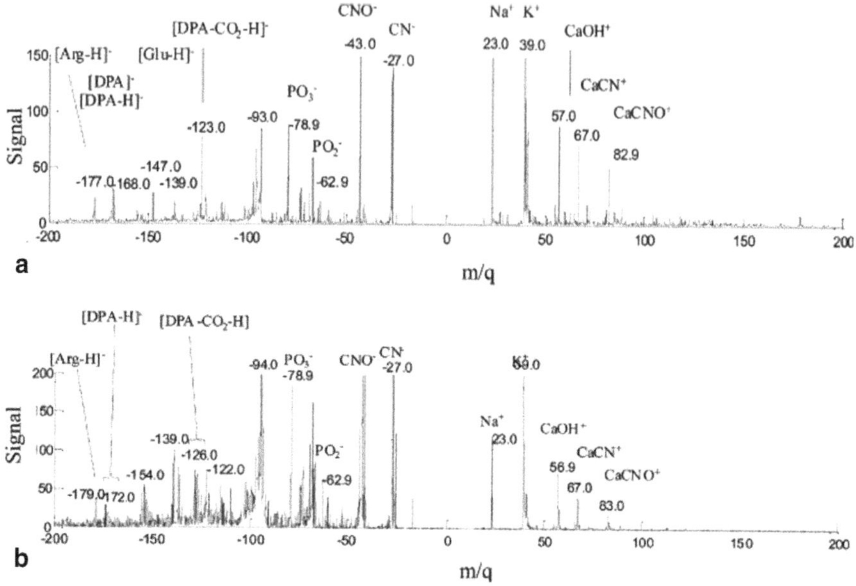

Fig. 9.12 (**a**) Average of 94 spectra of ^{15}N-labeled *B. atrophaeus* spores and (**b**) average of 28 spectra of ^{13}C-labeled spores using a dual-polarity TOF mass spectrometer. (Reprinted with permission from Czerwieniec et al. [6]. Copyright 2005 American Chemical Society)

in the $[DPA\text{-}H]^-$ and $[DPA\text{-}CO_2]^-$ peaks. The aerosol generation, spectral recording, and bacterial identity analysis occurred in the seconds' time frame.

A comprehensive identification of all major mass spectral peaks continued in a comparison of unlabeled BG aerosol spectra with that of the bacteria grown in ^{15}N- and ^{13}C-enriched media [58]. Fragments were also elucidated and standard biochemical aerosols were included to provide further confirmation of peaks. The compounds identified in the spectrum included metal ions, DPA, amino acids, trimethylglycine (betaine) and phosphate species, and the purine nucleobases adenine, guanosine diphosphate, and cytidine. These identified compounds from the BAMS system provide credibility that the pyrolyzed mass peaks seen at the <200 Da range are real and not artifacts of the pyrolysis event. These compounds are generally observed over a wide range of growth conditions since they are expected to be present in growing cells.

Sensitivity Parameter

Studies continued where an increase in sensitivity was addressed [46]. An updated BAMS system was constructed from the ion source, and ion guide wires were installed from the earlier version (Fig. 9.9). Initial testing used off-line MALDI by premixing matrix and analyte and impacting the particles on a plate for subsequent laser ionization. DHB was the matrix, and biochemicals such as gramicidin S, cyclodextrin, and carbohydrate molecules were used. Signals in the zeptomole concentration range were recorded.

Early in the BAMS technology it was observed that the laser fluence occupied a central position in the uniform ionization of particles and sensitivity issues. A concern was that the cross-section of the laser beam was inhomogeneous with respect to fluence or power [59]. Efforts were then made to provide a more uniform, flattened laser cross-section profile [60]. It was also observed that there was a minimum amount of power or fluence necessary to ablate, desorb, and ionize a particle. Many particles that passed through the laser cross-section were not ionized [59]. The explanation was that low-fluence regions in the laser cross-section existed, and particles did not experience enough energy for ionization. This in turn affected the reproducibility issue for a series of particles. Imperfectly focused aerosol particles are routine for an aerosol TOF-MS experiment and as such, different particles experienced different regions of a laser cross-section. This also affected the ability to identify a group of particles from a sample and resulted in differing amounts of energy absorbed by aerosol particles. BG and BT spores were investigated, and the hit rate was compared to laser pulse energy. The hit rate is the ratio of the number of particles producing spectra to the total number of particles. Laser energy vs. hit rate curve shows a 0.15 hit rate for 0.1 mJ laser pulse energy; a 0.125 mJ laser power yielded a 0.225 hit rate; and a >0.15 mJ laser power yielded a constant high hit rate of 0.25. A maximum of only 0.25 or 25 % of the particles produced spectra. Particles that were tracked by their time delay between two laser beams may have passed to the side of the 320 μm diameter of the downfield ionization laser beam and thus not experience the laser energy. The work stresses that proper laser fluence will yield a relatively flatter laser profile that produces better statistical particle identification.

Single Particle Aerosol Mass Spectroscopy

The BAMS evolved into the single-particle aerosol MS (SPAMS) system [61]. The suite of biological substances originally investigated was expanded to include chemical, biological, radiological, nuclear, and explosive (CBRNE) materials as well as clandestine/illegal drug substances. The SPAMS system used either two or three continuous wave laser beams to produce particle sizing properties. When a particle crosses the HeNe beam a light burst occurs that is captured by a detector. This allows for the particle's position, velocity, and aerodynamic diameter to be discovered. Three tracking lasers allowed for the evaluation of these three elements with high precision. The SPAMS could track up to 10,000 particles per second. The particle's position and velocity were used to predict when it passed through subsequent regions of the instrument. After the sizing region, the particle is interrogated by a laser-induced fluorescence region to determine the presence of UV fluorescence, which indicates biological nature. If the particle produced UV fluorescence, it was then ionized by a 266 nm laser beam. Identification of a particle occurred by mass spectral pattern matching with a database. Subsequent steps included selected ions (individual elements on a vector) which followed a rule tree. The angle between a database vector (mass spectrum) and the experimental vector (a mass spectrum) was determined. Ubiquitous ions such as Na^+ and K^+ were removed. The closest angle match within certain limits was determined as the database vector identity.

The various substances produced significantly different experimental mass spectra, and no false identification or false alarms were observed with sequential challenges of the CBRNE materials. In addition to the analytes, there was a constant background of ambient outdoor particles such that the background particles dominated or were equivalent in temporal signal responses to the particular CBRNE challenge signals. Individual components of the mixture were RDX + cobalt to simulate a dirty bomb, RDX, and BG with no false alarms.

The SPAMS system was tested at the San Francisco, CA airport (SFO) and experienced varying environmental conditions over a 7-week time frame. The internal ambient atmosphere of the airport was sampled every minute, and the spectra were recorded. Approximately 1 million particles were tracked and recorded. After the recording and storage of the aerosol data, it was analyzed in the laboratory. No real-time analyses or decisions were made in the SFO field. In any 1-min interrogation, no more than two particles were identified as BG or pentaerythritol tetranitrate (PETN) explosive.

Off-Line Aerosol Analysis of Bacterial Biomarkers

Detection of Presence of Gram-Positive and -Negative Bacteria in Dust From Inherent Biochemicals

A series of papers by Fox and Larsson that are discussed in this section exemplify the difficulties and challenges that bioaerosols present. Their experiments attempted to find specific biochemical species in bacterial aerosols that allowed different stages of discrimination. These include trigger, detection, discrimination, and identification. However, these series of investigations dealt with absolute, not relative, determinations of specific biochemicals that are directly related to bacterial species. The aerosol samples were collected by an impactor in many outdoor and indoor scenarios. Off-line analyses took place in a laboratory where the fundamental systems were GC-MS and GC-tandem MS.

Organic dusts are found in the ambient environment and contain trace to ultratrace amounts of biological species including bacteria [64]. Agricultural environments contain organic dusts where a complex mixture of components includes feed components, dander, feathers, fecal material, pollen grains, insect parts, bacteria, and other bioactive materials. Bacteria in environmental locales can cause disease such as bronchitis, asthma, and airway hyperactivity of the lung regions. It is well known that a very low percentage (0.1–10%) of bacteria in the environment can be observed by culture methods and that visual microscopy is limiting in identification information. These analytical methods were developed to search for chemical markers that are unequivocal in origin in batch collected environmental aerosols. Analyses were performed off-line. Chemical markers constitute monomeric substances usually found covalently linked in the macromolecular architecture in a typical bacterium. Sample processing is such that the collected aerosol can remain on the filter. The entire sample was hydrolyzed to break up the macromolecular architecture into individual units. Purification by extraction was the next step and the selected monomers are derivatized to reduce the polarity of the functional groups such that the resulting compounds can pass through an analytical system such as a GC column with minimal loss.

Dust samples were collected in a room in a home and in a swine confinement building [47, 64]. Three distinct biochemical compounds were sought in the dust samples. Lipopolysaccharides (LPS) constitute a portion of the outer membrane of gram-negative bacteria and they contain unique chemical components. LPS causes challenges and insults to the human body, and they activate the immune system. Lipid A is the endotoxin component of LPS. It is responsible for endotoxicity and contains 3-hydroxy fatty acids (3-OH-FA). A comprehensive set of chemical reactions was performed to isolate the 3-OH-FA of the LPS material, and both dust samples were processed. The chemical steps culminate in the derivatization of the isolated 3-OH-FA species to reduce the polarity. Methyl esters are created and they are further derivatized with a trimethylsilyl reagent. GC-tandem MS was used to analyze for the derivatized 3-OH-FA. The mass spectra and tandem mass spectra of select precursor ions along with the GC retention times were matched to standard FA compounds. Unique swine and house dust GC patterns were observed, indicating a different gram-negative bacteria constitution at each locale. Specific FA and 3-OH-FA species were observed at the different places. The 3-OH-C14 FA was the most intense peak in the swine dust chromatogram, and the 3-OH-C16 FA was the most intense GC peak in the house dust chromatogram. These analyses showed the presence of gram-negative bacteria endotoxin (LPS) at a level of 0.01–0.094 nmol/mg of dust sample. Both dust samples were processed using another protocol to isolate ergosterol, which is found in the membranes of fungi and molds. GC-tandem MS showed very little of the chemical in the house dust, but a significant amount was observed in the swine dust. The low amount in the home environment indicates a relatively dry condition since molds thrive in damp conditions.

A third compound, muramic acid (Mur), is a nitrogen-containing carbohydrate found in the macromolecular peptidoglycan architecture [47, 64]. Peptidoglycan is found in gram-positive bacteria cell walls and is about 10× less in amount in gram-negative bacteria. Peptidoglycan is an immune system stimulant. A different procedure from the aforementioned methods was used to isolate Mur. Isolated Mur was derivatized and subjected to GC-MS analysis, and it was found in significant amounts in both dust samples.

A study was conducted by Lau et al. [37] in which sites in the greater Hong Kong area were subjected to sampling outdoor atmospheric aerosols. The objective was to investigate the presence of the fungal ergosterol chemical marker in the collected dust samples over a fall to spring 8-month time frame on a roughly weekly schedule. Thus, one sample represented a week collection phase. The dust sample was processed to isolate ergosterol and it was derivatized for GC-MS analysis. Ergosterol was noted in amounts significantly more than the background levels in the October to December time region than that in the winter and spring months. The fall time frame saw amounts of ergosterol in the 200–400 pg/m^3 concentrations, while those in the winter and spring months averaged 100 pg/m^3.

Research led by Fox et al. [20] pioneered the concept of trace analysis of bacteria in ambient aerosol environments such as hospitals, farms, home locales, indoor company buildings, schoolrooms, and air duct vents in buildings. The key was to collect raw ambient aerosol samples and systematically use chemical and biochemical reactions to isolate specific and defined chemical monomers or components of the macromolecular architecture of gram-positive and gram-negative bacteria vegetative and spore cells as well as fungal/mold species. Specificity is relegated to the presence of either gram-positive or -negative organisms since that is what the procedure relates.

A very important concept in this research pioneered by Larsson was that samples were not subjected to liquid or solid growth media in flasks or Petri dishes. Almost 90% of ambient aerosols contain bacteria that do not grow on conventional microbiology media under controlled laboratory conditions. Thus, a vast amount of the bacterial load in the ambient atmosphere goes unaccounted. LPS includes L-, D-heptose, and 3-OH FA species. 3-OH dodecanoic acid (3-OH C12:0) and 3-OH tetradecanoic (3-OH C14:0) acids are the usual FA species in LPS which is exclusive to gram-negative cells. These two biochemicals, along with Mur from gram-positive peptidoglycan macromolecules, were investigated for their presence in the aerosol load in hospital air-conditioning systems.

The L-, D-heptose and Mur were analyzed by the alditol acetate technique [18, 19]. Further chemical reaction and processing methods were performed in order to isolate the derivatized compounds [20]. The 3-OH FAs were processed from the duct samples and resulted in methyltrimethylsilyl derivatives. All three isolated and derivatized biochemical species were analyzed by GC-MS. Results showed the presence of the biochemicals, which points to a contribution of gram-positive and gram-negative bacteria in the dust samples. Culturing techniques capture a small portion of the bacterial burden in an aerosol collection. Thus, biochemical and chemical methods were utilized directly on dust samples to extract and derivatize these three key biomarker compounds. Hospital air filters were investigated [21]. Air passed through a primary set of filters and was directed to a secondary set of filters found in the hospital rooms and hallways. The hospital and home filters were processed for the presence of gram-positive and gram-negative bacteria. The alditol acetate derivatization method was used to process the dust with GC-MS detection. The alditol acetate processing of Mur also removes interfering compounds [22]. Three ions characteristic of Mur were used, m/z 168, 404, and 446. Dust from the home and primary hospital filters were found to contain 26–31 ng of Mur per milligram of dust. The hospital secondary filters had about 5 ng/mg dust.

Shahgholi et al. [49] applied trace detection of chemical markers in environmentally collected barn dust aerosol samples with LC, ESI, tandem MS to tackle the problem of Mur detection and identification. GC-MS detection necessitates a time-consuming alditol acetate derivatization procedure, which results in 53 h of processing. About 135 ng of Mur per milligram of hydrolyzed dust was observed with the liquid analysis method.

A study was performed by Krahmer et al. to investigate the Mur and 3-OH-FA levels in dust particulate in horse stable and dairy agricultural farm buildings by GC-tandem MS. It was shown that as the collected amount of dust increased, the amount of Mur and 3-OH-FA also increased. However, the least squares linear fit value for graphs on the amount of Mur in dust in a private horse stable and Mur in dust versus the dairy barn scenario were r^2=0.98 and 0.60, respectively. Product ion mass spectra of the precursor ion confirmed the presence of Mur. This study provided specific biomolecule measures that correlated with constant dynamic activity in the dairy scenario versus a relatively limited movement in the horse/animal stable.

Agar plate media only magnifies the viable bacterial portion of an aerosol; however, the Mur procedure measures both the viable and nonviable bacterial components. Impaction of aerosols onto an agar plate takes several seconds, and the plates were incubated for 24–48 h. In addition, the process of collecting dust on filters for biochemical analysis of Mur and 3-OH-FA requires 24–48 h. The total bacterial load in the aerosol from Mur determinations was much higher than that from agar plate counting techniques (Krahmer et al. 1998). This was in agreement with general knowledge in the area of bioaerosol analysis. Therefore, amounts of agricultural dust directly correlated with the amounts of Mur and 3-OH FA species. However, it is not straightforward to suggest that measuring the amount of dust in a bioaerosol would directly relate to the biological burden in the air. Rather, it may be viewed as complementary for an overall measure of the air bioburden.

A report by Pashynska et al. [43] collected ambient air particulate on quartz fiber filters in Ghent, Belgium. The carbohydrate-related compound levoglucosan was monitored. Levoglucosan is an anhydrous monosaccharide and is produced when cellulose-containing biomass substances are pyrolyzed, heated, or burned. Levoglucosan is a stable heating product. This compound was detected by isolation from a dust sample with trimethylsilylation for derivatization purposes. Other anhydrous monosaccharide compounds isolated from biomass burning were galactosan and mannosan. The glucose, arabitol, mannitol, fructose, and inositol sugars were also monitored. GC-ion trap tandem MS was used for detection and analysis. Dichloromethane (DCM), methanol (MeOH), and an 80:20 mixture of DCM-MeOH were used to test the extraction efficiency of the carbohydrate compounds from the untreated aerosol dust. The compounds are formed from burning biomass and may have adhered to the ambient particulate load. DCM and MeOH were determined to be effective solvents to extract the levoglucosan anhydrous monosaccharide from aerosol particulates. The MeOH and the DCM-MeOH solvent mixtures were effective in the extraction of levoglucosan and glucose. Galactosan and mannosan displayed relatively poor extraction efficiencies from all three solvents. All the extracted carbohydrates were then derivatized with trimethylsilyl reagent and analyzed by GC-tandem MS. The compounds were clearly observed with comparisons to the respective standard compounds by GC retention time and mass spectral matching. Therefore, the collection of aerosols in winter and summer months [43] showed that the total amount of the anhydrous monosaccharide (levoglucosan) load in atmospheric particulates in the winter was greater than that in the summer. This correlates with the burning of wood. For the other carbohydrates (arabitol, mannitol, glucose, fructose, and inositol sugars), their amounts are relatively higher in summer than winter, because they arise from abundant vegetation particulates in the atmosphere.

A study was done to correlate the presence and levels of Mur, 3-OH FA, and CO_2 in two schools with different air-conditioning systems. A previous study [54] found that the amount of biomass, from the two biochemical markers, correlated with CO_2 levels in a direct manner. Thus, it was inferred that the organism load in the air in the school emanated from the children. The current work attempted to assess the chemical markers from the collected aerosol particulates and the CO_2 levels in occupied and unoccupied schoolrooms. Processing and analysis of the dust aerosol samples were similar to earlier studies, and GC-MS was used for detection. It was found that the chemical markers were considerably higher in the children-occupied versus unoccupied schoolrooms over defined periods of time. Mur was 6× higher in occupied versus unoccupied rooms, indicating differing levels of gram-positive organisms. The correlation of FA was more complex. C12 and C14 3-OH FAs were higher in occupied classrooms; C16 3-OH FA was lower in occupied versus unoccupied rooms; and the C10 3-OH FA was about the same in both types of rooms. This variability of 3-OH FA indicates that they originate from different types of gram-positive bacteria. In the earlier study [54], CO_2 levels mirrored that of the overall biomass. In this study, the airborne dust load, chemical marker amounts, and CO_2 levels yielded relatively higher amounts in occupied versus unoccupied classrooms. It was also suggested that the overall amount of gram-positive bacteria was higher than gram-negative bacteria in the classrooms based on the fact that Mur is found in both gram-positive and gram-negative bacteria, but that Mur is approximately 10× more abundant in gram-positive than gram-negative organisms. Another conclusion was that when children were not present a survey study of classrooms would underestimate the microbiological aerosol load.

Fox et al. continued investigations of the classroom indoor air quality with respect to the bacterial aerosol burden. Evidence strongly suggested that the presence of bacteria in the air in indoor schoolroom locales originated from the indoor environment and the building occupants. Analysis was then taken to a more sophisticated level. Instead of the determinations of bacterial compounds such as Mur and 3-OH FA species, the protein content of the collected aerosols was investigated. This takes the form of proteomics. Aerosol collection was performed as previously described previously. Total proteins were extracted from 50 to 100 mg dust and were processed. The purified protein mixture was separated using two-dimensional (2D) gel electrophoresis. Each protein spot was excised from the gel with subsequent isolation and purification procedures. These included dithreitol, iodoacetamide, trypsin digestion, and metal ion removal. The purified peptide mixture was mixed with a MALDI matrix and MALDI-TOF-tandem MS was performed. Dust samples originating from occupied and unoccupied rooms were collected. The most abundant proteins from the gel-staining procedure were identified. One spot was more intense on the gels in both occupied and unoccupied rooms. MALDI-MS and MALDI-tandem MS identified the protein as human epithelial keratin K10. A standard sample also directly correlated with the experimental spot in gel electrophoresis comparisons and MALDI-MS procedures. This study showed that aerosol collection of indoor air indeed has a measurable quantity of human shed epithelial skin cells. Keratin was found to be the most abundant protein in airborne dust, and this was the first study to provide that information. The objective of bacterial protein studies evolved into an analysis of human proteins. These events took place due to the fact that human proteins dominated the gel electrophoresis studies.

Conclusions

MS techniques are providing applications and improvements to the analysis of bioaerosols in different settings. Improvements started after the initial utilization of pyrolysis MS, because versatility was added to the analysis of a wide range of cellular components in bioaerosol analyses. This versatility occurred by the advancement of desorption ionization MS techniques, which increased the detection capability for relatively large cellular components such as proteins and deoxyribonucleic acid (DNA). Desorption MS techniques faced challenges such as the need for a relatively large sample size to establish satisfactory detection limits and mass resolution. The greater the amount of sample, the higher the mass signal becomes for better figures of merit for a mass spectrum. However, the introduction of a reflectron TOF-MS system diminished those issues and enhanced the mass resolution. Alternatively, the sensitivity and bioaerosol collection efficiency were further improved with the introduction of an efficient aerodynamic lens and ion trap MS system, respectively. The latter technique provided attomole-level sensitivity for certain biochemicals in bacteria contained in 1 μm bioaerosol particles. Such developments were vital to bioaerosol analysis, because ion trap MS systems are portable, fieldable, and provide detection and identification in real-time scenarios.

Advancement in the field of MS analysis of bioaerosols was the introduction of an online TOF-MS system hybridized with spectroscopy techniques. This hybridization of optical fluorescence and mass determination technologies served as trigger and detector in one system. The first generation was scrutinized for mass resolution performance in single particle versus bulk analyses of bioaerosols. Engineering improvements in the hybridized TOF-MS system led to the development of the BAMS system. BAMS expanded the type of mass spectral information that could be obtained from biological samples in a single analysis by simultaneously collecting positive and negative ion mass spectra. Another improvement was the simultaneous recording of positive and negative ion mass spectra from the same bioaerosol particle for the first time. Moreover, BAMS eliminated the need for an organic desorption matrix and provided analysis performance that matched conventional MALDI-MS techniques. Evolution of the BAMS led to the development of the SPAMS system. The SPAMS refined the BAMS technique and enhanced reproducibility through minimization of several hurdles such as the distribution of particle sizes and their effect on the laser beam triggers. The SPAMS expanded the application repertoire to the analysis of CBRNE aerosol materials.

Other significant studies that addressed the utilization of MS techniques in bioaerosol analysis were focused on finding biomarkers that were capable of discrimination of microbes present in indoor and outdoor environments in the literature of Larsson and Fox (*vide supra*). The importance of those studies relied on their ability to utilize mass spectral information to establish a comparative microbial profile to discriminate the origin of bioaerosols in environments such as in hospital, classroom, farm stable, or swine enclosures.

Overall, utilization of MS techniques in bioaerosol analysis has witnessed a steady development in the hardware, data processing tools, MS detectors, and range of applicability in real-time analysis scenarios. Such improvements and expansion in the type of information and diverse applications of MS techniques to bioaerosol detection and identification are advantageous. Those features position MS techniques as a forefront technology of choice for various bioaerosol applications. The continued improvement in the sensitivity of TOF-MS detectors combined with the cost-effectiveness of fieldable MS instruments is a factor that promises to provide value to the MS technology. The inherent complexity of bioaerosol contents and the emergence of environmental issues can be addressed with MS technologies by providing rapid detection and accurate identification of agents of interest to the public and private sectors.

Acknowledgements The authors wish to thank Mrs. Cynthia Swim and Mr. Alan Zulich and Dr. Vicky Bevilacqua for administrative and technical support.

References

1. Basile F, Voorhees KJ, Hadfield TL (1995) Microorganism Gram-type differentiation based on pyrolysis-mass spectrometry of bacterial fatty acid methyl ester extracts. Appl Environ Microbiol 61: 1534–1539
2. Basile F, Beverly MB, Abbas-hawks C, Mowry CD, Voorhees KJ, Hadfield TL (1998) Direct mass spectrometric evidence of in situ thermally hydrolyzed and methylated lipids from whole bacterial cells. Anal Chem 70: 1555–1562
3. Beeson MD, Murray KK, Russell DH (1995) Aerosol matrix-assisted laser desorption ionization: effect of analyte concentration and matrix-to-analyte ratio. Anal Chem 67:1981–1986
4. Bente M, Sklorz M, Streibel T, Zimmermann R (2009) Online laser desorption-multiphoton positionization mass spectrometry of individual aerosol particles: Molecular source indicators for particles emitted from different traffic-related and wood combustion sources. Anal Chem 81:4456–4467
5. Breed RS, Murray EGD, Smith NR (eds) (1957) Bergey's Manual of Determinative Bacteriology. The Williams and Wilkins Co., Baltimore, MD, 7th ed, pp 295–305.
6. Czerwieniec GA, Russell SC, Tobias HJ, Pitesky ME, Fergenson DP, Steele P, Srivastava A, Horn JM, Frank M, Gard EE, LeBrilla CB (2005) Stable isotope labeling of entire

Bacillus atrophaeus spores and vegetative cells using bioaerosol mass spectrometry. Anal Chem 77:1081–1087

7. de B Harrington P, Street TE, Voorhees KJ, Radicati di Brozolo F, Odom RW (1989) Rule-building expert system for classification of mass spectra. Anal Chem 61: 715–719
8. DeLuca S, Sarver EW, de B Harrington P, Voorhees KJ (1990) Direct analysis of bacterial fatty acids by Curie-point pyrolysis tandem mass spectrometry. Anal Chem 62: 1465–1472
9. DeLuca SJ, Voorhees KJ (1993) A comparison of products from an air atmosphere tube furnace with a vacuum Curie-point pyrolyzer: implications for biodetection. J Anal Appl Pyrolysis 24: 211–225
10. Dixon TC, Meselson M, Guillemin J, Hanna PC (1999) Medical progress: anthrax. N Engl J Med 341:815–826
11. Dongre AR, Eng JK, Yates III JR (1997) Emerging tandem-mass-spectrometry techniques for the rapid identification of proteins. Tibtech 15: 418–425
12. Dworzanski JP, Berwald L, Meuzelaar HLC (1990) Pyrolytic methylation-gas chromatography of whole bacterial cells for rapid profiling of cellular fatty acids. Appl Environ Microbiol 56: 1717–1724
13. Dworzanski JP, Berwald L, McClennen WH, Meuzelaar HLC (1991) Mechanistic aspects of pyrolytic methylation and transesterification of bacterial cell wall lipids. J Anal Appl Pyrolysis 21:221–232
14. Dworzanski JP, Snyder AP (2005) Classification and identification of bacteria using mass spectrometry-based proteomics. Expert Rev Proteomics 2: 863–878
15. Esposito AP, Talley CE, Huser T, Hollars CW, Schaldach CM, Lane SM (2003) Analysis of single bacterial spores by micro-Raman spectroscopy. Appl Spectrosc 57:868–871
16. Fei X, Wei G, Murray KK (1996) Aerosol MALDI with a reflectron time-of-flight mass spectrometer. Anal Chem 68:1143–1147
17. Fergenson DP, Pitesky ME, Tobias HJ, Steele PT, Czerwieniec GA, Russell SC, Lebrilla CB, Horn JM, Coffee, KR, Srivastava A, Pillai SP, Shih MTP, Hall HL, Ramponi AJ, Chang JT, Langlois RG, Estacio PL, Hadley RT, Frank M, Gard EE (2004) Reagentless detection and classification of individual bioaerosol particles in seconds. Anal Chem 76:373–378
18. Fox A, Morgan SL, Gilbart J (1989) Preparation of alditol acetates and their analysis by gas chromatography and mass spectrometry. In: Biermann CJ, McGinnis GD (eds) Analysis of Carbohydrates by GLC and MS. CRC Press, Boca Raton, FL, pages 87–117
19. Fox A, Black GE (1993) Identification and detection of carbohydrate markers for bacteria. Derivatization and gas chromatography-mass spectrometry. In: Fenselau C (ed) Mass Spectrometry for the Characterization of Microorganisms. American Chemical Society, Washington, DC, pages 107–131
20. Fox A, Rosario RMT, Larsson L (1993) Monitoring of bacterial sugars and hydroxyl fatty acids in dust form air conditioners by gas chromatography-mass spectrometry. Appl Environ Microbiol 59:4354–4360
21. Fox A, Rosario RMT (1994) Quantification of muramic acid, a marker for bacterial peptidoglycan, in dust collected from hospital and home air-conditioning filters using gas chromatography-mass spectrometry. Indoor Air 4:239–247
22. Fox A, Wright L, Fox K (1995) Gas chromatography-tandem mass spectrometry for trace detection of muramic acid, a peptidoglycan chemical marker, in organic dust. J Microbiol Methods 22:11–26
23. Gard E, Mayer JE, Morricai BD, Dienes T, Fergenson DP, Prather KA (1997) Real-time analysis of individual atmospheric aerosol particles: Design and performance of a portable ATOFMS. Anal Chem 69:4083–4091
24. Gieray RA, Reilly PTA, Yang M, Whitten WB, Ramsey JM (1997) Real-time detection of individual airborne bacteria. J Microbiol Methods 29:191–199
25. Griest WH, Wise MB, Hart KJ, Lammert SA, Thompson CV, Vass AA (2001) Biological agent detection and identification by the Block II chemical biological mass spectrometer. Field Anal Chem Technol 5: 177–184

26. Harris WA, Reilly PTA, Whitten WB (2005) MALDI of individual biomolecule-containing airborne particles in an ion trap mass spectrometer. Anal Chem 77:4042–4050
27. Hart KJ, Wise MB, Griest WH, Lammert SA (2000) Design, development, and performance of a fieldable chemical and biological agent detector. Field Anal Chem Technol 4: 93–110
28. Havey CD, Basile F, Mowry C, Voorhees KJ (2004) Evaluation of a micro-fabricated pyrolyzer for the detection of bacillus anthracis spores. J Anal Appl Pyrolysis 72: 55–61
29. Huang S-s, Chen D, Pelczar PL, Vepachedu VR, Setlow P, Li Y-q (2007) Levels of Ca^{2+}-dipicolinic acid in individual *Bacillus* spores determined using microfluidic Raman tweezers. J Bacteriol 189:4681–4687
30. Jackson SN, Murray KK (2002) Matrix addition by condensation for matrix-assisted laser desorption/ionization of collected aerosol particles. Anal Chem 74:4841–4844
31. Jackson SN, Mishra S, Murray KK (2004) On-line laser desorption/ionization mass spectrometry of matrix-coated aerosols. Rapid Commun Mass Spectrom 18:2041–2045
32. Kaufmann AF, Meltzer MI, Schmid GP (1997) The economic impact of a bioterrorist attack: are prevention and postattack intervention programs justifiable? Emerg Infect Diseases 3:83–94
33. Kim J-K, Jackson SN, Murray KK (2005) Matrix-assisted laser desorption/ionization mass spectrometry of collected bioaerosol particles. Rapid Commun Mass Spectrom 19:1725–1729
34. Kleefsman I, Stowers MA, Verhrijen PJT, van Wuijckhuijse AL, Kientz Ch E, Marijnissen JCM (2007) Bioaerosol analysis by single particle mass spectrometry. Part Part Syst Charact 24:85–90
35. Liu D-Y, Rutherford D, Kinsey M, Prather KA (1997) Realtime monitoring of pyrotechnically derived aerosol particles in the troposphere. Anal Chem 69:1808–1814
36. Liu H, Lin D, Yates III JR (2002) Multidimensional separations for protein/peptide analysis in the post-genomic era. BioTechniques 32: 898–911
37. Lau APS, Lee AKY, Chan CK, Fang M (2006) Ergosterol as a biomarker for the quantification of the fungi biomass in atmospheric aerosols. Atmospheric Environ 40:249–259
38. Mansoori BA, Johnston MV, Wexler AS (1996) Matrix-assisted laser desorption/ionization of size- and composition-selected aerosol particles. Anal Chem 68:3595–2601
39. Meuzelaar HLC, Windig W, Harper AM, Huff SM, McClennen WH, Richards JM (1984) Pyrolysis mass spectrometry of complex organic materials. Nature 226: 268–274
39. Meuzelaar HLC, Windig W, Harper AM, Huff SM, McClennen WH, Richards JM (1984) Pyrolysis mass spectrometry of complex organic materials. Nature 226: 268–274
40. Murray KK, Lewis TM, Beeson MD, Russell DH (1994) Aerosol matrix-assisted laser desorption ionization for liquid chromatography time-of-flight mass spectrometry. Anal Chem 66:1601–1609
41. Murray KK, Russell DH (1994a) Aerosol matrix-assisted laser desorption ionization mass spectrometry. J Am Soc Mass Spectrom 5:1–9
42. Murray KK, Russell DH (1994b) Liquid sample introduction for matrix-assisted laser desorption ionization. Anal Chem 65:2534–2537
43. Pashynska V, Vermeylen R, Vas G, Maenhaut W, Claeys M (2002) Development of a gas chromatographic/ion trap mass spectrometric method for the determination of levoglucosan and saccharidic compounds in atmospheric aerosols. Applications to urban aerosols. J Mass Spectrom 37:1249–1257
44. Ramponi AJ, Chang JT, Langlois RG, Estacio PL, Hadley RT, Frank M, Gard EE (2004) Reagentless detection and classification of individual bioaerosol particles in second. Anal Chem 7:373–378
45. Russell SC, Czerwieniec G, Lebrilla C, Tobias H, Fergenson DP, Steele P, Pitesky M, Horn J, Srivastava A, Frank M, Gard EE (2004) Toward understanding the ionization of biomarkers from micrometer particles by bio-aerosol mass spectrometry. J Am Soc Mass Spectrom 15:900–909
46. Russell SC, Czerwieniec G, Lebrilla C, Steele P, Riot V, Coffee K, Frank M, Gard EE (2005) Achieving high detection sensitivity (14 zmol) of biomolecular ions in bioaerosol mass spectrometry. Anal Chem 77:4734–4741

47. Sebastian A, Larsson L (2003) Characterization of the microbial community in indoor environments: a chemical-analytical approach. Appl Environ Microbiol 69:3103–3109
48. Setlow P J (2006) Spores of *Bacillus subtilis*: their resistance to radiation, heat and chemicals. Appl Microbiol 101:514–525
49. Shahgholi M, Ohorodnik S, Callahan JH, Fox A (1997) Trace detection of underivatized muramic acid in environmental dust samples by microcolumn liquid chromatography-electrospray tandem mass spectrometry. Anal Chem 69:1956–1980
50. Silva PJ, Prather KA (1997) On-line characterization of individual particles from automobile emissions. Environ Sci Technol 31:3074–3080
51. Sinha MP, Giffen CE, Norris DD, Estes TJ, Vilker VL, Friedlander SK (1982) Particle analysis by mass spectrometry. J Colloid Interface Sci 87:140–153
52. Sinha MP, Platz RM, Vilker VL, Friedlander SK (1984) Analysis of individual biological particles by mass spectrometry. Intl J Mass Spectrom Ion Processes 57:125–133
53. Sinha MP, Platz RM, Friedlander SK, Vilker VL (1985) Characterization of bacteria by particle beam mass spectrometry. Appl Environ Microbiol 49:1366–1375
54. Smedje G, Norback D (2001) Incidence of asthma diagnosis and self-reported allergy in relation to the school environment: A four-year follow-up study in schoolchildren. Intl J Tuberculosis Lung Dis. 5:1059–1066
55. Smith PB, Snyder AP (1993) Characterization of bacteria by oxidative and non-oxidative pyrolysis-gas chromatography/ion trap mass spectrometry. J Anal Appl Pyrolysis 24: 199–210
56. Snyder AP, Eiceman GA, Windig W (1988) Influence of pyrolytic techniques on mass spectra of biopolymers with chemical ionization at atmospheric pressure. J Anal Appl Pyrolysis 13: 243–257
57. Spurny KR (1994) On the chemical detection of bioaerosols. J Aerosol Sci 25:1533–1547
58. Srivastava A, Pitesky ME, Steele PT, Tobias HJ, Fergenson DP, Horn JM, Russell SC, Czerwieniec GA, Lebrilla CB, Gard EE, Frank RM (2005) Comprehensive assignment of mass spectral signatures from individual *Bacillus atrophaeus* spores in matrix-free laser desorption/ionization bioaerosol mass spectrometry. Anal Chem 77:3315–3332
59. Steele PT, Tobias HJ, Fergenson DP, Pitesky ME, Horn JM, Czerwieniec GA, Russell SC, LeBrilla CB, Gard EE, Frank M (2003) Laser power dependence of mass spectral signatures from individual bacterial spores in bioaerosol mass spectrometry. Anal Chem 75:5480–5487
60. Steele PT, Srivastava A, Pitesky ME, Fergenson DP, Tobias HJ, Gard EE, Frank M (2005) Desorption/ionization fluence thresholds and improved mass spectral consistency measured using a flattop laser profile in the bioaerosol mass spectrometry of single *Bacillus* endospores. Anal Chem 77:7448–7454
61. Steele PT, Farquar GR, Martin AN, Coffee KR, Riot VJ, Martin SI, Fergenson DP, Gard EE, Frank M (2008) Autonomous, broad-spectrum detection of hazardous aerosols in seconds. Anal Chem 80:4583–4589
62. Stowers MA, van Wuijckhuijse AL, Scarlett B, van Baar BLM, Klentz Ch E (2000) Application of matrix-assisted laser desorption ionization to on-line aerosol time-of-flight mass spectrometry. Rapid Commun Mass Spectrom 14:829–833
63. Su Y, Sipin MF, Furutani H, Prather KA (2004) Development and characterization of an aerosol time- of-flight mass spectrometer with increased detection efficiency. Anal Chem 76:712–719
64. Szponar B, Larsson L (2001) Use of mass spectrometry for characterizing microbial communities in bioaerosols. Ann Agric Environ Med 8:111–117
65. van Wuijckhuijse AL, Stowers MA, Kleefsman WA, van Baar BLM, Kientz Ch E, Marijnissen JCM (2005) Matrix-assisted laser desorption/ionisation aerosol time-of-flight mass spectrometry for the analysis of bioaerosols: development of a fast detector for airborne biological pathogens. J Aerosol Sci 36:677–687
66. Voorhees KJ, Harrington PB, Street TE, Hoffman S, Durfee SL, Bonelli JE, Firnhaber CS (1990) Approached to pyrolysis/mass spectrometry data analysis of biological materials. In: Meuzelaar HLC (ed) Computer-enhanced analytical spectroscopy. Modern analytical chemistry, vol 2. Plenum Press, New York, pp 259–275

67. Wilson KR, Jimenez-Cruz M, Nicolas C, Belau L, Leone SR, Ahmed M (2006) Thermal vaporization of biological nanoparticles: Fragment-free vacuum ultraviolet photoionization mass spectra of tryptophan, phenylalanine-glycine-glycine, and beta-carotene. J Phys Chem A 110:2106–2113
68. Windig W, Jakab E, Richards JM, Meuzelaar HLC (1987) Self-modeling curve resolution by factor analysis of a continuous series of pyrolysis mass spectra. Anal Chem 59: 317–323
69. Yates III JR (1998) Mass spectrometry and the age of the proteome. J Mass Spectrom 33: 1–19
70. Yates III JR (2004) Mass spectral analysis in proteomics. Annu Rev Biophys Biomed Struct 33: 297–316

Chapter 10
Detection of Bioaerosols Using Raman Spectroscopy

Hilsamar Félix-Rivera and Samuel P. Hernández-Rivera

Introduction to Raman Spectroscopy

Historical Perspective and Basic Principles

In 1923, Austrian professor of physics Adolf Smekal predicted on theoretical grounds the inelastic scattering of light from a gas, liquid, or solid, with a corresponding shift in wavelength from that of the usually monochromatic incident radiation [1]. A year later, the Indian physicist C. V. Raman, working in collaboration with K. S. Krishnan, discovered the eponymous effect [2] virtually simultaneously with the Russian physicists G. S. Landsberg and L. I. Mandelstam [3]. Two years after the experimental verification of the Smekal–Raman Effect, Sir Chandrasekhara Venkata Raman was awarded the Nobel Prize in Physics for his work on the scattering of light and the discovery of the Raman Effect. In the 50 years that followed its prediction and discovery, Raman scattering spectroscopy became established as an excellent tool for both materials characterization and biophysical studies. Today, Raman spectroscopy (RS), also called normal or spontaneous RS, remains a powerful spectroscopic technique that is widely used for steady state and time-resolved studies in many areas of science, engineering, forensic investigation, and even art restoration. Raman scattering has come of age in the last 20 years due to technological advances that have significantly decreased the cost of the instrumentation and made the technique available to a much wider audience. This important spectroscopic tool is routinely used in vibrational analysis, including low-frequency modes in crystalline solid systems. Raman is a high-resolution photonic technique that can provide chemical and structural information on almost any Raman-active material

S. P. Hernández-Rivera (✉) · H. Félix-Rivera
Center for Chemical Sensors Development/Chemical Imaging Center (CCSD/CIC),
ALERT DHS—Center of Excellence for Explosives Research, Department of Chemistry,
University Puerto Rico-Mayagüez, PO Box 9000,
Mayagüez, PR 00681, USA
e-mail: samuel.hernandez3@upr.edu

P. Jonsson et al. (eds.), *Bioaerosol Detection Technologies*, Integrated Analytical Systems,
DOI 10.1007/978-1-4419-5582-1_10, © Springer-Verlag New York 2014

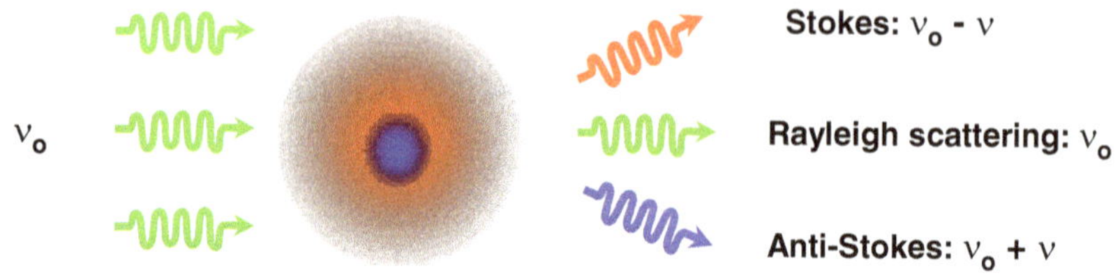

Fig. 10.1 Representation of collision phenomena in which photons with frequency ν_0 interact with the electron cloud of a molecule of the sample. The photon–molecule collision events result in scattering of the incident radiation with photons with the same frequency (Rayleigh radiation) and photons with shifted frequencies producing the Raman radiation

(organic, inorganic and biological compounds), allowing for their identification and characterization within a few seconds.

For a molecule to participate in the Smekal–Raman Effect, it has to undergo a change in one or more of its molecular polarizability components while the molecule vibrates. The extent of this change will determine the signal intensity of the Raman scattering associated with the vibrational transition. This phenomenon is based on the inelastic scattering of light by the material under study, and its analysis is simpler when a single frequency/wavelength source of light interacts with the molecules in the system. The radiation is usually, but not necessarily, generated by a laser source in the UV, VIS or NIR regions of the electromagnetic spectrum. When a sample is illuminated with a monochromatic light beam of frequency ν_0, photons in the beam interact with the electron density surrounding the bonds of the molecules under investigation (Fig. 10.1).

Most of the radiation is elastically scattered (Rayleigh scattering), and the emerging photons have the same frequency as the incident photons. This type of scattering event does not provide information about the sample composition, but it can be used, for example, to determine the average molecular size in a collection of high molecular weight polymers in a technique called dynamic light scattering. The Rayleigh scattering event is at least 10 million times more probable than the Raman scattering phenomenon. A small fraction of the incident photons (as low as 1 in 10^7) are inelastically scattered in discrete amounts of lower and higher frequencies (Raman scattering). Thus, in order to observe the Raman scattering a high power source that generates large quantities of photons is typically required.

If the incident photons excite one of the electrons into a virtual electronic state, the molecule will spontaneously decay to an allowed vibrational state of the ground electronic state of the molecule, resulting in a fast emission of photons having different energies than those of the exciting photons. This produces the inelastic event called the Spontaneous or Normal Raman Effect (Fig. 10.2).

The small frequency changes ($\Delta\nu_i$) associated with the inelastically scattered photons correspond to differences in energy of the vibrational modes of the molecules, providing information about the vibrational structure of the system under study. According to the Plank–Einstein energy conservation principle, inelastically scattered photons have energies given by:

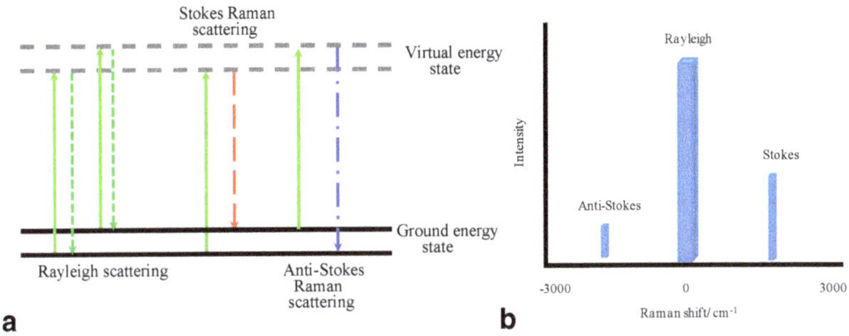

Fig. 10.2 **a** Energy level diagram showing the states involves in photon scattering. Excitation energies: *light green lines/solid lines*; Rayleigh scattering (no energy/frequency change): *dark green lines/square dot lines*; inelastic (energy loss) Stokes scattering: *red line/long dash line*; superelastic scattering (energy/frequency gain): *blue line/long dash dot line*. **b** Schematic of the intensity distribution in scattering types: Anti-Stokes region, Rayleigh, Stokes region

$$\Delta\varepsilon_i = h\Delta\nu_i = h\ \nu_o \pm h\ \nu_i \tag{10.1}$$

Comparison with Infrared Spectroscopy

Modern infrared spectroscopy (IRS) and RS are the two main vibrational spectroscopic analysis techniques. Infrared is an inherently strong process due to its photonic enabling mechanism: resonant absorption; Raman is a weak scattering event that requires specialized equipment to enable the implementation of the technique. Aside from the complementary nature of these two techniques, there are fundamental differences between them:

- RS requires a minimum sample preparation, both for macro Raman and micro Raman.
- Water is the favored solvent for Raman studies due to its extremely low Raman scattering probability (cross section). In IR spectroscopy, water absorbs profusely and limits the observation of its solutes, even at high concentrations.
- Ordinary glass cells and capillary tubes can be used for liquid sampling in visible Raman excitation; quartz cells are required for UV-Raman studies. IRS requires cells and windows that allow for a high transmission of infrared radiation.
- Raman lines (signals) arising from partially to totally symmetric vibrations are much sharper than the corresponding asymmetric vibrations favored by IRS. These features are more amenable to peak location and quantification.
- The possibility of measuring the equivalent of the full IR spectrum, from the far-IR to the near-IR, in a single scan, without changing the excitation source, detector, transfer optics, sample holder, etc. That is, the spectral range can be $\varpi \leq 100$–$10000\ \text{cm}^{-1}$.

- In RS, the measurements of parallel and perpendicular polarization spectra are relatively easy, and the subsequent calculation of depolarization ratios allows for the assignment of the normal vibrational mode symmetries. The excitation sources in RS are typically polarized sources of electromagnetic radiation.
- The use of tunable excitation laser sources to search for resonantly enhanced Raman spectra (Resonance Raman Spectroscopy, RRS), in which it is possible to selectively increase vibrations associated with chromophore groups in molecules by several orders of magnitude (important in biophysical and biomedical studies).
- There are excellent techniques for coupling with optical microscopy sampling. Noble gas ion lasers typically have very small, round beams (at TEM_{00} mode). Microscopy studies are diffraction limited to $\geq\lambda/2$ of the excitation source. For the current state of the art, this represents 125–300 nm spatial and optical resolution under a microscope objective with high magnification.
- The technique is well suited for fiber optics coupling for measurements outside the sample compartment.
- Stand-off detection using several RS modalities has been successfully achieved.
- Hyperspectral RS studies, including the generation of vibrational maps (2-dimensions) and images (3-dimensions), are becoming much more popular.

Some clear disadvantages of RS relative to IRS (particularly to Fourier transform IR are:

- A highly monochromatic excitation source is required for RS. A laser source is preferred because of its many attributes: it is monochromatic, coherent, polarized and collimated.
- Fluorescence is a very strong resonance absorption process that can mask the vibrational features of a Raman Shift spectrum. It is a persistent obstacle in molecular systems with low energy electronic states that can participate in fluorescence (endogenous fluorescence) or in systems with contaminants that can fluoresce (exogenous fluorescence).
- The highest spectral resolution achievable ($\Delta\varpi$) is potentially lower than that of IRS. Raman normally operates in the UV to NIR regions of the electromagnetic spectrum where the resolving power ($RP=\varpi/\Delta\varpi$) of most monochromators is limited and wavelength-dependent. Typical high resolution RS studies are characterized by $1\geq\Delta\varpi\geq0.1\ cm^{-1}$. In IRS, the resolution can be as high as $\Delta\varpi\sim0.0001\ cm^{-1}$ in gaseous molecular systems.
- The cost is typically lower for IRS instrumentation. However, new breeds of low cost fiber optics-coupled Raman spectrometers are as cost effective as entry level FTIR interferometers.

Scattered Raman Signal and Sample Requirements

The intensities that can be observed in a Raman scattering experiment (following McCreery [4]) are given by:

$$I_R = I_o\sigma_j N\,dz = I_o\sigma_j{}^{\circ}(v_o \pm v_j)^4 N\,dz \tag{10.2}$$

where I_0 and I_R are the incident and scattered intensities (W), respectively, N is the number density of scatterers (molecules per cm^3), σj is the integrated cross section over all scattering directions (transition probability, cm^2/molecule) for vibrational mode "j", σ_j^o is the frequency-independent cross section of vibrational mode j (cm^6 molecule^{-1}), v_0 and v_j are the exciting laser line and the Raman scattered light frequencies, respectively, and dz is the path length (cm) of the laser in the sample [4–6]. For modern systems that count the number of incident and scattered photons per second (P_0 and P_R, counts/s, respectively) this equation can be recast to:

$$P_R = P_o \sigma_j' v_o (v_o \pm v_j)^3 N dz \tag{10.3}$$

The integrated cross sections: σ_j (W) and σ_j' (photons·s^{-1}) represent the integration over all scattering angles and over the bandwidth of a vibrational line. It is often more practical to define a differential scattering cross section (per unit solid angle) as:

$$\beta = d\sigma_j / d\Omega \ [\text{cm}^2 \text{molec} \cdot \text{sr}^{-1}], \tag{10.4}$$

or per unit solid angle and per unit Raman Shift (wavenumber) as:

$$\beta' = \delta^2 \sigma_j \ / \ \delta\Omega \, \delta(\Delta n) \ [\text{cm}^2 \text{molec} \cdot \text{sr}^{-1} \cdot \text{wavenumber}^{-1}]. \tag{10.5}$$

Differential scattering cross sections can be used in qualitative form to compare line strengths of samples to those reported in the literature. However, the most truthful comparison for quantitative measurements can be obtained when Raman signal intensities are compared within the same instrument and scattering geometry.

In general terms, minimum sample preparation and handling is required to obtain qualitative and quantitative information from Raman spectra. This contrasts markedly with IRS, for which sample preparation is typically a dominant factor for obtaining good quality data. Solids can be examined on plates (stainless steel, aluminum, etc.) or in capillary tubes in the case of Raman spectrometers coupled to optical microscopes. Neat liquid compounds and liquid solutions can be studied in glass cells or capillary tubes for visible light excitation or quartz cells for UV light excitation. Unless a specialized Raman Effect technique is used (as will be discussed later), special cells are needed for gas phase measurements because the low density of the rarefied molecular systems requires multiple excitation light reflections though the sample to compensate for the combination of low scattering cross sections and low densities in these samples.

Type of Information Obtained from Raman Spectroscopy

The Raman shift spectrum of a molecule is an intrinsic fingerprint of the sample that is complementary to the infrared spectrum. They jointly provide useful information about molecular structure (chemical bonds, angles between bonds, and symmetry of molecules), molecular interactions, both intramolecular and intermolecular (through studies of vibrational mode locations and signal widths), and composition

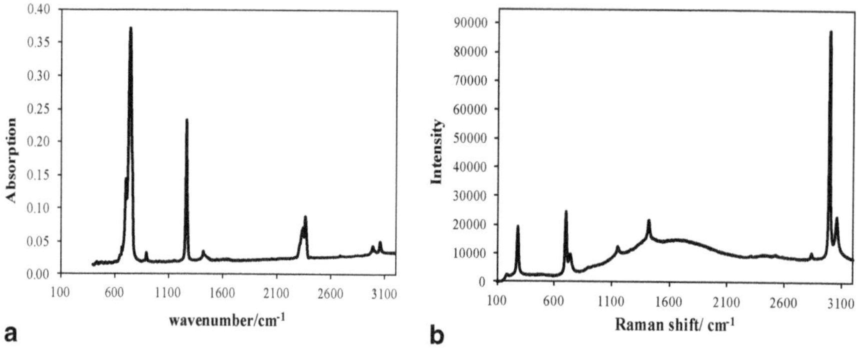

Fig. 10.3 Spectra of dichloromethane 99.9%. (**a**) IR and (**b**) Raman: 532 nm laser, 10 mW, 10 s

(characterization and physical properties). Although it is possible to obtain pure rotational and rotational–vibrational information from samples, these studies are limited to simple diatomic or small polyatomic molecules in the gas phase at low densities. Therefore, the foci of most typical Raman investigations are systems in the solid or liquid phases and their solutions. Table 10.1 contains some of the group frequencies that are important in the detection of energetic materials, chemical agents and their simulants, biological agents and their simulants, and toxic industrial compounds and environmental contaminants. Groups of atoms bonded as part of a molecular structure retain the location of their vibrational frequencies in much the same way as a family of organic compounds retains its chemical characteristics.

A good example of the latter is the family of aliphatic alcohols, which absorb in the IR (scatters in Raman) at approximately 3600 cm^{-1}, nearly independent of the number of carbon atoms in the linear or branched chain structures. The persistency in wavenumber absorbance of strong functional groups renders vibrational spectroscopy an excellent tool for the chemical characterization of samples. An atom other than H or C adjacent to the C atom bearing the functional group provides a secondary chemical environment (electron density withdrawal or donation) that shifts the group frequency slightly, but significantly, and providing more physical information. Figure 10.3 shows IR and Raman spectra of dichloromethane, used as example. Table 10.2 summarizes the spectral details. Vibrational information occurs at nearly the same wavenumber location but with different intensities.

Applications

Spectroscopic methods have been successfully applied to the detection, identification, and characterization of energetic compounds, chemical agents, and biological threats. RS has a wide range of applications in these fields [8–10]. Chemical agents, more commonly used as liquids and highly energetic materials are successfully de-

Table 10.1 Group frequencies and wavenumber ranges in infrared and Raman spectroscopies [7]

Family of compounds	Functional group	Wavenumber range/cm^{-1}	IR intensity	RS intensity
Alkanes	C–H stretch	2950–2800	m-s	s
	CH_2 bend	~1465	s	w
Alkenes	=CH stretch	3100–3010	m	s
	C=C stretch (isolated)	1690–1630	m	s
	C=C stretch (conjugated)	1640–1610	m	s
	C–H in-plane bend	1430–1290	m-s	w
Alkynes	Acetylenic C–H stretch	~3300	m	vw
	C, C triple bond stretch	~2150	w	s
	acetylenic C–H bend	650–600	s	
Aromatics	C–H stretch	3020–3000	s	s
	C=C stretch	~1600–1475	m	w
Alcohols	O–H stretch	3400–3650 or	br	w
	C–O stretch	1260–1000	s	w
Ethers	C–O–C stretch (dialkyl)	1300–800	m	w
	C–O–C stretch (diaryl)	~1250–1120	m	s
Aldehydes	C–H aldehyde stretch	~2850–2750	m	m
	C=O stretch	~1725	s	m
Ketones	C=O stretch	~1715	s	m
	C–C stretch	1300–1100	m	m
Carboxylic acids	O–H stretch	3400–2400	br	vw
	C=O stretch	1730–1700	s	m
	C–O stretch	1320–1210	s	w
	O–H bend	1440–1400	br	w
Esters	C=O stretch	1750–1735	v	m
Acid chlorides	C=O stretch	1810–1775	m	m
	C–Cl stretch	730–550	w-m	s
Anhydrides	C=O stretch	1830–1800	m	m
		1775–1740	m	m
	C–O stretch	1300–900	s	w
Amines	N–H stretch (1°)	3500–3300	s	m
	N–H bend	1640–1500	m-s	w
	C–N stretch (alkyl)	1200–1025	m	m
	C–N stretch (aryl)	1360–1250	m	m
	N–H bend	~800	v	w
Amides	N–H stretch	3500–3180	m	m
	C=O stretch	1680–1630	m	m
	N–H bend	1640–1550	m	m
	N–H bend (1°)	1570–1515	s	m
Alkyl halides	C–F stretch	1400–1000	s, br	s
	C–Cl stretch	785–540	w-m	s
	C–Br stretch	650–510	m-s	s
	C–I stretch	600–485	m-s	s
Nitriles	C, N triple bond stretch	~2250	m	m
Isocyanates	–N=C=O stretch	~2270	m	w
Isothiocyanates	–N=C=S stretch	~2125	m	m
Imines	R_2C=N–R stretch	1690–1640	m	s

Table 10.1 (continued)

Family of compounds	Functional group	Wavenumber range/cm^{-1}	IR intensity	RS intensity
Nitro groups	$-NO_2$ (aliphatic)	1600–1530	s & w	w & s
		1390–1300	s & w	w & s
	$-NO_2$ (aromatic)	1550–1490	w	s-m
		1355–1315	w	s-m
Mercaptans	S–H stretch	~2550	w	m
Sulfoxides	S=O stretch	~1050	s	w
Sulfones	S=O stretch	~1300 & ~1150	s	w
Sulfonates	S=O stretch	~1350 & ~1750	s	w
	S–O stretch	1000–750	s	w
Phosphines	P–H stretch	2320–2270	m	w
	PH bend	1090–810	w	s
Phosphine oxides	P=O	1210–1140	s	w

s strong, *m* medium, *w* weak, *vw* very weak, *br* broad, *v* variable

Table 10.2 IR and Raman frequencies of the vibrational modes of dichloromethane

Tentative assignment	IR frequency		Raman frequency	
	ν (cm^{-1})	Intensity	ν (cm^{-1})	Intensity
			283.3	m
	705.8	m	702.2	m
ν (C–Cl)	744.4	s	740.2	w
	896.7	w	1154.4	w
	1265.0	s		
	1421.3	w	1423.6	w
	2341.2	m	2341.5	vw
	2360.4	m	2361.0	vw
ν (–C–H) symmetric	2821	vw	2833.5	vw
ν (–C–H) asymmetric	2987.2	w	2993.9	s
	3054.7	w	3056.7	m

s strong, *m* medium, *w* weak, *vw* very weak; unfilled frequencies/intensities: no IR or RS activity

tected and identified by RS. The remote detection of highly energetic [11, 12] and biological materials [13] have been successfully achieved at lower limits of detection useful for defense and security applications and for monitoring environmental pollution.

Recently, Spontaneous RS and several of its variations have been applied in the detection and spectroscopic analysis of microorganisms and other airborne biological matter. Most of these organisms have Raman active constituents. Bioaerosol particles such as pollen, bacteria, and viruses are of particular interest in this book. Bioparticles of interest in the aerosol size range usually have characteristic dimensions (diameter for spheres and long axis for ellipsoidal rods) between 0.002 μm

(2 nm) and 100 μm in size. These bioaerosols can be detected in air or collected on a solid or liquid and analyzed in bulk or as single particles.

Properties and Preparations of Aerosols

Light absorption and scattering of aerosol particles depend on the shape and size of the microorganisms. The absorption of spherical particles can be calculated using Lorenz–Mie theory. They can be characterized by their efficiency of absorption and the cross section of the scattering process. For nonspherical particles, light interaction can be approximated using the Mie theory but requires more complicated numerical methods for its application [14]. Several studies have applied this theory for a variety of detection research interests using aerosol particles. The application of the Mie theory is very useful in RS applications of direct aerosol detection.

Several bioaerosol sample preparation methods have been used in detection experiments. Other contributions in this book contain more information about the subject. The reader is referred to the corresponding sections for further information on the topic. However, NR spectroscopy, in general, does not require extensive procedures for sample preparation.

Technologies for Raman Spectroscopy

In the last 25–30 years RS has transitioned from the specialized research laboratory to the chemist's benchtop for routine analysis [4–6]. A series of slowly developing stages were responsible for the current status of the 80-year-old technique. The Raman Effect can occur in virtually any region of the electromagnetic spectrum, from the ultraviolet (UV) and deep UV regions (200–400 nm) to the visible (VIS) frequencies (400–700 nm) and near infrared (NIR) region (700–1200 nm). There are specific advantages and limitations inherent to each of the wavelength regions for Raman studies. However, in all of these regions there are five basic components required to assemble a Raman system. The current state of the art RS can be understood by discussing the technological advances in each of these five areas:

- Monochromatic excitation source (generally a laser system)
- Sample holder
- Means of separating Rayleigh scattered photons from Raman scattered photons
- Spectrograph to separate the Raman scattered light waves into a frequency spectrum
- An efficient detection system

During the first 30 years after its discovery, RS remained an esoteric research tool, in much the same way as X-ray analysis was treated until several decades ago. The stringent requirement of a powerful monochromatic source of energy limited the popular use of the technique. Adding to that requirement, the only effective means of separating the Raman scattered light from the Rayleigh scattering was to use a pre-monochromator or a double pre-monochromator equipped with diffraction gratings, forming double or triple monochromator assemblies when the final filtering stage, the spectrograph, was added to the picture. This led to a high laser line rejection to reduce the unwanted stray light and allowed for the examination of frequency spectral components very close to the laser line. Thus, the first 30 years were characterized by expensive, bulky, and cumbersome instrumentation. A mercury arc lamp became the principal light source, first with photographic detection and then with spectrophotometric detection.

In the 25 years after the 1960s, RS experienced its first revolutionary and defining change. In the early 1960s, with the coupling of a laser source for Raman excitation, the technique experienced a quantum leap in its development. The effects of switching from an atomic arc source to a photonic source are still emerging and variations on normal RS are still being developed as a result of using a highly collimated, coherent, phase-retaining and polarized radiation beam. Over the next 25 years lasers became more powerful, smaller and more energy efficient. UV and NIR lasers became available and affordable. This important event was followed by the incorporation of photon counting techniques for photomultiplier detectors of the time. The sample holder was also an important piece of equipment because this carefully designed chamber needed to collect every possible Raman scattered photon. The excitation laser beam had to be transported from the laser head and ultimately focused on the focal point of the optically dark sample chamber.

In the post-1985 era, RS experienced a second revolutionary stage in the development of technological and instrument components. Multichannel detectors were introduced in RS in 1982 by the incorporation of intensified photodiode arrays (IPDAs), but the real breakthrough came in 1985 when charge coupled devices (CCDs) were first used as Raman detectors. Multichannel detectors significantly increase the signal-to-noise ratios (SNR), add the possibility of detecting a wide range of Raman shifted frequencies, and cut down the acquisition (signal integration) time. Today, CCDs operating in the visible region can have quantum efficiencies as high as 95% at their maximum operating wavelengths [5].

A second important development early in the new RS era was the incorporation of a new breed of RS systems: Fourier transform Raman spectroscopy (FTRS) [15, 16]. FT Raman was able to overcome some of the limitations of dispersive RS with the proper selection of conditions. In FT Raman, the time domain spectrum is recorded and subsequently Fourier-transformed to the frequency domain spectrum. All the frequency components in a spectrum are recorded simultaneously faster (Fellgett or multiplexing advantage). In theory, all NIR energy is used simultaneously, suggesting the possibility of achieving much higher signal-to-noise ratios (SNR) by lowering the inherent noise levels introduced by the aperture slits of dispersive instruments (Jacquinot advantage). In practice, a lower sensitivity than dispersive RS operating in the VIS region is achieved, because at 1064 nm (Nd:YAG laser source) the Raman signals are of significantly lower intensity owing to the ν^4 disadvantage in the NIR. An anticipated gain did materialize, however: the significant reduction in interference from the endogenous or exogenous florescence of the samples analyzed. An internal laser (He:Ne) is used to calibrate the interference information, providing very high wavenumber accuracy and reproducibility.

Another important innovation that revolutionized RS was the use of new filter designs based on holograph optics, improved dielectric filters, and semiconducting absorbers that effectively reject the intense Rayleigh light. These rejection filters can be used in lieu of premonochromators and are basically composed of two types of functionality: a rejection or notch filter that allows the passage of light at all wavelengths except the central (laser) wavelength ($\lambda_o \pm \delta\lambda$) and an edge filter (high wavelength pass filter). These filters are typically small and simple and permit the use of a small, single-stage spectrograph for wavelength dispersion. A notch (or edge) filter and a single spectrograph can be much smaller and more efficient than a traditional double or triple spectrometer, reducing both the size and cost of the spectrometer. The use of multichannel detectors (CCDs) for multiplexing the optical information enabled the reengineering of compact and fast spectrographs, significantly increasing the instrument throughput.

Sampling also received a boost by the addition of an optical (mineralogy) microscope that enabled Raman Microspectroscopy (or Raman Microscopy). With the mindset of taking the RS to the sample, rather than taking the sample to the instrument, fiber optics sampling enabled the conduction of the laser beam through a highly transmisive medium (99.99+% T) to excite the sample, collect the scattered light, filter the Rayleigh component, and transfer the Raman scattered photons through a fiber thread or bundle to the wavelength sorter and detector. When all of these technological improvements were coupled together, new breeds of dispersive and nondispersive RS flourished, resulting in the current stage of development.

Bioaerosol Detection Using Raman Spectroscopy

Bioaerosol particles, though considered some of the most complex aerodispersed particles, can be detected and successfully characterized by RS. RS (RS), also called Normal or Spontaneous, including some of its modalities, can potentially demonstrate advantages in the vibrational spectroscopy detection of aerosols. As an analytical method, RS is a fast, specific, and noninvasive technique with which fingerprinting information on the chemical composition of the samples can be obtained with high reliability. Part of the information presented on the detection of bioaerosols by Raman and its modalities has been recently reviewed by the authors [17].

Spontaneous Raman Spectroscopy

Spontaneous or normal RS (RS) is the inelastic scattering process that occurs by a change in photon frequency upon interaction with matter (Fig. 10.1). Bioaerosol

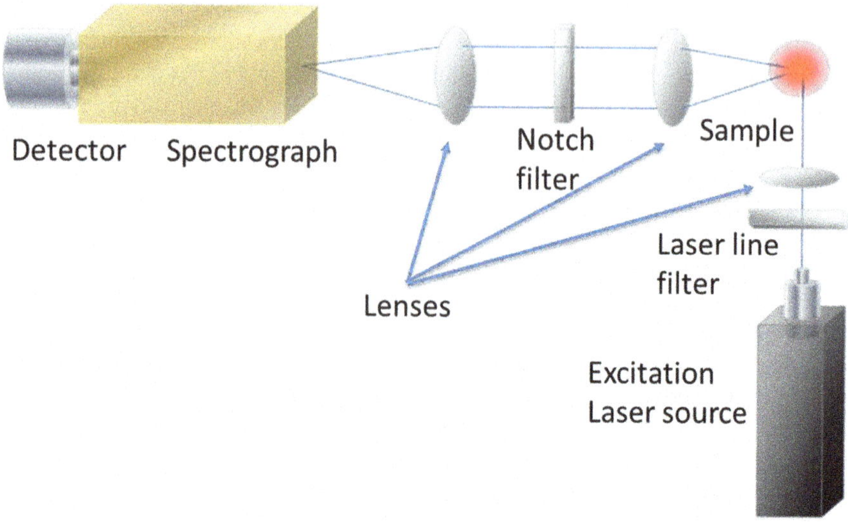

Fig. 10.4 General schematic diagram of normal or spontaneous RS conventional detection geometry (90° scattering). This arrangement is typically used as one of the simplest experimental setups to implement

samples have been directly detected by RS (on the fly) and the results of these studies have been compared with the spectroscopic information obtained by other RS techniques such as Raman Microscopy analysis and Raman vibrational chemical mapping and imaging. A general Raman system for RS detection is shown in Fig. 10.4. This conventional sampling system shows a conventional 90° scattering geometry. The monochromatic laser source arrives at the sample and the scattered Raman signal is directed towards the detector. This is one of the simplest experimental setups to implement.

Bioaerosol Detection by RS

Essential components of bacteria and bacterial spores, including dipicolinic acid (DPA) and other components of biological samples have been detected by RS. The solid phase and liquid solution/suspension spectra allow for the identification of the characteristic vibrational modes. Raman Microspectroscopy has enabled the detection and characterization of individual bacterial endospores such as *Bacillus cereus, B. megaterium, B. subtilis, B. thuringiensis* [18, 19] and complex samples containing mixtures of microorganisms [20]. The vibrational bands can be compared to those detected by other techniques and the bands can almost always be observed at low to very low intensity levels [21, 22]. RS can also be used to obtain images of biological samples as detection and characterization techniques providing a full spectral assessment of bioaerosol samples [23]. The fluorescent compounds found in bioaerosol samples such as proteins and amino acids present problems in obtain-

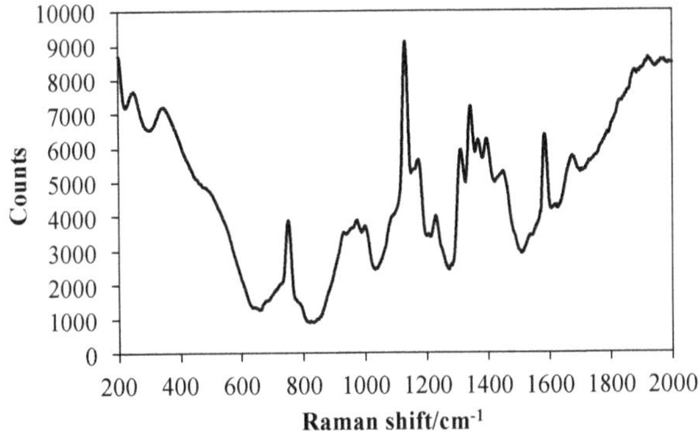

Fig. 10.5 Normal or spontaneous Raman spectrum of *E. coli* bacteria. The spectral information can be efficiently used to identify and characterize the biosample. However, the spectral definition is generally poor except for a few RS molecular signatures (authors' unpublished results)

ing their corresponding RS spectra because of endogenous fluorescence. Many of these compounds resonantly absorb the incident Raman excitation frequency, leaving the molecule in an upper electronic state rather than in a virtual state. Due to this absorbance, the Raman spectra of the target chemicals present a broad fluorescence emission that overlaps the Raman signals [24].

Figure 10.5 presents the vibrational signals for an *Escherichia coli* bacterial sample using RS [25]. As can be observed, the RS spectrum can be efficiently used to identify and characterize some types of biosamples. However, the spectral definition is generally poor except for a few RS molecular signatures.

Spontaneous Raman imaging is also possible and very attractive because, contrary to other imaging techniques, sample images are free of additives (only the sample is present). The limitation of this technique is the time required to obtain the complete Raman image and the restriction to live cell applications. Modifying the scanning mode and detection approach can reduce this limiting factor. Laser line and point scanning are discussed by Chan et al. [23]. The scanning in one axis reduces the acquisition time required to obtain the vibrational image and allows the imaging of living cells.

Due to those disadvantages, bioaerosol detection and characterization has been very commonly achieved by Raman variations such as Resonance Raman (RR), UV Raman Spectroscopy, Coherent Anti-Stokes Raman Spectroscopy (CARS) and Surface-Enhanced Raman Spectroscopy (SERS).

Resonance and UV Raman Spectroscopy

The resonance Raman (RR) effect is obtained when the incident laser frequency used is near the half width of a spectral absorption band of the scattering sample. Figure 10.6 shows a schematic energy diagram of the process. Pre-resonance and post-resonance frequencies are often used to designate the intensity of the events based on the incident photon energy equivalent wavelength location with respect to the resonance condition.

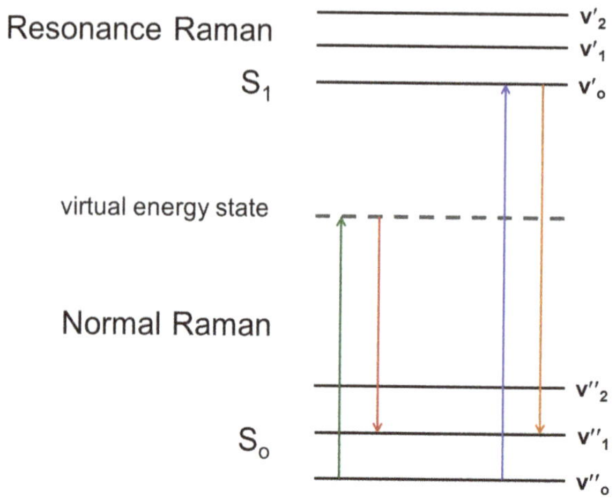

Fig. 10.6 Resonance Raman spectroscopy (RSS) uses the excitation frequency in the range that resonates with the electronic excited state frequency of the target providing enough energy to excite the electrons to a higher electronic state as show the schematic energy diagram involved in RRS

The RR experimental setup is generally similar to that of spontaneous RS, but it uses the excitation frequency in the range that is resonant with the electronic excited state frequency of the target. Resolution and sensitivity problems can be improved using this technique, the wavelength of the incident light used is near the electronic transition of the molecule and the resonance Raman bands excited are strengthened. The resonance Raman spectrum is obtained by the electronic and vibrational transitions combination [26]. Some variations of the sensor system can be used with positive results to detect biochemical warfare agents, stand-off of chemical, biological and explosives samples, and water and surface contamination monitoring as Hug et al. demonstrate in several works [27–30]. Frequently, biological samples do not exhibit fluorescence effects in the UV range around two hundred nanometers but other biomolecules absorb well in this region [31].

Bioaerosol Detection by RR in the UV Region

The UV absorption bands of the samples show that the resonance frequencies in the ultraviolet region of the electromagnetic spectrum can be used to obtain RR spectra of the samples. Because the spectra are collected in a liquid solvent, the absorption maxima of the target analytes obtained by UV–VIS spectrophotometry do not necessarily correspond to the solid state, neat sample values. Resonance Raman restricts the resonance scattering to ± 1 nm of the band absorption maximum. Thus, a tunable laser source such as a frequency doubled dye laser system must be scanned to obtain the resonance scattering condition. It allows the selection of the excitation wavelength depending on the chromophore in the biological sample of interest. The RRS spectrum contains highly sensitive and selective vibrational information about the target. Figure 10.7 presents the Raman signals for *E. coli* using RRS. As can be seen, the RR spectrum is used efficiently to identify and characterize biosamples

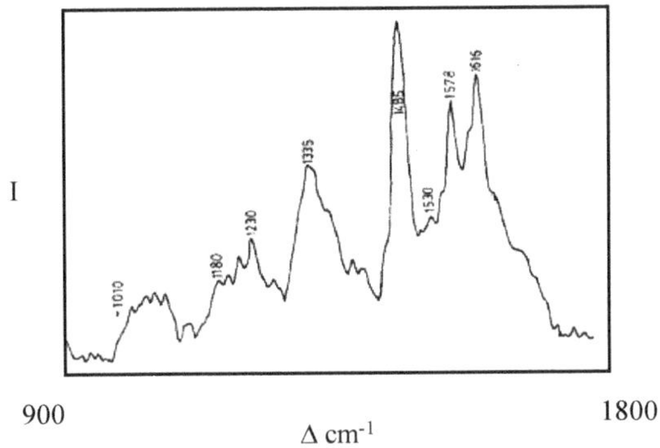

Fig. 10.7 *E. coli* RR spectrum using an excitation source at 242 nm. (Reproduced with permission from Society for Applied Spectroscopy, [40])

in the UV region. The frequency signals can be compared to the signals from the normal RS (Fig. 10.5) spectrum.

Resonance Raman spectra allow for more efficient identification and characterization of the peaks detected than can be achieved by normal RS, with high intensity vibrational signals observed. A large enhancement in the spectra of crystalline particles and droplet samples was first demonstrated by Fung [32, 33]. Studies on the anharmonicity of molecular vibrations are difficult to perform because of the weakness of the overtones, but RR has enough sensitivity to achieve this detection using droplets. RR also has the capacity to minimize Mie resonance scattering and is therefore a good analytical technique for studying aerosol droplets [32, 33]. Chromophoric aerosol species are successfully analyzed by RR and Ultra Violet Resonance Raman (UVRR) and single-cell spectra in mixtures [26] was easy observed using resonance RS. The quantitative UV resonance Raman study of viruses and bacteria has shown significant information regarding the structure, interactions, and dynamics of nucleic acids [34]. Components of biosamples that have been detected by UV and RRS are described in Table 10.3.

The detection of spores using UV resonance Raman is "especially attractive because of the potential single-spore detection limits [18] and because of the potential for remote detection using laser optical methods" [41]. Dipicolinic acid (DPA) , a common constituent of bacterial spores, has a strong vibrational signature that is detectable by UV Raman. It exhibits the n-π* transitions in the UV region that are attributed to carboxylate groups and pyridine rings. DPA is the principal component of pollen and flavonoid compounds. The Raman spectrum of bacterial spores occurs at 242 nm and the bacterial peaks that can be observed at this excitation frequency are attributed to rRNA material. Thus, the presence of pollen will not typically obstruct the detection of bacteria [36, 40]. An enhancement of the Raman signal for bacterial spore detection is observed when using a probe pulse in the UV region with femtosecond-CARS [42]. Although UVRR spectroscopy shows great potential for the detection and identification of biosamples, it also requires a large amount of material to work properly because the high energy density of the UV light can burn the samples through local heating [39]. Two dimensional (2D) resonant Raman bacterial spectra also allowed to obtained microbial signatures

Table 10.3 Bioaerosol components detected by UV Resonance Raman Spectroscopy

Biosample	Laser radiation (nm)	Observation	Reference
Bacillus subtilis (Bs)	242	Spectra at different grow stages	[35]
	222	Near growth-rate and media independent	[35]
	222	Weak scattering	[36]
	222, 230, 250	Spectral range: 900–1800 and 1300–1800 cm^{-1}	[26]
	228, 244, 248	Lag, log and stationary grow stages (10^8 cells/mL)	[37]
	257	Cells attached to a quartz place	[38]
Bacillus megaterium	222	Near growth-rate and media independent	[35]
	251, 242, 231, 222	242 nm selective excitation line	[36]
	242	Spectral range: 1300–1800 cm^{-1}	[26]
	242	FAST-CARS	[39]
Enterobacter aerogenes	222	Near growth-rate and media independent	[35]
Aeromonas hydrophila	222	Near growth-rate and media independent	[35]
Flawbacterium capsdatum	257	~20 cells sufficient to obtain a quality spectra	[38]
Escherichia coli (*E. coli*)	222	Near growth-rate and media independent	[35]
	242	Spectra at different grow stages	[35]
	242	Spectral range: 900–1800 cm^{-1}	[26]
	228, 244, 248	Lag, log and stationary grow stages (10^8 cells/mL)	[37]
	242	Contributions primarily from ribosomal RNA	[40]
	251	Log growth phase (10^6 cells/mL)	[34]
Bacillus cereus spores	242	Spectra of vegetative cells, spores, and germinated spores	[36]
	231, 222	Weak scattering	[36]
	242	Spectral range: 900–1800 cm^{-1}	[26]
	242	FAST-CARS	[39]
Whole apricot pollen Whole grass pollen Whole ragweed pollen	222, 231, 242	Detection in complex mix possible	[40]

of *B. thuringiensis, B. cereus, E. coli* and *Staphylococcus epidermidis* in mixture microbes and contaminants [43]. Several studies have demonstrated the viability of bioaerosol detection by RR and UV RRS by taking advantage of the fact that some biomolecules resonantly absorb UV laser light radiation. This fact enhances the Raman scattered signal intensities and improves the sensitivity and selectivity of biomolecule detection.

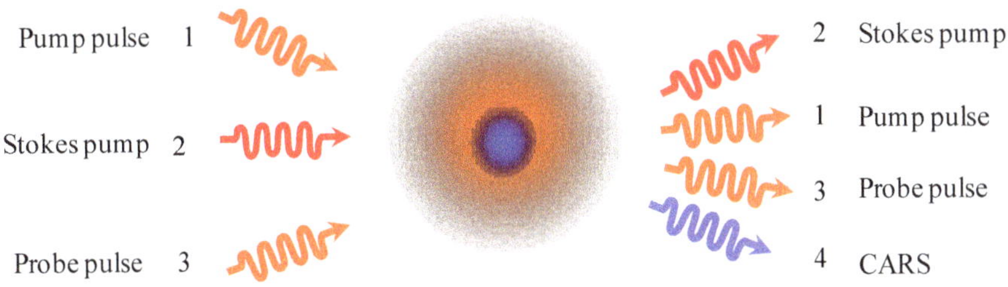

Fig. 10.8 Schematic photonics in CARS experiment: three waves, two at the pump frequency and one at the Stokes tunable frequency, reach the sample and produce the anti-Stokes beam of frequency. CARS photons (*4*) are produced after the pump pulse (*1*), the Stokes pump pulse (*2*), and the probe pulse (*3*) interact with the sample

Fig. 10.9 Two laser radiation pulses of different frequencies force the vibration of the molecule to form a coherent Raman scattering by phase matching as shown in the energy diagram for the generation of a CARS signal

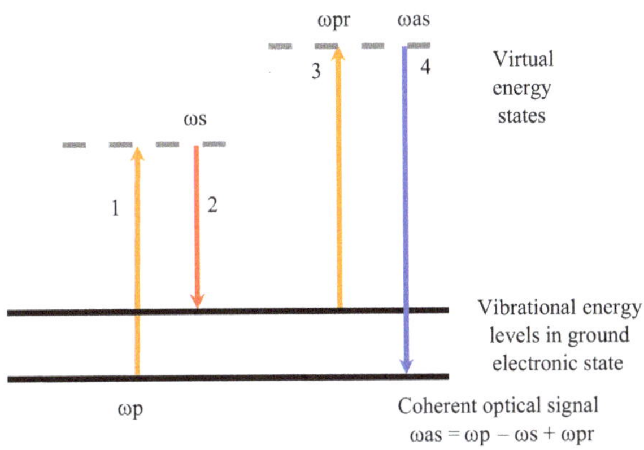

Coherent Anti-Stokes Raman Spectroscopy

Stimulated RS using coherence scattering has several advantages for bioaerosol detection applications. Coherent Anti-Stokes Raman Spectroscopy (CARS) is a third order nonlinear optical process based on a 4 wave mixing process that increases Raman cross sections through resonance enhancement. CARS uses three pulsed beams (pump, Stokes, and the probe) that reach the sample to produce the fourth beam (Anti-Stokes pulse), which is known as the CARS signal. (Fig. 10.8)

The pump and the probe pulses commonly overlap and have the same frequency. Three waves, two at the pump frequency ω_p and one at the Stokes tunable frequency ω_s, reach the sample and produce the Anti-Stokes beam of frequency ω_{pr}. Two laser radiation pulses of different frequencies force the vibration of the molecule by phase matching to form a coherent Raman scatter (Fig. 10.9).

A general setup for Coherent Anti-Stokes Raman in microscopy and spectroscopy is shown in Figs. 10.10a and b (used for noncollinear phase matching experiments), respectively.

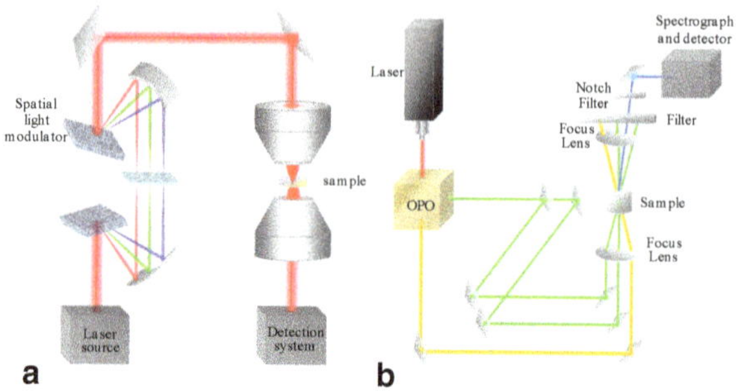

Fig. 10.10 General schematic for CARS setup: **a** Microscopy and imaging. **b** Spectroscopy experiments using noncollinear phase matching

This coherent Raman process has been employed in spectroscopy, microscopy, and imaging as a tool for characterization and detection purposes. Bioaerosol samples can be detected using CARS and its several variations. A CARS spectrum allows for more efficient identification and characterization of peaks detected in aerosol samples than does normal RS. Large peak intensities in the CARS spectrum can typically be observed using this stimulated process. The background fluorescence from the sample does not interfere with the CARS detection because the scattered Raman signals are observed in the Anti-Stokes region of the spectrum, corresponding to a higher energy loss as shown in Fig. 10.9. However, CARS presents interference problems with the nonresonant background. To reduce or avoid the nonresonant background CARS variations can be incorporated.

CARS presents advantages in the real-time detection of aerosols, in both micro and macro spectroscopies of chemical and biological systems. The spectroscopic experiments in macro mode generally use an optical parametric oscillator (OPO) to obtain a broadly tunable laser source that increases the scan range in the Stokes pulse. The OPO converts the input laser wave into two output waves of lower frequency by means of a nonlinear optical interaction. The CARS-microscopy setup is used as a chemical imaging tool for biological systems such as living cells, molecules, and other unstained samples that have strong vibrational modes. DNA and proteins still remain great challenges for CARS [23, 44–50]. Applications in both areas using CARS microscopy and macro mode CARS experiments show high quality results as compared to those obtained from spontaneous Raman microscopy. CARS imaging of biosamples can show mapping distribution, orientation, and diffusion of specific molecules in dynamic samples [46].

The CARS spectra display the vibrational information of the target with high sensitivity. CARS vibrational imaging surpasses the normal Raman levels with lower acquisition times compared to those of spontaneous RS. At the same time, CARS takes advantage of the C-H vibrational stretching in the biological samples to obtain a CARS image [23]. Figure 10.11 presents the Raman signals for *B. subtilis* using CARS. As can be seen, the CAR spectrum can be efficiently used to identify and characterize biosamples. Frequency signals can be compared to the signals obtained in the normal RS spectrum.

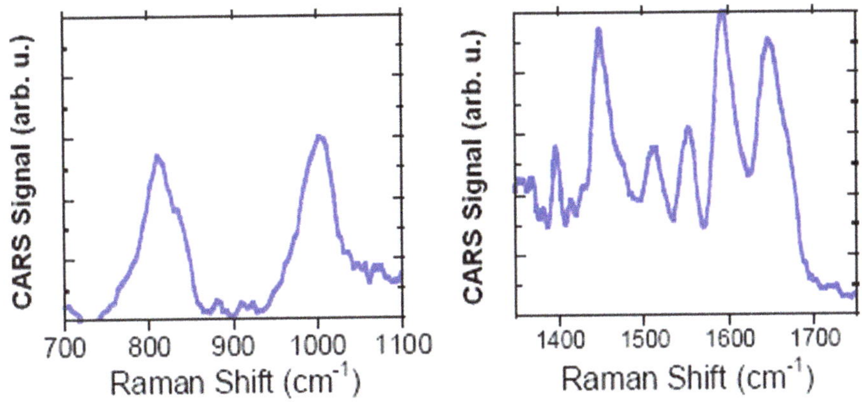

Fig. 10.11 CARS spectra of a solution of *B. subtilis* bacterial spores. CARS spectrum can be efficiently used to identify and characterize biosamples and vibrational signatures can be readily compared with signals obtained in normal RS experiments. (Reproduced with permission from Optical Society of America, [51])

Bioaerosol Detection by CARS

CARS microscopy techniques have been successfully employed to detect the endospores of bacteria. Bacterial endospores such as *B. subtilis, B. anthracis* [51–53] and biological constituents such as dipicolinic acid (DPA) , phospholipids, glycogen, ribonuclease A, and ribonucleic acid (RNA) [50] have been detected by CARS and its variations. Techniques such as pulse shaping, single pulse multiplex [47], time-gated approaches [54], frequency resolved, femtosecond techniques [39, 51–53], and resonant enhancement [42, 55] have been used. These techniques are commonly used in the detection of biomolecules and can enhance the coherence Raman signal. Femtosecond Adaptive Spectroscopic Techniques for CARS (FAST CARS) excite the molecule into vibrational modes and create coherence between the vibrational levels of the molecule. This technique combines quantum coherence and femtosecond lasers capabilities. Scully and collaborators proposed a possible FAST CARS protocol to identify and characterize particular bacterial spores in real-time, stand-off detection [39].

To demonstrate the potential of FAST CARS, the coherent and spontaneous Raman shift peaks for *B. subtilis* spores were compared and showed a good spectral match. CARS results could be obtained in a shorter integration time and the technique suppressed the nonresonant background as was also the case for time-resolved CARS with high sensitivity [52]. Table 10.4 shows some examples of biosamples that have been detected by CARS.

Recently, CARS spectroscopy has been widely used in the investigations of a broad variety of samples in the liquid, gas, and condensed phases. It has also been used in stand-off detection [38] and is two or more orders of magnitude more efficient than spontaneous RS [23, 39, 54].

Table 10.4 Biological content detected by CARS

Biological sample	CARS variation	Pulse properties	Observation	Reference
NaDPA powder	Frequency and time resolved CARS	Ultra short ~ 50 fs	Probe pulse delay to eliminate NR FWM signal	[52]
Bacillus subtilis spores	Frequency and time resolved CARS	Ultra short ~ 50 fs	Probe pulse delay to eliminate NR FWM signal	[52]
	FAST-CARS	Ultra short ~ 50 fs	Energy dependence on CARS signal demonstration	[52]
	CARS	Broadband continuous wave	Aqueous solution	[51]
	Hybrid CARS: broadband preparation and narrowband time delay of probe pulse	~ 50 fs laser shot	Single laser shot detecting ~ 10^4 spores	[53]

The Lorenz-Mie theory was applied to the femtosecond dynamics of the nonlinear processes associated with the stand-off detection of bacterial spores. In bioaerosol detection, the Mie theory is applied to study the variations of intensity in CARS spectra, taking into account the various laser parameters and directions of detection [46]. For example, the angular dependence of the CARS signal has been shown [56]. This theory can be used for variations in CARS techniques, including FAST CARS [38, 56] and other RS experiments in which the scattering of electromagnetic radiation by spherical particles occurs. The theoretical model was applied to model the backscatter signal of anthrax spores and resulted in the identification of a large number of detectable photons by FAST CARS, making this technique a good approach to the stand-off detection of bioaerosol samples [56]. All of these results indicate the great potential for the remote sensing detection of biosamples using CARS.

Surface-Enhanced Raman Spectroscopy

Surface Enhanced Raman Spectroscopy (SERS) has been more successful in the detection of bioaerosols than spontaneous RS. SERS is a powerful spectroscopic technique that is routinely used to detect analytes at very low concentrations. Based on SERS, the intensities of vibrational signatures can be enhanced by factors of 10^6 to 10^{14} or even higher under well controlled conditions.

SERS is a powerful spectroscopic technique for detecting very low concentrations of analytes [4, 57, 58]. It was first observed by Fleischmann, Hendra and McQuillan in 1974 but they attributed their anomalous Raman enhancement of pyridine in aqueous solution in close proximity to a roughened silver electrode to the increased surface area of the electrode to which pyridine had adsorbed [57, 58].

In 1977, two independent groups: Jeanmaire and Van Duyne [59] and Albrecht and Creighton [60] discovered and correctly attributed the huge signal enhancements to a surface effect. After these events, SERS has been steadily developing motivated by the overcoming the extremely low cross sections (10^{-31} cm^2) of conventional RS3, which is based on a weak scattering process.

As its name indicates SERS requires a supporting substrate. The most commonly used substrates are gold, silver or copper nanoparticles, because these metals absorb visible light between wide 400 and 700 nm range. This relatively wide extension of the optical absorption region is very attractive for practical SERS measurements, which in the case of silver and gold colloids are believed to depend on their optical absorption properties.

According to classical theory of Raman scattering [61], when a molecule is irradiated by light, an electric dipole moment P proportional to the electric field (E) and to the molecular polarizability (α) is established as shown in Eq. (10.6):

$$P = \alpha E \tag{10.6}$$

It is clearly noted that any enhancement can come from the molecular polarizability tensor through any one or more of its 9 components or from the electrical field. Thus, the SERS phenomenon is mainly due to a contribution of two effects: the electromagnetic effect and the chemical effect. The electromagnetic effect is produced by the interaction between the electric field from the incident radiation and the surface metal electrons which results in an increase of the electric field at the molecule. This enhancement occurs because the local electromagnetic field at the surface of the metal changes when the incident radiation arrives at the surface. This optical excitation at the surface stimulates conduction electrons and generates a surface plasmon resonance. This causes the roughness features of the metal to be polarized and the electromagnetic field in the interior of the particle at the surface to increase considerably from the applied field. There is an increase of the electromagnetic field incident on an adsorbed molecule at a metallic surface.

The other enhancement (chemical) involves the incident radiation striking the roughened metallic surface resulting in a photon being excited within the metal to higher energy level. From this excited state, a charge transfer (C-T) process to a vibrational level of the same energy within the target analyte takes place. A bond formation also can result between the metal and the adsorbate, which may increase the molecular polarizability components.

Gold, silver and copper nanoparticles absorb visible light in a 400–700 nm range, which makes them ideal substrates for this type of light scattering phenomenon [62]. They are very attractive for SERS applications in several fields. Molecules adsorbed on an appropriate metal surface can have large enhancements of the Raman signals in a small sample volume with detection limits in the range of picomoles to femtomoles.

The first studies of SERS were obtained from works based on absorbed species on electrode surfaces. Electrochemical works in RS were focused on the absorption of pyridine and other nitrogen heterocycles and amines on a silver metal electrode

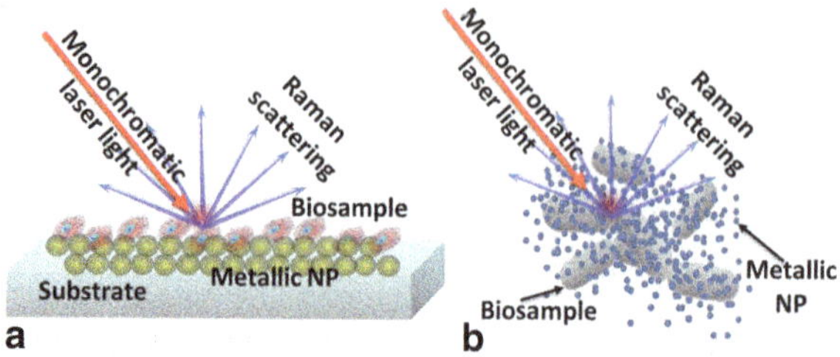

Fig. 10.12 Signal enhancement effect of metallic NP in SERS is produced by closely spaced interacting particles generated at the location of adsorbed molecules on **a** the metal surface and **b** in colloidal suspension. So-called "hot spots" seem to provide extra electromagnetic field signal amplification

resulting in a surface enhanced Raman signals with details about the double layer and the interaction between species used. These studies demonstrate the basis of enhancement factors of absorbed molecules using SERS [62–65]. The enhancement in the Raman scattering due to the presence of suspended metallic colloidal nanoparticles or a roughened metal surface has been explained using two complementary theories: electromagnetic field enhancement and chemical field enhancement [7, 62]. The electromagnetic field theory is based on the change in the electric field of the laser light in the vicinity of a roughened or discontinuous metal surface, leading to an interaction between the analyte and the surface plasmon of the metallic nanoparticle surface. The surface plasmon on the surface of the substrate transfers energy through the electric field to the molecule and it became polarized. The rough surface can provide enhancements of 10^4–10^7. The enhancement depends on the electrical properties of the metal, the distance of the molecule from the surface, the orientation of the molecule in the surface, the energy of the incident radiation, and the morphology, size and geometry of the rough surface. The chemical enhancement theory is based on the chemical bonds that can be produced by the transfer of electron density between the analyte and the surface. The effect increased the polarizability of the molecule due to the interaction with the metallic electrons; however this effect contributes only a small enhancement in SERS (10^1–10^2). Figure 10.12 illustrates SERS for biosample detection using a metallic solid substrate and colloidal suspension.

The complete elucidation of the strong electromagnetic field enhancement produced is still a matter of intense research and interpretation. Highly active spots ("hot spots") contribute to the large vibrational signal enhancement and are generated at the location of adsorbed molecules on the metal surface, where the local electric field is large due to the collective excitation of conducting electrons within the small metallic structure (surface plasmon) [7, 24]. Substrates like zero valence gold, silver, and copper nanoparticles are the most commonly used for SERS experiments. These metals absorb visible light between 400 and 700 nm, making them excellent substrates for the light scattering phenomenon. The method of preparation of silver nanoparticles reported by Lee and Meisel is the mostly widely used [63].

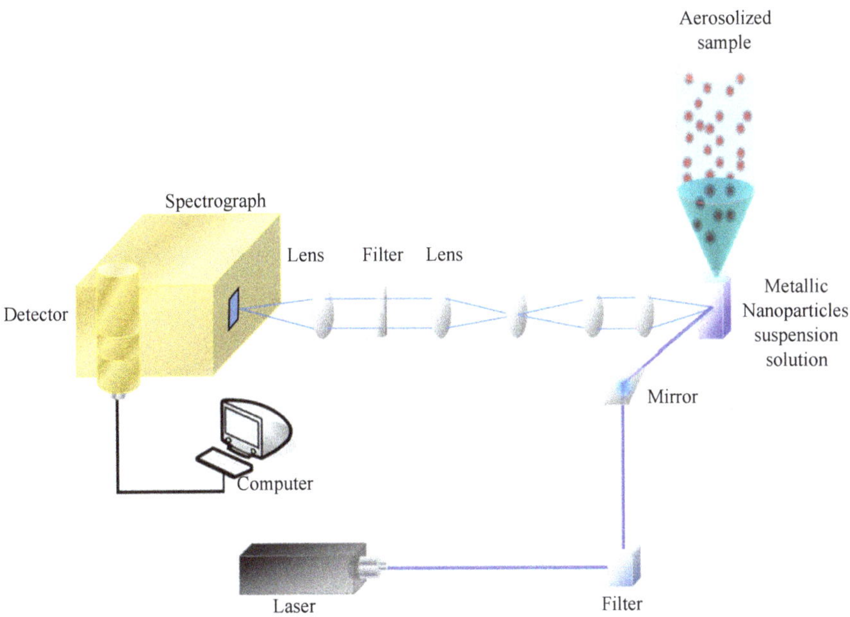

Fig. 10.13 General experimental Raman system for SERS experiments using aerosolized samples

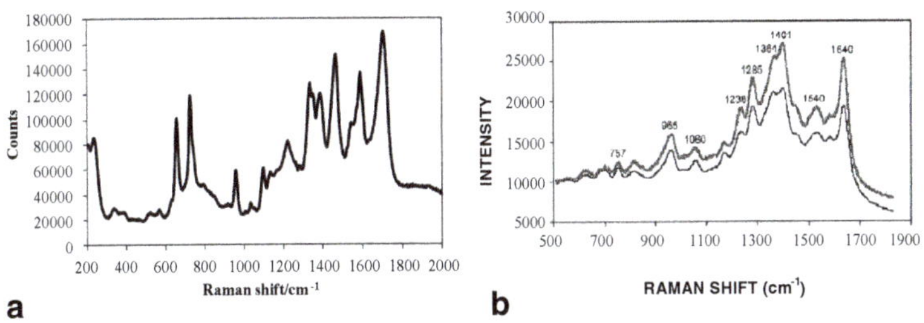

Fig. 10.14 **a** SERS spectrum of *E. coli* bacteria (authors' unpublished results); **b** *E. coli* aerosolized bacteria SERS spectra of two identical samples. (Reproduced with permission from Elsevier, [70])

The dynamic charge transfer between the molecule and the metallic nanoparticle allows for the enhancement of the detected signal. The experimental Raman system used in SERS experiments is shown in Fig. 10.13. The sample is aerosolized and then falls into the metallic nanoparticle suspension to be detected.

Figure 10.14a presents the Raman signals for *E. coli* bacteria using SERS. As can be seen, the SER spectrum can be used efficiently to identify and characterize biosamples. The enhanced frequency signals can be differentiated from the signals observed in the spontaneous Raman spectrum. The Surface Enhanced Raman spectrum allows for the identification and characterization of the peaks detected in bioaerosol samples more efficiently than does RS. In general, large and more defined peak intensity can be observed in the SER spectrum for suspension solution (Fig. 10.14a) and aerosolized samples (Fig. 10.14b) if it is compared to the SR spectrum (Fig. 10.5). The Raman spectra of the biosamples can be different or slightly different depending on the components of the biosample present at the exact location of the spectral reading. The identification in SERS experiments varies from the interior to the exterior of the bacterial cell [64]. Most Raman measurements identify biosamples using the exterior membrane components for detection. SERS allows for the analysis and identification of biosamples from the detection of either extracellular or intracellular features of the bacterial cell [65].

Bioaerosol Detection by SERS

Table 10.5 presents biosamples that were detected, identified, and characterized successfully by SERS. The detection of aerosol samples by SERS can be influenced by the colloidal substrate and the laser source used. Specific procedures were used to obtain the optimum SERS signal.

SERS has some restrictions on analyte properties in order to ensure a highly enhanced Raman signals and consequently, low limits of detection. A particular bioaerosol sampling system was prepared by Sengupta et al. [70], in which the biological matter was transferred to a Raman spectrometer by deposition onto a surface to maintain an aqueous suspension for the analysis. Another way to generate aerosols for SERS experiments in an aqueous suspension was presented by Ayora et al. [93]. Important assessments based on the requirements for handling sample devices for bioaerosol generation including dry powder bioaerosols have been summarized in the review by Griffiths et al. [31, 105]. SERS has also been used for kinetic process monitoring of endospore germination in real time. The electromagnetic enhancement provided by the metallic NP allowed for the identification of endospores at different concentrations [73] and temperatures [106]. Fluorescence quenching or suppression of fluorescence is caused by absorption of the stimulating radiation by biosample chromophores when using SERS metallic NP because of the independence of the energy transfer process [24]. Quenching of the endogenous fluorescence is necessary to obtain a good signal with Raman detection.

Other Raman Variations

Raman microspectroscopy with optical trapping, laser trapping Raman spectroscopy (LTRS), is a new approach that enables the investigator to consistently monitor the changes in a single cell of a homogeneous or heterogeneous sample. The method relies on trapping the sample in the focus of the laser beam, which enhances the collection of the Raman scattering and decreases background interference (Fig. 10.15) [80]. Bacterial cells and spores at various growth conditions in aqueous solutions require biosample sensors for detection, identification, and discrimination from the complex media. The molecular structure and chemical composition can be ascer-

Table 10.5 Biological samples detected by SERS under certain optimum conditions for detection

Biological sample	SERS substrate	Laser wave-length (nm)	Observation	References
Bacillus globigii	Ag oxide film	633	UV irradiation before SERS analysis	[66]
	Ag colloids	830	Portable Raman used for ppm detection	[8]
Erwinia herbicola	Ag oxide film	633	UV irradiation before SERS analysis	[66]
Bacillus thuringiensis	Ag oxide film	633	UV irradiation before SERS analysis	[66]
	Au nanoparticles covered SiO_2	785	No fluorescence; enhancements > 10^4	[24]
	Ag colloids	830	Portable Raman used for ppm detection	[8]
	Ag colloids	785	Citrate reduced nanoparticles	[67]
	Ag colloids	532	Activation of "hot spots" and surface charge modification of hydroxylamine, and citrate capped borohydride NP	[68]
Adenovirus	Ag nanorods	785	best binding to substrate: 1 hr/RT	[69]
Rhinovirus	Ag nanorods	785	best binding to substrate: 1 hr/RT	[69]
Human immuno-deficiency virus (HIV)	Ag nanorods	785	best binding to substrate: 1 hr/RT (room temp.)	[69]
Escherichia coli (*E. coli*)	Ag colloids	514	Aerosolized bacteria spectra similar to stock suspension	[70, 71]
	Au-SiO_2	785	Similar spectra to Ag colloids	[24]
	Ag colloids	532	Citrate reduced Ag favored identification	[72]
	Ag colloids	514	Borohydride reduced Ag; LOD 10^3 cfu/mL	[73]
	Ag colloids	785	Volume ratio, time and pH variations	[74]
	Ag-nanocrystals and Ag-NS	633/785	10^2–10^6 cfu/mL bacterial concentrations	[75]
Salmonella typhimurium	Ag colloids	514	Aerosolized bacteria spectra similar to stock suspension	[70]
	Au nanoparticles covered SiO_2	785	No fluorescence; enhancements > 10^4	[24]
	Ag colloids	785	Volume ratio, time and pH	[74]
	Ag colloids	633	Various colloids tested and optimized	[76]

Table 10.5 (continued)

Biological sample	SERS substrate	Laser wavelength (nm)	Observation	References
	Ag-nanocrystals and Ag-NS	633/785	10^2–10^6 cfu/mL bacterial concentrations	[75]
Pseudomonas aeruginosa	Ag colloids	514	Aerosolized bacteria spectra similar to stock suspension	[70]
	Ag colloids	785	Volume ratio, time and pH	[74]
Populus deltoides (cottonwood)	Ag colloids	514	Aerosolized pollen	[71]
Sequoia sempervirens (redwood)	Ag colloids	514	Aerosolized pollen	[71]
Staphylococcus aureus	Ag-nanocrystals and Ag-NS	633/785 nm	10^2–10^6 cfu/mL bacterial concentrations	[75]
Bacillus cereus	Au nanoparticles covered SiO_2	785	No fluorescence; enhancements > 10^4	[24]
Bacillus subtilis	Au nanoparticles covered SiO_2	785	No fluorescence; enhancements > 10^4	[24]
	Ag nanoparticles	532	Citrate reduced Ag favored identification	[72]
	Silver film over nanospheres	785	LOD: 10^3 endospores	[77]
Bacillus anthracis Sterne (nonvirulent strain)	Au NP/SiO_2	785	No fluorescence; enhancements > 10^4	[24]
	Ag films on Ag NS substrates	785	Detection of endospores	[77]
DPA	Au-coated silicon nanowires	633	High reproducibility	[78]
	Au NP immobilized on Au surfaces	633	Equilibrium for adsorption on the substrate	[79]
Erwinia herbicola	Ag colloids	830	Portable Raman: ppm detection	[8]
Legionella pneumophila	Ag colloids	633	Several sols tested and optimized	[76]

tained by micromanipulating the size of the biological samples in the gas or liquid media. One advantage attributed to this technique is simple substrate preparation [23, 81–83]. This tool can be combined with the Raman variations described earlier for aerosol detection applications [83–86].

Cavity enhanced Raman Spectroscopy (CERS) can be used to determine the droplet size and composition to within 20–50 μm of the sample radius. The spectral fingerprint can differentiate the aerosol particles from binary aqueous droplets and determine the droplet size, cross-section, sample concentration, and laser intensity dependence [87, 88].

Spatially offset Raman Spectroscopy (SORS) is a simple method based on the collection of Raman signals from spatial regions offset from the point of illumina-

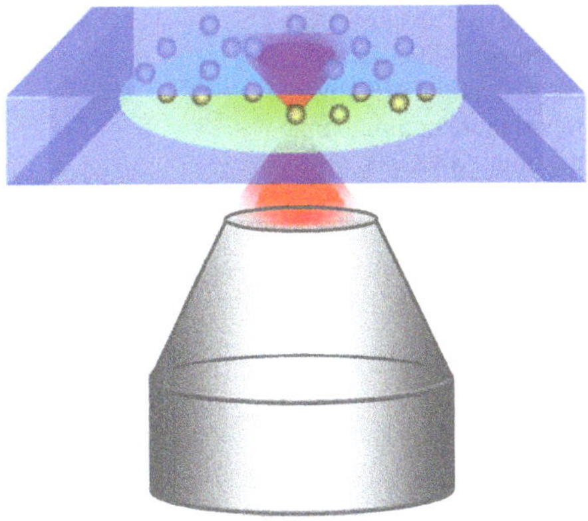

Fig. 10.15 General schematic showing the optical trap for individual particles in LTRS. It relies on trapping the sample at the focus of the laser beam, which enhances the collection of the Raman scattered signal from the sample and discriminates against background scattering

tion on the sample surface. It can be applicable only for nonabsorbing and weakly absorbing samples. The advantage of this technology for biological fluorescent materials is based on the extra suppression of the surface. It allows the recovery of weaker signals from the sublayers [89]. SORS exceeds other spectroscopic methods able to obtained Raman signals from depths and can be used to obtain spatial mapping of chemical components that interferes superficial layer [90]. The sensitivity and accuracy of SORS allows studies based on security screening and forensics, biomedical sciences and pharmaceutical process [91].

Optimal Control in Raman Analysis

Raman analysis requires optimal control on the experimental setups for the particular Raman technique used. This control depends markedly on the type of Raman experiment performed: incoherent or coherent and spontaneous or stimulated processes.

Spontaneous Raman Scattering has poor efficiency with reduced scatterers, low concentrations and high fluorescence samples. The type of the monochromator limits the spectral resolution. There are significant problems for aerosol of biological samples that can be avoided using others Raman variations.

Resonance Raman and UV Resonance Raman, as nonlinear coherent process, can overcome the problems mentioned, enhancing the Raman scattering cross section by orders of magnitude. No special Raman instrumentation is needed can be used conventional Raman spectrometers with appropriate excitation wavelength. The quantification measurement using RR becomes difficult and can be affected by [62]:

- Competition between scattering and absorption process
- Thermal decomposition of the sample
- Photolysis

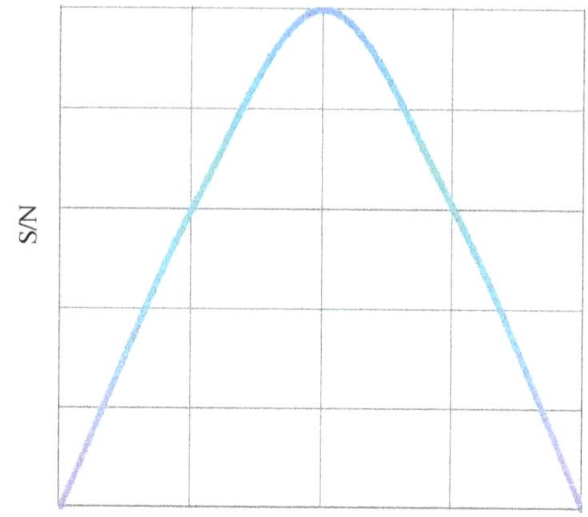

Fig. 10.16 Effect of colloid volume on SERS signal-to-noise ratio (SNR). It is important to optimize for maximum SNR as a function of colloid volume for increasing the scattered radiation collected and decreasing contributions from interferences

The use of a continuous-wave UV laser, a high duty-cycle picosecond UV laser source, and a sensitive CCD detector allow a fast collection of the spectral intensity data for quantification, achieving sensitivity and reproducibility in the analysis of biological samples. Effects of hypochromism (reduction of Raman intensity), and hyperchromism (enhancements of Raman intensity) can be observed with biological samples using this technique [34, 37].

A mix of the fundamental and third harmonics of a Ti-sapphire laser pumped with by the second harmonic of a Q-switched Nd-YLF laser provides a deep UV excitation allows the study of biocomponents. The tunability, linewidth, high average power of the laser and repetition rate promise a good UV resonance Raman instrument for biological applications [92].

SERS, which is an incoherent technique, has become one of the most used RS modalities for obtaining large signal enhancement and attaining low limits of detection in a very wide variety of samples. Bioaerosol detection by SERS has certain parameters that need to be optimized to obtain optimum signal enhancement without masking the Raman signals. Selectivity, sensitivity, reproducibility, intensification of signal, response, and analysis time in SERS experiments depend on the following:

- *Ratio of the biosample to metallic nanoparticle concentration*: The amount of sample used for detection must be controlled in order to obtain a good Raman signal. Low and high concentrations of the biosample in the mixture affect the SERS experiments because of very weak Raman signals or strong background scattering (background can also be Raman enhanced). Figure 10.16 shows the general behavior for the scenario, in which the optimum volume of the colloid increases the scattered radiation, reaching a maximum in intensity, and then decreases the detected SERS signal. For high metal nanoparticles concentrations, the vibrational SERS signals cannot be clearly observed, and there is a decrease in the intensity that is inversely proportional to the nanoparticle concentration. In the case of SERS experiments for the detection of bacterial aerosols, concentrations of 10^2 cfu/mL and 10^6 cfu/mL (cfu: colony forming units) are considered the low and high concentrations, respectively [71]. Determining the optimum volume of nanoparticle used is also important for increasing the scattered radiation collected [93] but a minimum amount of the sample is required [82].

- *The thermodynamics of the analyte adsorption to the surface of interest*: This property is related to both the time required to obtain the effective binding of the analyte and metallic nanoparticle as well as the type and strength of the adsorbate–substrate bond. It also depends on the effective binding to Raman active sites involved in the signal enhancement ("hot spots"). The binding kinetics can be fast (i.e., seconds to a few minutes) or slow (i.e., several minutes to hours) depending on the interaction between the analyzed system and the NP surface [66]. It has been clearly demonstrated that factors such as pH, density of bacteria in solution, and type of bacteria greatly influence the time-dependent behavior of colloidal/bacterial suspensions or the adsorption rate of the metallic colloidal NP and the bacteria analyzed [74, 76, 101]. Modifying the surface charge of the NP can increase the electrostatic interaction between the NP surfaces and the bacterial cell wall. This in turn depends on the reducing agent of the metallic ions solutions [102]. Ag-borohydride NP surface charges were modified by changing the pH of the colloid to obtain a more intimate interaction with the bacterial cell wall in which at low and high pH values (3.25 and 9.74), but did not result in improved detection of *E. coli* [73–74]. Improved SERS signals at pH 5 and higher were obtained for *Bacillus* Gram-positive species with citrate reduced NP [95] and for biological molecules with amine groups using hydroxylamine hydrochloride reduced NP [101]. The changes in NP surfaces increase the formation of hot spots and consequently further augment SERS effect observed. To induce aggregation, pH changes as well as adding halide anions to the NP can also be used. The use of aggregation agents in SERS of bacterial samples was studied to allow the NP induced aggregation and to generate highly SERS-active inter-particle spaces [76]. *B. thuringiensis* detection has been studied by SERS using silver oxide films [66], Au NP covered by SiO_2 [24], Ag colloids reduced by citrate [67] and by hydroxylamine hydrochloride [8].
- *The shape and size of the sample*: Raman light scattering in SERS analysis results in intensity variations that depend on the morphological properties of the biosample. The biological samples have a wide range of particle shapes and sizes and the dependence on the Raman scattered light has been theoretically and experimentally studied. For example, Vehring et al. [94] studied the size dependence of homogeneous spheres. The shape, size, and orientation of metallic nanoparticles affect the selection of the optimal excitation wavelength. Heterogeneously roughened surface limit the surface enhancement, but it is difficult to obtain reproducible SERS-active substrates because the density and location of "hot spots" are very hard to control.

Kahraman et al. has studied the SERS sensitivity in bacterial detection [95]. The study included comparisons between types of colloid and excitation wavelengths in two types of bacteria for reproducibility purposes. The reproducibility and consistency of enhancement of vibrational signatures of microorganisms and their stability and dependence on morphology of SERS active substrates has also been studied in depth [24]. It is well known that the electromagnetic and chemical mechanisms of SERS produced by NP upon interaction with analytes is in general determined by the intimacy of the interaction of the molecules to the surface plasmon field and the charge transfer at the "hot spots" sites. The optimization procedure for achieving maximum signals plays an important role in the attainment of high Raman enhancements for *B. thuringiensis*. Several previous reports concentrated on the experimental parameters that had to be optimized when studies focused on biological samples detection and discrimination was carried on.

However, due to the coherent nature of CARS the optimum control parameters have to include several parameters that incoherent techniques do not require. The signal intensity of CARS depends strongly on the bioaerosol properties and experimental parameters. For example, a few of the parameters are:

- N^2: where N is the number of molecules participating in the process [23]
- Laser line width: laser line width determines the spectral resolution of the technique defining the multiple or single Raman vibration wavelength [23, 46, 47, 82]
- Synchronization of excitation pulses [46]
- Beam geometry: control of CARS excitation lasers and detection of scattered signals depending on the physical properties of the sample [51]
- Laser energy/power: control required to avoid photo-damage of the sample
- Raman instrumentation for nonlinear spectroscopy [96]

Moore recently discussed the optimal conditions for improving selectivity and sensitivity in the coherent spectrochemical analysis [96]. The constructed model shows a closed loop relationship with the laser bandwidth of the source chosen, producing an optical waveform using a pulse shaper to obtain the sample signal. To select the initial pulse shape to interact with the sample, an algorithm was modified until the optimal pulse shape and a strong sample signal were achieved. This optimization procedure illustrates the way to reduce the nonresonant background problems of coherent spectra [96]. Scully et al. also applied the model using an ultra-fast laser system [39]. Optimizing the control parameters allows obtaining reproducible and strong Raman signals for rapid identification, characterization, and detection of bioaerosols.

Data Analysis

The spectral data obtained by RS measurements show the vibrational modes of the biosample constituents. The vibrational signals allow for the detection of biosamples and the subsequent identification and characterization of the biological matter according to frequency ranges and vibrational mode assignments. Tentative band assignments obtained can be compared to those reported in the literature to assure the correct identification. In addition to this, the presence of additional peaks or

Table 10.6 Assignment of bands frequently found in Raman spectra of biological specimens [97]

Tentative assignment	Frequency (cm^{-1})
(C=C–H) aromatic str	3059
CH_3 str	2975
CH_3 & CH_2 str	2935
CH_2 str	2870–2890
C=O ester str	1735
Amide I	1650–1680
Tyrosine	1614
Phenylalanine	1606
Guanine, Adenine ring str	1575
C–H_2 def	1440–1460
CH_2 def	1295
Amide III	1230–1295
C–N and C–C str	1129
PO_2^- sym str	1102
CC skeletal and COC str from glycosidic link	1098
C-O str	1085
C–N and C–C str	1061
Phenylalanine	1004
COC str	897
CC str, COC 1,4-glycosidic link	858
Tyrosine "buried"	852
Tyrosine "exposed"	829
Cytosine and Uracil ring str	785
Adenine	720
Guanine	665
Tyrosine skeletal	640
Phenylalanine skeletal	620
COC glycosidic ring def	540
S–S str	520–540

str stretching, *def* deformation, *sym* symmetric

the lack of peaks in the spectra indicates the formation or breakdown of chemical bonds. Table 10.6 includes the vibrational assignments of signals frequently found in Raman spectra of biological specimens [97].

The detection of aerosol and suspension samples by SERS can be influenced by the colloidal substrate and the laser radiation of the source used. Specific procedures must be used to obtain the best SERS signal.

Chemometrics statistical routines such as partial least squares (PLS) principal component analysis (PCA) , discriminant analysis (DA), partial least squares-discriminant analysis (PLS-DA), discriminant factor analysis (DFA), hierarchical cluster analysis (HCA), linear discriminant analysis (LDA), quadratic discriminant analysis (QDA), K-nearest neighbor classifier (kNN), support vector machines (SVMs), Gaussian mixture discriminant analysis (MDA), and spectral angle mapping are some of the currently used methods to find patterns in spectral data. These routines have been used to identify microbial cells from the single cell level to the

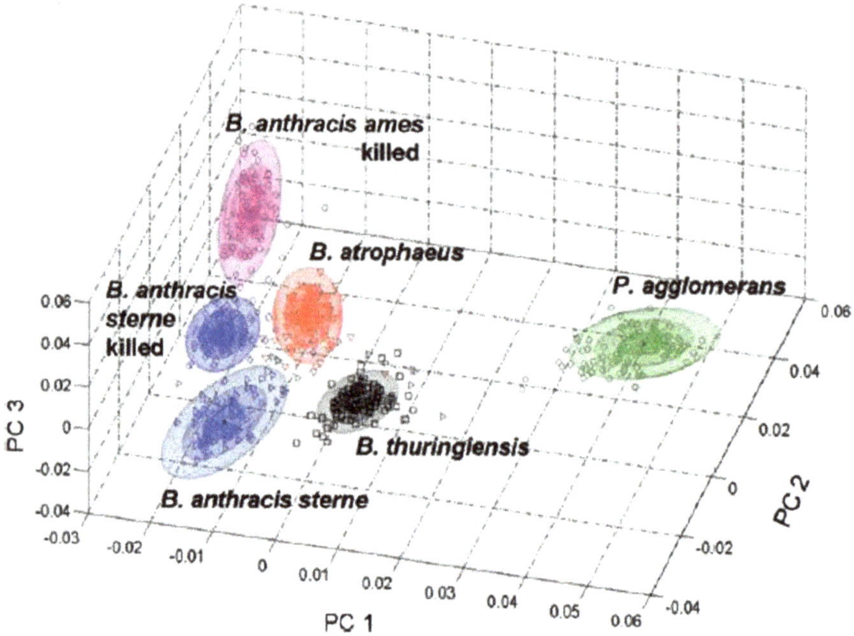

Fig. 10.17 PCA discrimination of five *Bacillus* spore samples and *Pantoea agglomerans*. (Reproduced with permission from Society of Applied Spectroscopy, [67])

bulk analysis level, discriminating between bacterial strains in mixtures of aerosol particles, different cultivation parameters, and taxonomic discrimination between microorganisms, among others. An example of applications of multivariate analysis in discrimination of 5 *Bacillus* spore samples and *pantoea agglomerans* can be observed in Fig. 10.17. Spectral angle mapping can be used under the microscope objective to compare an experimental spectrum with a library spectrum. The accurate interpretation of models can provide robustness and reliability to the correlative analysis of Raman data [19, 20, 55, 67, 98–100].

Data analysis has been used to validate the applicability of qualitative SERS spectroscopy for analysis of bacterial species by utilizing PCA to show discrimination of biological threat stimulants. Different types of bacteria have been detected and discriminated after important optimization experiments to obtain good SERS results. For example, PCA of extracted SERS spectra from single bacterial cells coated with silver NP such as *B. cereus* vegetative cells, *B. anthracis* spore and *B. thuringiensis* vegetative cells were sufficient to allow for discrimination between the three single bacteria [103]. Also, SERS spectra have been used to discriminate between different species and the possible differentiation of strains of *Bacillus* spores [67]. Gram-negative and Gram-positive species as well as spores and vegetative cells of *B. globiggi* were discriminated by PCA with spectral differences mainly attributed to lipid layer component of the cell walls and membranes and C–C skeletal in proteins at endospores samples. However, SERS spectral results that contain fluorescence can limit the chemical information obtained and the reliability of the sample identification. A multivariate analysis was performed on multiple sets of measurements, wavelength, samples and data sets [104].

Outlook

RS technique and its variations have been successfully used for bioaerosol detection, identification, and characterization with high sensitivity, selectivity, and reproducibility with fast and real-time detection schemes. The results presented show several advantages over the conventional methods of bioaerosol detection, with significant improvements in sensitivity and selectivity, however the techniques have some limitations. Special interest in stand-off detection of bioaerosol using RS represents a great technological challenge that is being continuously improved. The coupling of RS with others detection and characterization analytical tools can be used to overcome many of the challenges that current detection boundaries pose and will undoubtedly contribute to reach the goals of improved selectivity, sensitivity, lower duty cycles, lower biosample concentrations, and real-time analysis. Kahraman [95] and Sengupta [74] proposed that a standard protocol must be developed depending on the goal of the study in order to optimize the SERS detection system by controlling interaction parameters.

Acknowledgments This work was made possible by funding from the U.S. Department of Defense, Proposal Number: 58949-PH-REP, Agreement Number: W911NF-11-1-0152. The authors also acknowledge contributions from Dr. Richard T. Hammond from the Army Research Office, DoD. Support from the U.S. Department of Homeland Security and the Awareness and Localization of Explosives Related Threats (ALERT) Program of the DHS Center of Excellence for Explosives under Award Number 2008-ST-061-ED0001 is also acknowledged. However, the views and conclusions contained in this document are those of the authors and should not be interpreted as necessarily representing the official policies, either expressed or implied, of the U.S. Department of Homeland Security. Thanks are due to Dr. Luis F. de la Torre-Quintana for collaboration in the design and preparation of several of the figures and to Luis A. Echevarría for his help and support.

References

1. Smekal A (1923) Zur quantentheorie der dispersion. Naturwissenschaften 43:873–875
2. Raman CV, Krishnan KS (1928) A new type of secondary radiation. Nature 121:501–502
3. Landsberg GS, Mandelstam LI (1928). Naturwissenschaften 16:557–558
4. McCreery RL (2000) Raman Spectroscopy for Chemical Analysis. John Wiley & Sons, Inc., New York, NY
5. Long DA (1977) Raman Spectroscopy. Mc-Graw-Hill, New York, NY
6. Pelletier MJ (ed) (1999) Analytical Applications of Raman Spectroscopy. Blackwell Science Ltd., London, UK
7. Smith E, Dent G (2005) Modern Raman Spectroscopy—A Practical Approach. J. Wiley & Sons, Ltd., Hoboken, NJ

8. Yan F, Vo-Dinh T (2007) Surface-enhanced Raman scattering detection of chemical and biological agents using a portable Raman integrated tunable sensor. Sensors Actuat B-Chem 121 (1):61–66. doi:10.1016/j.snb.2006.09.032
9. Kiefer W (2007) Recent Advances in linear and nonlinear Raman spectroscopy I. J Raman Spectrosc 38 (12):1538–1553. doi:10.1002/jrs.1902
10. Kiefer W (2008) Recent advances in linear and nonlinear Raman spectroscopy II. J Raman Spectrosc 39 (12):1710–1725. doi:10.1002/jrs.2171
11. Pacheco-Londoño LC, Ortiz-Rivera W, Primera-Pedrozo OM, Hernandez-Rivera SP (2009) Vibrational spectroscopy standoff detection of explosives. Anal Bioanal Chem 395 (2):323–335. doi:10.1007/s00216-009-2954-y
12. Wallin S, Pettersson A, Ostmark H, Hobro A (2009) Laser-based standoff detection of explosives: a critical review. Anal Bioanal Chem 395 (2):259–274. doi:10.1007/s00216-009-2844-3
13. Sharma S, Misra A (2008) Remote Raman Spectroscopic Detection of Inorganic, Organic and Biological Materials to 100 m and More. Proc ICOPVS, 1: 3–7
14. Moosmuller H, Chakrabarty RK, Arnott WP (2009) Aerosol light absorption and its measurement: A review. J Quant Spectrosc Ra 110:844–878. doi:10.1016/j.jqsrt.2009.02.035
15. Chase DB (1986) Fourier Raman transform spectroscopy. J Am Chem Soc 108:7485
16. Chase DB, Rabolt J-F (1994) Fourier Transform Raman Spectroscopy: From Concept to Experiment. Academic Press, New York, NY
17. Félix-Rivera H, Hernández-Rivera S (2012) Raman Spectroscopy Techniques for the Detection of Biological Samples in Suspensions and as Aerosol Particles: A Review. Sens Imaging 13 (1):1–25. doi:10.1007/s11220-011-0067-0
18. Esposito AP, Talley CE, Huser T, Hollars CW, Schaldach CM, Lane SM (2003) Analysis of single bacterial spores by micro-Raman spectroscopy. Appl Spectrosc 57 (7):868–871
19. Tripathi A, Jabbour RE, Guicheteau JA, Christesen SD, Emge DK, Fountain AW, Bottiger JR, Emmons ED, Snyder AP (2009) Bioaerosol Analysis with Raman Chemical Imaging Microspectroscopy. Anal Chem 81 (16):6981–6990. doi:10.1021/ac901074c
20. Rŏsch P, Harz M, Peschke K-D, Ronneberger O, Burkhardt H, Schüle A, Schmauz G, Lankers M, Hofer S, Thiele H, Motzkus H-W, Popp J (2006) On-Line Monitoring and Identification of Bioaerosols. Anal Chem, vol 78.
21. Carmona P (1980) Vibrational-spectra and structure of crystalline dipicolinic acid and calcium dipicolinate trihydrate. Spectrochim Acta A 36 (7):705–712
22. Kolomenskii AA, Jerebtsov SN, Opatrny T, Schuessler HA, Scully MO (2003) Spontaneous Raman spectra of dipicolinic acid in microcrystalline form. J Mod Optic 50 (15–17):2369–2374. doi:10.1080/09500340320001208O3
23. Chan J, Fore S, Wachsman-Hogiu S, Huser T (2008) Raman spectroscopy and microscopy of individual cells and cellular components. Laser Photonics Rev 2 (5):325–349. doi:10.1002/lpor.200810012
24. Premasiri WR, Moir DT, Klempner MS, Krieger N, Jones G, Ziegler LD (2005) Characterization of the Surface Enhanced Raman Scattering (SERS) of bacteria. J Phys Chem B 109 (1):312–320. doi:10.1021/jp040442n
25. Luna-Pineda T, Soto-Feliciano K, De La Cruz-Montoya E, Londono LCP, Rios-Velazquez C, Hernandez-Rivera SP (2007) Spectroscopic characterization of biological agents using FTIR, Normal Raman and surface enhanced Raman scattering. Proc SPIE 6554, 1–11. doi: 10.1117/12.720338
26. Manoharan R, Ghiamati E, Dalterio RA, Britton KA, Nelson WH, Sperry JF (1990) UV resonance Raman spectra of bacteria, bacterial spores, protoplasts and calcium dipicolinate. J Microbiol Methods 11 (1):1–15. doi:http://dx.doi.org/10.1016/0167-7012(90)90042-5
27. Hug WF, Bhartia R, Taspin A, Lane A, Conrad P, Sijapati K, Reid RD (2005) Status of miniature integrated UV resonance fluorescence and Raman sensors for detection and identification of biochemical warfare agents. Proc SPIE 59940J. doi:10.1117/12.628923
28. Hug WF, Reid RD, Bhartia R, Lane AL (2008) A new miniature hand-held solar-blind reagentless standoff chemical, biological, and explosives (CBE) sensor. Proc SPIE 69540I–69540I

29. Hug WF, Reid RD, Bhartia R, Lane AL (2009) Performance status of a small robot-mounted or hand-held, solar-blind, standoff chemical, biological, and explosives (CBE) sensor. Proc SPIE 73040Z–73040Z
30. Hug WF, Bhartia R, Tsapin A, Lane A, Conrad P, Sijapati K, Reid RD (2006) Water and surface contamination monitoring using deep UV laser induced native fluorescence and Raman spectroscopy. Proc SPIE 63780S–63780S
31. Chadha S, Nelson WH, Sperry JF (1993) Ultraviolet Micro-Raman spectrograph for the detection of small numbers of bacterial-cells. Rev Sci Instrum 64 (11):3088–3093
32. Fung KH, Tang IN (1992) Analysis of Aerosol-Particles by Resonance Raman-Scattering Technique. Appl Spectrosc 46 (1):159–162
33. Fung KH, Tang IN (1992) Aerosol-particle analysis by Resonance Raman-Spectroscopy. J Aerosol Sci 23 (3):301–307
34. Wu Q, Nelson WH, Elliot S, Sperry JF, Feld M, Dasari R, Manoharan R (2000) Intensities of *E. coli* Nucleic Acid Raman Spectra Excited Selectively from Whole Cells with 251-nm Light. Anal Chem 72 (13):2981–2986. doi:10.1021/ac990932p
35. Manoharan R, Ghiamati E, Chadha S, Nelson WH, Sperry JF (1993) Effect of cultural conditions on deep UV Resonance Raman-spectra of bacteria. Appl Spectrosc 47 (12):2145–2150
36. Ghiamati E, Manoharan R, Nelson WH, Sperry JF (1992) UV Resonance Raman-Spectra of *Bacillus* Spores. Appl Spectrosc 46 (2):357–364
37. Wu Q, Hamilton T, Nelson WH, Elliott S, Sperry JF, Wu M (2001) UV Raman Spectral Intensities of *E. Coli* and Other Bacteria Excited at 228. 9, 244.0, and 248.2 nm. Anal Chem 73 (14):3432–3440. doi:10.1021/ac001268b
38. Ooi CHR, Beadie G, Kattawar GW, Reintjes JF, Rostovtsev Y, Zubairy MS, Scully MO (2005) Theory of femtosecond coherent anti-Stokes Raman backscattering enhanced by quantum coherence for standoff detection of bacterial spores. Phys Rev A 72 (2). doi:02380710.1103/PhysRevA.72.023807
39. Scully MO, Kattawar GW, Lucht RP, Opatrny T, Pilloff H, Rebane A, Sokolov AV, Zubairy MS (2002) FAST CARS: Engineering a laser spectroscopic technique for rapid identification of bacterial spores. P Natl Acad Sci USA 99 (17):10994–11001. doi:10.1073/pnas.172290899
40. Manoharan R, Ghiamati E, Britton KA, Nelson WH, Sperry JF (1991) Resonance Raman-spectra of aqueous pollen suspensions with 222.5–242.4 nm pulsed laser excitation. Appl Spectrosc 45 (2):307–311
41. Nelson WH, Dasari R, Feld M, Sperry JF (2004) Intensities of calcium dipicolinate and *Bacillus subtilis* spore Raman spectra excited with 244 nm light. Appl Spectrosc 58 (12):1408–1412
42. Pestov D, Zhi MC, Sariyanni ZE, Kalugin NG, Kolomenskii AA, Murawski R, Paulus GG, Sautenkov VA, Schuessler H, Sokolov AV, Welch GR, Rostovtsev YV, Siebert T, Akimov DA, Graefe S, Kiefer W, Scully MO (2005) Visible and UV coherent Raman spectroscopy of dipicolinic acid. P Natl Acad Sci USA 102 (42):14976–14981. doi:10.1073/pnas.0506529102
43. Grun J, Manka CK, Nikitin S, Zabetakis D, Comanescu G, Gillis D, Bowles J (2007) Identification of Bacteria from Two-Dimensional Resonant-Raman Spectra. Anal Chem 79 (14):5489–5493. doi:10.1021/ac070681h
44. Müller M, Zumbusch A (2007) Coherent anti-stokes Raman scattering microscopy. ChemPhysChem 8 (15):2157–2170. doi:10.1002/cphc.200700202
45. Kalasinsky KS, Hadfield T, Shea AA, Kalasinsky VF, Nelson MP, Neiss J, Drauch AJ, Vanni GS, Treado PJ (2007) Raman chemical imaging spectroscopy reagentless detection and identification of pathogens: Signature development and evaluation. Anal Chem 79 (7):2658–2673. doi:10.1021/ac0700575
46. Cheng JX, Xie XS (2004) Coherent anti-Stokes Raman scattering microscopy: Instrumentation, theory, and applications. J Phys Chem B 108 (3):827–840. doi:10.1021/jp035693v
47. Silberberg Y (2009) Quantum Coherent Control for Nonlinear Spectroscopy and Microscopy. Annu Rev Phys Chem 60:277–292. doi:10.1146/annurev.physchem.040808.090427
48. Downes A, Mouras R, Elfick A (2009) A versatile CARS microscope for biological imaging. J Raman Spectrosc 40 (7): 757–762. doi:10.1002/jrs.2249

49. Rodriguez LG, Lockett SJ, Holtom GR (2006) Coherent anti-stokes Raman scattering microscopy: A biological review. Cytom Part A 69A (8):779–791. doi:10.1002/cyto.a.20299
50. Naumann D (2001) FT-infrared and FT-Raman Spectroscopy in biomedical research. In: Gremlich H-U, Yang B (eds) Infrared and Raman spectroscopy of biological material.
51. Petrov GI, Yakovlev VV, Sokolov AV, Scully MO (2005) Detection of *Bacillus subtilis* spores in water by means of broadband coherent anti-Stokes Raman spectroscopy. Opt Express 13 (23):9537–9542
52. Pestov D, Murawski RK, Ariunbold GO, Wang X, Zhi MC, Sokolov AV, Sautenkov VA, Rostovtsev YV, Dogariu A, Huang Y, Scully MO (2007) Optimizing the laser-pulse configuration for coherent Raman spectroscopy. Science 316 (5822):265–268. doi:10.1126/science.1139055
53. Pestov D, Wang X, Ariunbold GO, Murawski RK, Sautenkov VA, Dogariu A, Sokolov AV, Scully MO (2008) Single-shot detection of bacterial endospores via coherent Raman spectroscopy. P Natl Acad Sci USA 105 (2):422–427. doi:10.1073/pnas.0710427105
54. Petrov GI, Arora R, Yakovlev VV, Wang X, Sokolov AV, Scully MO (2007) Comparison of coherent and spontaneous Raman microspectroscopies for noninvasive detection of single bacterial endospores. P Natl Acad Sci USA 104 (19):7776–7779. doi:10.1073/pnas.0702107104
55. Harz A, Rosch P, Popp J (2009) Vibrational Spectroscopy-A Powerful Tool for the Rapid Identification of Microbial Cells at the Single-Cell Level. Cytom Part A 75A (2):104–113. doi:10.1002/cyto.a.20682
56. Ooi CHR (2009) Theory of coherent anti-Stokes Raman scattering for mesoscopic particle with complex molecules: angular-dependent spectrum. J Raman Spectrosc 40 (7):714–725.
57. Aroca R, Rodriguez-Llorente S, (1997) Surface-enhanced vibrational spectroscopy. J Molec Struct 408–409: 17–22.
58. Fleischmann M, Hendra P, McQuillan A (1974) Raman spectra of pyridine adsorbed at a silver electrode. Chem Phys Lett 26 (2):163–166
59. Jeanmaire D, Van Duyne R (1977) Surface Raman Spectroelectrochemistry. Part I Heterocyclic, aromatic, and aliphatic amines adsorbed on the anodized silver electrode. J Electroanal Chem 84:1–20.
60. Albrecht M G, Creighton JA (1977) Anomalously intense Raman spectra of pyridine at a silver electrode. J Amer Chem Soc 99 (15), 5215–5217.
61. Moskovits M, (1985) Surface-enhanced spectroscopy. Rev Mod Phys 57 (3).
62. Schrader B (ed) (1995) Infrared and Raman spectroscopy: Methods and applications. VCH, New York, NY.
63. Lee PC, Meisel D (1982) Adsorption and Surface-Enhanced Raman of Dyes on Silver and Gold Sols. J Phys Chem 86:3391
64. Efrima S, Zeiri L (2009) Understanding SERS of bacteria. J Raman Spectrosc 40 (3):277–288. doi:10.1002/jrs.2121
65. Jarvis RM, Law N, Shadi LT, O'Brien P, Lloyd JR, Goodacre R (2008) Surface-enhanced Raman scattering from intracellular and extracellular bacterial locations. Anal Chem 80 (17):6741–6746. doi:10.1021/ac800838v
66. Yan F, Wabuyele MB, Griffin GD, Vass AA, Vo-Dinh T (2005) Surface-enhanced Raman scattering, detection of chemical and biological agent simulants. IEEE Sens J 5 (4):665–670. doi:10.1109/jsen.2005.850993
67. Guicheteau J, Argue L, Emge D, Hyre A, Jacobson M, Christesen S (2008) *Bacillus* spore classification via surface-enhanced Raman spectroscopy and principal component analysis. Appl Spectrosc 62 (3):267–272
68. Félix-Rivera H, González R, Rodríguez G, Primera-Pedrozo OM, Ríos-Velázquez C, Hernández-Rivera SP (2011) Improving SERS Detection of *Bacillus thuringiensis* using Silver Nanoparticles Reduced with Hydroxylamine and with Citrate Capped Borohydride. International Journal of Spectroscopy 2011. doi:10.1155/2011/989504
69. Shanmukh S, Jones L, Driskell J, Zhao YP, Dluhy R, Tripp RA (2006) Rapid and sensitive detection of respiratory virus molecular signatures using a silver nanorod array SERS substrate. Nano Lett 6 (11):2630–2636. doi:10.1021/nl061666f

70. Sengupta A, Laucks ML, Dildine N, Drapala E, Davis EJ (2005) Bioaerosol characterization by surface-enhanced Raman spectroscopy (SERS). J Aerosol Sci 36 (5–6):651–664. doi:10.1016/j.jaerosci.2004.11.001
71. Sengupta A, Brar N, Davis EJ (2007) Bioaerosol detection and characterization by surface-enhanced Raman spectroscopy. J Colloid Interf Sci 309 (1):36–43. doi:10.1016/j.jcis.2007.02.015
72. Jarvis RM, Goodacre R (2008) Characterisation and identification of bacteria using SERS. Chem Soc Rev 37 (5):931–936. doi:10.1039/b705973f
73. Sengupta A, Mujacic M, Davis EJ (2006) Detection of bacteria by surface-enhanced Raman spectroscopy. Anal and Bioanal Chem 386 (5):1379–1386. doi:10.1007/s00216-006-0711-z
74. Sengupta A, Laucks ML, Davis EJ (2005) Surface-enhanced Raman spectroscopy of bacteria and pollen. Appl Spectrosc 59 (8):1016–1023
75. Wang YL, Lee K, Irudayaraj J (2010) Silver Nanosphere SERS Probes for Sensitive Identification of Pathogens. J Phys Chem C 114 (39):16122–16128. doi:10.1021/jp1015406
76. Knauer M, Ivleva NP, Niessner R, Haisch C (2010) Optimized Surface-enhanced Raman Scattering (SERS) Colloids for the Characterization of Microorganisms. Anal Sci 26 (7):761–766
77. Zhang XY, Young MA, Lyandres O, Van Duyne RP (2005) Rapid detection of an anthrax biomarker by surface-enhanced Raman spectroscopy. J Am Chem Soc 127 (12):4484–4489. doi:10.1021/ja0436623b0b
78. Dhawan A, Du Y, Yan F, Gerhold MD, Misra V, Vo-Dinh T (2010) Methodologies for Developing Surface-Enhanced Raman Scattering (SERS) Substrates for Detection of Chemical and Biological Molecules. IEEE Sens J 10 (3):608–616. doi:10.1109/jsen.2009.2038634
79. Cheng HW, Luo WQ, Wen GL, Huan SY, Shen GL, Yu RQ (2010) Surface-enhanced Raman scattering based detection of bacterial biomarker and potential surface reaction species. Analyst 135 (11):2993–3001. doi:10.1039/c0an00421a
80. Chan JW, Esposito AP, Talley CE, Hollars CW, Lane SM, Huser T (2004) Reagentless identification of single bacterial spores in aqueous solution by confocal laser tweezers Raman spectroscopy. Anal Chem 76 (3):599–603. doi:10.1021/ac0350155
81. Butler JR, Wills JB, Mitchem L, Burnham DR, McGloin D, Reid JP (2009) Spectroscopic characterisation and manipulation of arrays of sub-picolitre aerosol droplets. Lab Chip 9 (4):521–528. doi:10.1039/b814545h
82. Schweiger G (1990) Raman-Scattering on Single Aerosol-Particles and on Flowing Aerosols—A Review. J Aerosol Sci 21 (4):483–509
83. Petrov DV (2007) Raman spectroscopy of optically trapped particles. J Opt A-Pure Appl Op 9 (8):S139–S156. doi:10.1088/1464-4258/9/8/s06
84. Hopkins RJ, Mitchem L, Ward AD, Reid JP (2004) Control and characterization of a single aerosol droplet in a single-beam gradient-force optical trap. Phys Chem Chem Phys 6 (21):4924–4927. doi:10.1039/b414459 g
85. Lübben JF, Mund C, Schrader B, Zellner R (1999) Uncertainties in temperature measurements of optically levitated single aerosol particles by Raman spectroscopy. J Mol Struc 480–481:311–316
86. Alexander TA, Pellegrino PM, Gillespie JB (2003) Near-infrared surface-enhanced-Raman-scattering-mediated detection of single optically trapped bacterial spores. Appl Spectrosc 57 (11):1340–1345
87. Symes R, Gilham RJJ, Sayer RM, Reid JP (2005) An investigation of the factors influencing the detection sensitivity of cavity enhanced Raman scattering for probing aqueous binary aerosol droplets. Phys Chem Chem Phys 7 (7): 1414–1422. doi:10.1039/b500385 g
88. Symes R, Sayer RM, Reid JP (2004) Cavity enhanced droplet spectroscopy: Principles, perspectives and prospects. Phys Chem Chem Phys 6 (3):474–487. doi:10.1039/b313370b
89. Hargreaves M, Macleod N, Brewster V, Munshi T, Edwards H, Matousek P (2009) Application of portable Raman spectroscopy and benchtop spatially offset Raman spectroscopy to interrogate concealed biomaterials. J Raman Spectrosc 40:1875–1880. doi:10.1002/jrs.2335

90. Maher J, Berger A (2010) Determination of ideal offset for spatially offset Raman Spectroscopy. Appl Spectrosc 64 (1):61–65
91. Matousek P (2006) Inverse spatially offset Raman Spectroscopy for deep noninvasive probing of turbid media. Appl Spectrosc 60 (11):1341–1347
92. Balakrishnan G, Hu Y, Nielsen SB, Spiro TG (2005) Tunable kHz Deep Ultraviolet (193–210 nm) Laser for Raman Application. Appl Spectrosc 59 (6):776–781
93. Ayora MJ, Ballesteros L, Perez R, Ruperez A, Laserna JJ (1997) Detection of atmospheric contaminants in aerosols by surface-enhanced Raman spectrometry. Anal Chim Acta 355 (1):15–21
94. Vehring R, Aardahl CL, Schweiger G, Davis EJ (1998) The characterization of fine particles originating from an uncharged aerosol: Size dependence and detection limits for Raman analysis. J Aerosol Sci 29 (9):1045–1061
95. Kahraman M, Yazici MM, Sahin F, Culha M (2007) Experimental parameters influencing surface-enhanced Raman scattering of bacteria. J Biomed Opt 12 (5). doi:10.1117/1.2798640
96. Moore D (2009) Optimal coherent control of sensitivity and selectivity in spectrochemical analysis. Anal Bioanal Chem 393 (1):51–56. doi:10.1007/s00216-008-2318-z
97. Chalmers JL, Griffiths PR (eds) (2002) Handbook of Vibrational Spectroscopy, vol I. Theory and Instrumentation. John Wiley & Sons Ltd, Chichester
98. Xie C, Mace J, Dinno MA, Li YQ, Tang W, Newton RJ, Gemperline PJ (2005) Identification of single bacterial cells in aqueous solution using confocal laser tweezers Raman spectroscopy. Anal Chem 77 (14):4390–4397. doi:10.1021/ac0504971
99. Schmid U, Rosch P, Krause M, Harz M, Popp J, Baumann K (2009) Gaussian mixture discriminant analysis for the single-cell differentiation of bacteria using micro-Raman spectroscopy. Chemometr Intell Lab 96 (2):159–171. doi:10.1016/j.chemolab.2009.01.008
100. Jarvis RM, Brooker A, Goodacre R (2004) Surface-enhanced Raman spectroscopy for bacterial discrimination utilizing a scanning electron microscope with a Raman spectroscopy interface. Anal Chem 76 (17):5198–5202. doi:10.1021/ac049663f
101. Kazanci M, Schulte JP, Douglas C, Fratzl P, Pink D, Smith-Palmer T (2009) Tuning the Surface-Enhanced Raman Scattering Effect to Different Molecular Groups by Switching the Silver Colloid Solution pH. Appl Spectrosc 63 (2):214–223
102. Alvarez-Puebla RnA, Arceo E, Goulet PJG, Garrido JnJ, Aroca RF (2005) Role of Nanoparticle Surface Charge in Surface-Enhanced Raman Scattering. The J Phys Chem B 109 (9):3787–3792. doi:10.1021/jp045015o
103. Guicheteau J, Christesen S, Emge D, Tripathi A (2010) Bacterial mixture identification using Raman and surface-enhanced Raman chemical imaging. J Raman Spectrosc 41 (12):1632–1637. doi:10.1002/jrs.2601
104. Guicheteau J (2006) Principal component analysis of bacteria using surface-enhanced Raman spectroscopy. Proceedings of SPIE–the international society for optical engineering 6218 (1):62180–62181
105. Griffiths WD, Decosemo GA L (1994) The assessment of bioaerosols-A critical-review, J Aerosol Sci 25, 1425–1458
106. Daniels JK, Caldwell TP, Christensen KA, Chumanov G (2006) Monitoring the kinetics of *Bacillus subtilis* endospore germination via surface-enhanced Raman scattering spectroscopy, Anal Chem 78, 1724–1729

Chapter 11
Biological Detection with Terahertz Spectroscopy

Tatiana Globus and Boris Gelmont

Basic Principles and Background Overview

Vibrational spectroscopy in the terahertz (THz) frequency range, 0.1–10 THz, or in the submillimeter-wave range, (3–300 cm^{-1}), is a fast emerging technique for fingerprinting biological molecules and species. (Fig. 11.1) For more than 20 years, theoretical studies predicted multiple resonances in absorption (or transmission) spectra of biological molecules in this range. THz radiation interacts with the low-frequency internal molecular vibrations by exciting these vibrations. Most of these vibrations are due to the torsion angles variations that produce collective motions of many atoms. These motions involve molecular groupings connected by weakest bonds: weak hydrogen bonds, van der Waals and/or other nonbonded interactions [1, 2]. Internal hydrogen bonds, like –C···H–N–, –C···H–O–, –N–H···O=C– (shown by dots), determine structure, function, and dynamics of biopolymers. Although hydrogen bonds are weak and have only ~5 % of the strength of covalent bonds, multiple hydrogen bonds stabilize the structure of bio-polymers, in particular hold the two strands of the DNA double helix together, or hold polypeptides together in different secondary structure conformations. The subTHz/THz regions of experimental absorption spectra of bio-molecules and species reveal these low frequency molecular motions. The resonant frequencies of such motions, low energy vibrational modes, are strongly dependent on a three-dimensional molecular structure. As such, these vibrational modes are sensitive to conformational change of molecules and to environment. It can be expected that each biomolecule possesses its unique resonant absorption characteristic, a THz fingerprint. THz vibrational spectroscopy technique for identifying and characterizing objects is based on the specificity of spectroscopic signature that reflects absorption of THz radiation at characteristic resonance frequencies.

Bacteria and spores are very complex biological objects. Since the absorption of biomaterials in the THz range is not very strong (absorption coefficient

T. Globus (✉) · B. Gelmont
University of Virginia, 104 Chaucer Rd., Charlottesville, VA 22904-1000, USA
e-mail: tg9a@virginia.edu

P. Jonsson et al. (eds.), *Bioaerosol Detection Technologies*, Integrated Analytical Systems, DOI 10.1007/978-1-4419-5582-1_11, © Springer-Verlag New York 2014

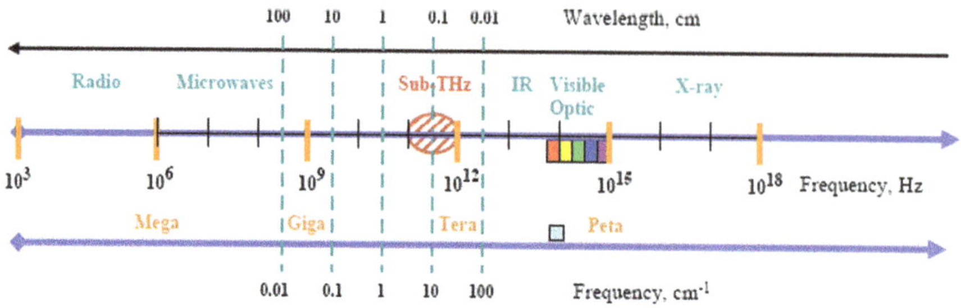

Fig. 11.1 Electromagnetic waves wavelengths and frequencies

~10–100 cm^{-1}), the radiation propagates through the entire object allowing genetic material (DNA and RNA), proteins, bacterial cell walls and membranes to contribute to the THz signature of bacterial cell/spore. Thus contributions from separate molecular components can provide us with an insight into the features of the spectrum from the whole bacterial cell.

The general need for faster and less expensive techniques that can provide useful molecular structure information related to the weakest bonds has naturally led to the development of spectroscopic methods that utilize the interaction of an applied electromagnetic field (EM) with the vibration modes of the material in the THz range. The capability of THz spectroscopy to detect directly low-frequency vibrations of weakest bonds is unique providing the information quite different from the visible or IR spectroscopic characterization. This uniqueness opens potentially many applications for THz vibrational spectroscopy including detection and identification of harmful biological species by using multiple resonances as distinctive spectral fingerprints. Optical spectral sensing of biological cells and spores seems to be an obvious solution for the problem in the nearest future.

Many benefits result from experimental observations of spectroscopic features in the sub-THz region. These include low absorption by water vapors and liquid water, at least two orders of magnitude less than in the far IR (100–1000 cm^{-1} or 10–100 μm). Thus, water does not mask absorption by biological materials and characterization of molecules in the solution is possible. THz radiation penetrates nonmetallic materials such as skin and clothing allowing detection of hidden bio-agents. The spectral range less than 1 THz (sub-THz) is especially attractive for practical applications because of low disturbance from liquid water. Although liquid water absorbs and contributes to background in the sub THz/THz spectral range, the level of water absorption in the low THz range is at least 2.5 orders of magnitude less compared to IR and far-IR. Because of less disturbance from water absorption lines, sensors in sub-THz range do not require evacuation or purging with dry nitrogen (see Fig. 11.2, less than 30 cm^{-1} at low vacuum). The atmosphere becomes more and more transparent with reducing frequencies and there are only few water vapor absorption lines less than 1 THz. A big advantage of THz spectroscopy is that it is an optical method and is nondestructive for living species [3]. The absence of problems with light-scattering that are often encountered with short wave-length

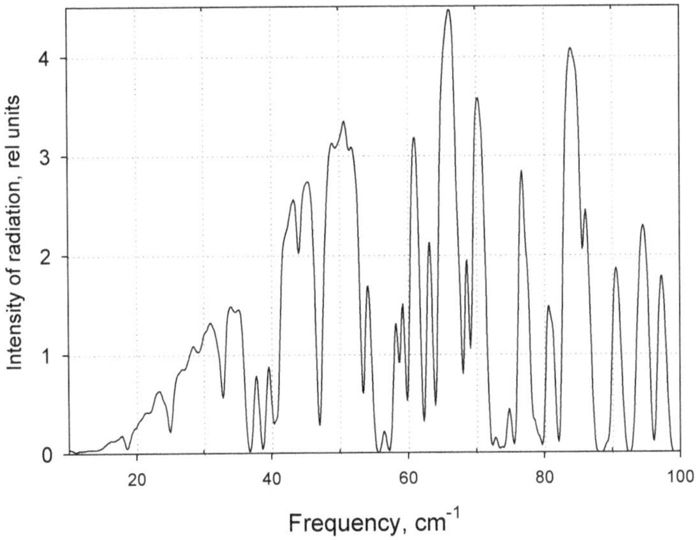

Fig. 11.2 Intensity of THz radiation (Bruker Spectrometer, low vacuum, spectral resolution 0.5 cm^{-1})

optical techniques is another advantage of THz spectroscopy. Some concerns existed regarding the scattering phenomena and mechanisms (Rayleigh or Mie) that can influence the results of measurements of biological materials at THz [4]. The scattering from molecules and very tiny particles (< 1/10 wavelength) is predominantly Rayleigh scattering. For particle sizes larger than a wavelength, Mie scattering predominates. This type of scattering from ensembles of many species is possible. However, although this scattering produces a pattern like an antenna lobe, with a sharper and more intense forward lobe for larger particles, it is not wavelength dependent and would not look like absorption resonances. Besides, for Mie scattering the difference in refractive index values for two medias (bioparticles and air) is important, and in the case of biological materials this difference is small.

The lowest vibrational frequencies of biological macromolecules start from 1to 2 cm^{-1} (30–60 GHz) and this limit shifts toward higher frequencies for smaller molecules. The smaller the molecule, the higher is the low frequency limit where vibrational modes can be observed. From experiments using Fourier transform (FT) spectroscopy with a liquid helium cooled bolometer, the width of spectral lines (or bands) for these vibrations in the sub-THz range is ~0.5 cm^{-1} [5, 6]. Thus, to be able to resolve vibrational modes from biological molecules, the spectra have to be taken with a relatively good spectral resolution of 0.2–0.3 cm^{-1}. On the contrary, many rather complex organic molecules, which however are much smaller compared to biological macromolecules, have broad absorption lines due to vibrations that happen in the range of 1–10 THz, with observed spectral line widths of 1–5 cm^{-1} [7]. Figure 11.3 demonstrates the difference between typical spectral features in the THz range for a plastic bonded RDX explosive (1,3,5-trinitro-1,3,5-triazacyclohexane)

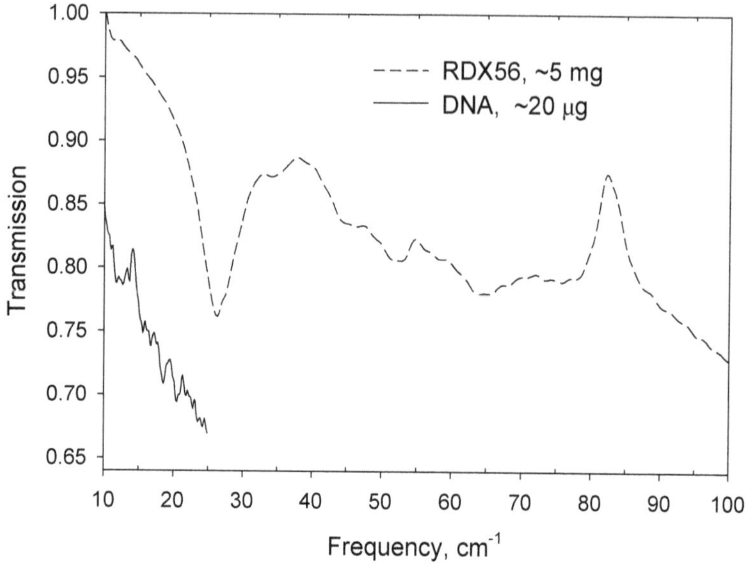

Fig. 11.3 Comparison between THz spectral features from explosive material (RDX) and biological macromolecule (salmon DNA), both in solid state (Brucker FTIR spectrometer)

and a DNA solid sample, both prepared as thin layers on very thin (~ 10 μm) plastic substrates. This figure illustrates the narrow width of low frequency resonances (features in the DNA curve) as compared to the broader RDX peaks at higher frequencies.

Many light molecules such as water, carbon monoxide, ozone, and chlorine monoxide have rotational transition energies in the sub-THz and THz frequency bands. The exact absorption spectra of molecules such as these have long been studied by chemists in an effort to improve their understanding of molecular structure and dynamics [8]. Radio astronomers have used the THz emission from clouds of molecules in interstellar space to identify the chemical constituents of these clouds and to investigate their structure, temperature, and internal motions [9, 10]. In a similar application, atmospheric scientists have used THz spectroscopy to study the trace constituents of the upper atmosphere, with recent emphasis on monitoring ozone concentrations and the distribution of molecules involved in ozone depletion such as chlorine monoxide [11]. In these studies, experimentators took advantages from the sharpness of rotational modes of small molecules that are easily separated on the frequency scale from artificial effects in the measurement systems. These differences justify specific approaches to be used in THz sensing of small and large biomolecules.

Detail analysis of THz science and technology applications in remote detection and identification is given in the review [12 and references therein]. The important results for the stand-point detection and identification have been the experimental measurement of absorption signatures on *Bacillus subtilis* (BG) in different forms including aerosol.

Progress in THz spectroscopy in the last several years (sample preparation techniques, repeatability and reliability of spectroscopic features, intensity of features) significantly enhanced experimental spectroscopic features, thus resulting in more reliable signatures from macromolecules and bacterial cells/spores. Nevertheless, there is still a widespread skepticism, as discussed in [13], that a large density of overlapping states contributing to the absorption bands of macromolecules might obscure vibrational resonances and yield essentially structureless spectra. This skepticism can be argued due to the fact that vibrational bands in spectra of, for example, proteins are observed and very well studied in the far IR region at much higher densities of states compared to THz. It was also demonstrated that not only biopolymers but also whole microorganisms can be characterized in the THz range [6, 14]. Multiple resonances due to low energy vibrational modes within biological macromolecules and biocells have been unambiguously demonstrated experimentally in the sub-THz frequency range in agreement with the theoretical prediction [15].

Biological materials in a form of water or buffer solution (gel) can be characterized as solids films [6]. Spectra of liquid samples have been measured and compared with spectra obtained from solid films. Very close similarities and also some different spectral features are demonstrated. There is almost no interference between spectral features of material and water background in spectral range of 10–25 cm^{-1} except for the band around 18.6 cm^{-1}. Much higher level of sensitivity and higher sharpness of vibrational modes for liquid sample characterization in comparison with solid films is observed in many cases. The mechanism of intensive vibration modes in liquid phase still has to be understood.

Simulations can help us to interpret the experimental data and even to predict spectral features from macromolecule components. Simultaneously with experimental characterization, computational modeling techniques have been developed using the energy minimization, normal mode analysis, and molecular dynamics (MD) approaches to understand and predict low frequency vibrational absorption spectra of short artificial DNA and RNA [5, 15–18], large macromolecules of DNA [19, 20], and proteins [21, 22]. Direct comparison of experimental spectra with theoretical prediction for a short chain a-helix RNA fragment with known structure [5], transfer RNA [20], and protein thioredoxin [22] from *Escerichia coli* (*E. coli*) showed reasonably good correlation thus validating both experimental and theoretical results.

There is, however, other controversial problem in understanding interaction of THz radiation with biological materials. The spectral width and the intensity of vibrational modes depend on relaxation processes of intermolecular dynamics, however the entire mechanism that determines relaxation (dissipation) times in dynamics processes is still not completely understood. The length of dissipation time is one important problem concerning THz vibrational modes in biological molecules. For each particular mode, the relaxation time, τ, and the dissipation factor, γ, that is reciprocal to τ are parameters that determine the intensity, the spectral width of vibrational modes, and eventually the absorption coefficient spectra and discriminative capability of spectroscopy. It is surprisingly narrow width of spectral lines (or dissipation factor) in transmission (absorption) spectra of BG spores "that makes them detectable, in spite of their relatively low absorption cross section compared to polar molecules in the THz region" [12]. Till recently γ was obtained only from experimental width of spectral features and big uncertainty exists in both modeling and experimental results interpretation. Due to overlapping of modes occurring at close frequencies, only the upper limit for γ has been estimated from experiments on Bruker FT spectrometer as $\sim$0.5 cm^{-1}, which corresponds to the lower limit for relaxation time for observed features $\tau \sim 70$ ps. Very recently highly resolved spectroscopic instruments have been introduced, which permit observation even more narrow spectral lines (0.07–0.1 cm^{-1} or 2–3 GHz) from biological macromolecules and bacterial cells/spores [23, 24]. These new experimental results have also been supported by the analyses of intra-molecular dynamics that demonstrated the coexistence of multiple relaxation processes with different scale of relaxation times [25]. The resulting "THz biosignatures that are inherently narrow in spectral width ... may be strong enough for remote sensing" [12]. Overall, THz resonance spectroscopy is expected to become a useful tool for the biological detection in air.

Characterization Techniques

THz resonance spectroscopy was impeded in the past by the absence of THz sources of radiation, detectors and spectroscopic systems that utilize effectively coupling between the radiation and biological material. In the last 10 years, the situation was changing rapidly due to new sources and detectors becoming available [26]. However, most of experimental data were obtained at frequencies greater than 1 THz where the power of sources is growing fast with frequency. At the same time, the strongest THz signatures of large macromolecules and organisms occur in the sub-THz region.

In the past there have been a limited number of transmission/absorption spectroscopic studies in the sub-THz region due to the experimental difficulties that are presented there [27–35]. In the region between the microwave (~100 GHz) and the lower frequency end of the far IR (~1000 GHz), the output power of the available sources is limited, while the absolute absorption of the biological material is not very strong. In this spectral region where the wavelength of radiation is on the same order of magnitude as the components of the spectrometer, multiple reflections lead to interference or etalon effects in the system, and significant care must be taken not to misinterpret these types of artifacts as spectral features.

There are different sources of spectral artifacts. The measurement system layout can be one of the sources due to multiple reflections between system elements. Sample preparation technique, including sample geometry (thickness), sample cell windows, or sample substrate, can be the other source of spectral artifacts. The physical origin of these artifacts can be again caused by multiple reflections at interfaces. Proper choice of a sample preparation procedure and a sample cell geometry can eliminate or significantly reduce these effects. To achieve this goal, the thickness of cell windows or sample substrates, and refractive index of materials used in construction need to be reduced. Substrates or windows from plastic materials (PE, Teflon, PMMA) that are thinner than 100 μm and very transparent in the

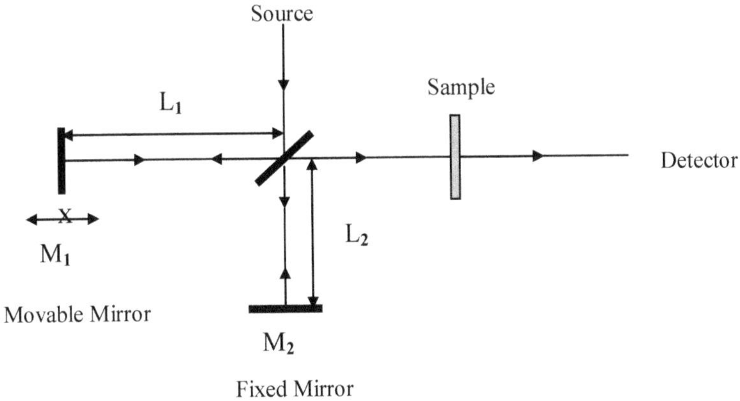

Fig. 11.4 FTIR: schematics of a Michelson interferometer

sub-THz range still produce interference fringes. However, these spectral fringes are stretched on the frequency scale and do not obscure spectroscopic signatures from bioparticles.

Environmental effects can be factors limiting the minimum detectable absorption change associated with a given spectroscopic instrument even in the case of a good signal-to-noise ratio. Water and gas absorption lines present in the spectral range of interest can make reliable measurements in the laboratory conditions extremely difficult because of mismatch between the reference (background) and sample measurements. The longer the acquisition time, the more difficult to obtain reproducible and reliable results even in the system with evacuated sample chamber.

Absorption spectra of water vapor and other gases in the spectral range of interest have to be included in data analysis to insure the reliability of spectral characteristics of material under the test. Artifacts are closely related to the spectral resolution and the width of spectral lines. The best choice for the spectral resolution depends on measurements objectives and for the broadband characterization a relatively poor spectral resolution has to be recommended to eliminate artifacts. On the contrary, for vibrational spectroscopic characterization of materials, the required spectral resolution is dictated by the spectral width of vibrational modes that can be used for fingerprinting and have to be detected.

It is not clear if the artifacts can be eliminated reliably using mathematical methods that can be applied to unknown materials and samples. All these difficulties have severely limited the direct identification of phonon modes in biological materials at submillimeter-wave frequencies in the past.

Fourier transform spectroscopy (FTS) and time domain spectroscopy (TDS) are mostly often used techniques for studying THz molecular resonances [36]. In an FTS, the radiation from a broad band source splits into two beams, reflects by two mirrors, one of which is fixed and the other moves back and forth, and recombined beam is directed through a sample (see schematic of Michelson Interferometer in Fig. 11.4).

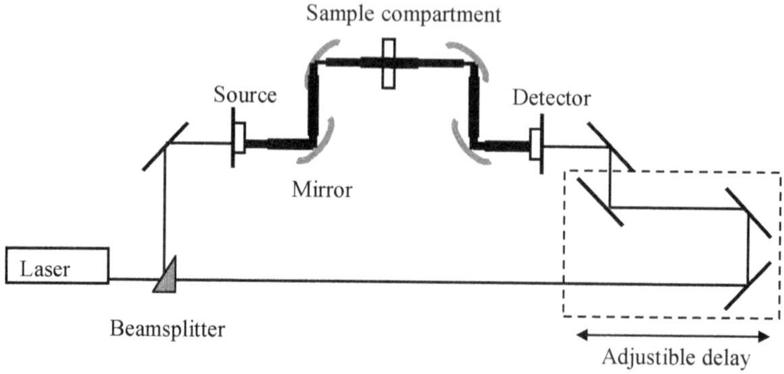

Fig. 11.5 Current terahertz time-domain spectrometers generally employ a Ti:Sapphire laser generating brief pulses of visible light which are used both to trigger a high-bandwidth terahertz pulse and, after a variable time delay, to trigger detection

The signal is measured as a periodic function of a movable mirror displacement (interferrogramm), since a path length difference has been introduced between two beams. The term *Fourier transform spectroscopy* reflects the fact that a mathematical procedure (a Fourier transform) is required to turn the raw data measured in a displacement (or time) domain into the spectrum in a frequency domain. Since sample is illuminated by all wavelengths at once, the measurements time is short and depends mainly on the time required to sweep the mirror. The spectral resolution is determined by the maximum traveling distance of a movable mirror.

In a terahertz time domain spectroscopy (TDS), the material is probed with the terahertz pulses (a few picoseconds) generated by an ultrashort pulsed laser (Fig. 11.5). A single pulse contains frequency components covering the whole terahertz range up to 4 THz. The signal is measured as a function of a time delay between two pulses, and an FT is again used to extract the frequency spectrum from the time-domain data [37]. Typical THz TDS have a spectral resolution of 50 GHz [36].

Raman spectroscopy can also give useful information on the far–IR molecular vibrations. However the THz spectroscopy is sensitive to the molecular polarizability while Raman is related to the derivative of this polarizability vs. local dimensions. In addition, different vibrational phonon modes are active in Raman experiments compare to absorption spectroscopy, and these two methods can provide complimentary information.

Till recently, Fourier transform (FT) transmission spectroscopy (Bruker IFS66v) with cooled Si bolometer operating at 1.7 K (see Fig. 11.6) provided the most detailed information on sub-THz vibrational spectral signatures of biological molecules [6, 15, 38].

Spectral resolution in these studies was 0.25 cm^{-1}. Mylar beam splitter of 125 μm thickness was used for the measurements in the spectral range of 10–25 cm^{-1}. Among other important spectrometer and software setting parameters are the phase resolution of 2 cm^{-1}, the scan velocity 80 kHz, the optical aperture of 12 mm, acquisition function of "single sided, fast return" and zero filling factor equal to 4. A cooled filter with a cut-off frequency of 35 cm^{-1}was used inside the

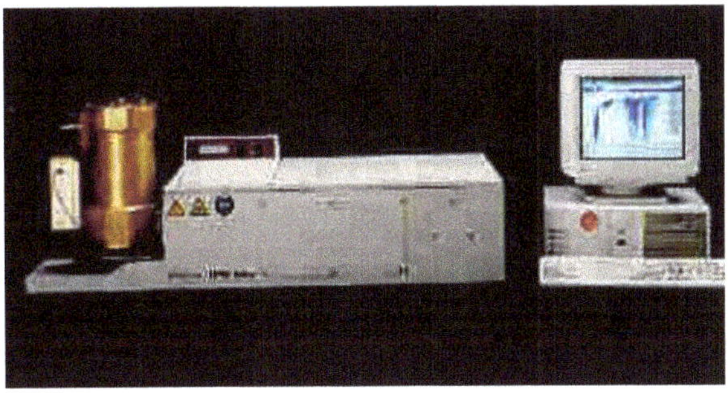

Fig. 11.6 Bruker IFS-66 spectrometer with an Hg-lamp source and He cooled Si-bolometer (1.7°K). Three vacuum systems are not shown

bolometer cryostat. Under these settings, the signal to noise ratio is >500 and the reproducibility of the instrument in transmission mode is better than 0.2–0.3% in almost entire spectral range, although it becoming worse at the both ends due to the rapid reducing of radiation intensity. The sample is usually placed in a standard Bruker sample holder in the focus of the beam inside the sample compartment. The whole instrument, except for the sample compartment, evacuated to 10 mbar. Measurements in air are possible since there are almost no absorption lines from water vapors or oxygen in the 10–25 cm^{-1} spectral range (see Fig. 11.2). A notable exception is a water vapor absorption band at around 18.6 cm^{-1}. Test measurements with a wire polarizer (25 μm wire diameter and 75 μm spacing between wires) were performed indicating 80% polarization of radiation in FTIR spectrometer in the spectral range of interest, and the electric field vector E oriented preferably in the vertical direction. Typically, 32 interferograms were averaged to reduce noise. With the "Blackman-Harris 3 terms" apodization function used for data processing, there were no standing wave problems in the system.

Time-Domain Spectroscopy has potentially many advantages in comparison with FT transmission spectroscopy. Some of them include the availability of more THz power by using ultra-fast laser sources (instead of a mercury lamp), effective THz emitters and sensitive room temperature detectors, the capability to extend characterization to the lower frequency limit, the capability to characterize simultaneously in one experiment both imaginary and real part of material complex dielectric constant, and most importantly, the absolute new capacity to study the dynamics of optical excitation and relaxation processes within materials. However, a very serious problem is the conflict between the good spectral resolution required for the resonance spectroscopy and the standing wave phenomenon [39, 40]. The samples used with time-domain spectroscopy are often prepared in the form of thick pellets from the mixture of biomaterials with polyethylene powder that causes multiple reflection effects on pellet surfaces. These geometry optic effects mask characteristic spectral features and prevent application of good spectral resolution required for resolving relatively narrow lines. Due to poor spectral resolution and sample preparation techniques that are not adequate to the problem, the investigators were able to observe only smooth broad absorption band less than 1 THz and the expected advantages of this promising technique, which employs a time-domain methodology [41] have not been fully realized.

Promising results with very good spectral resolution were obtained using a photomixing sweep oscillator as a source of a THz radiation [42–48]. In the earlier setups, very low power of radiation available through photomixing required enlarged time for data acquisition to provide large enough signal/noise ratio. However, later the coherent photomixing sweep oscillator [44] was combined with a coherent photomixing transceiver [45] both driven by the same pair of lasers with the frequency offset. The system produces the ratio of the power response to the noise of 60 dB at 1 THz and with the line width of ~ 100 MHz at room temperature operation. A high resolution terahertz spectrometer for gas phase chemical detection has been reported in [49, 50]. The instrument is based on CW heterodyne photomixer technology in the range of 0.1–1.2 THz and demonstrates frequency line widths and repeatability at the MHz level.

New compact frequency domain spectroscopic systems have been demonstrated for the sub-THz range [23, 24, 51]. Both systems utilize sets of frequency multipliers developed by the VDI Ltd for the THz source of radiation [26]. High spectral resolution transmission patterns were derived setting the angle between the incident beam and the sample surface close to the Brewster angle to eliminate artifacts [51].

For aerosol characterization, reflection mode can be more important than transmission. Vibrational spectra of biological macromolecules and species can be measured in reflection mode as well, and reflection data show spectral features correlated with transmission [52]. THz time domain spectroscopy and FT spectroscopy both were used to characterize materials in reflection mode. Fitting spectra from independent measurements of transmission and reflection to modeling results permits to extract the refractive index and absorption coefficient spectra of DNA and protein solid thin film samples [53].

Complex dielectric constant of liquid water was determined in the range from 0.1 to 2 THz [54].

Sub-THz Spectroscopy of Biological Macromolecules and Bioparticles. Sensitivity of THz Characterization and Data Analysis

In all cases, sub-THz transmission/absorption spectra of biological macromolecules measured in thin layers/films or in liquid environment with a good spectral resolution of 0.25 cm^{-1} or better reveal multiple resonances caused by fundamental physical mechanisms of interactions between radiation and biological material [15, 38, 55]. It has also been reported that using FTS, identification of whole-organism bacteria and other microorganisms to the species is possible in the near- and mid-IR [14, 56]. Transmission spectra of cells and spores have been investigated in the sub-THz range using photomixing and FTIR technologies with a relatively good spectral resolution, better than 0.25 cm^{-1} [42, 43, 57–59]. Spectra reveal detectable resonant features in the range less than 25 cm^{-1}. Our early results on spectroscopic characterization of microorganisms also detected multiple resonances in transmiss ion/absorption spectra of species [15, 58, 60–64] in the sub-THz frequency region. To enhance the sensitivity, we used a Fourier spectrometer model FTS-66v with a Si-bolometer operating at 1.7 K. The latest most detail results on sub-THz spectra of biological cells and spores can be found in [64].

B. thuringiensis (BT) , *B. cereus* (BC) and *B. subtillis* (BG) bacterial spores were characterized using submillimeter-wave Fourier Transform (FT) spectroscopy to investigate the capability of this technique to discriminate between different species [15, 60]. In this early work, solid films of controlled thickness and quantitative content have been prepared very reproducibly on membrane filters or Teflon IR cards. Samples in the form of powder spread between two teflon films (Polythetrafluoroethylene—PTFE) or two polyethylene (PE) films have been also prepared and characterized. Statistical analysis of data was performed to determine a standard deviation of measurement results for a BG-spore samples with a low level of signature signal. The error from these measurements was less than 0.5% in the entire spectral range from 10 $^{-1}$ to 25 cm^{-1}, and this error includes the deviation due to difference in samples. Spectra of BG spores from different sources are very close. Spectra are rich with structure and many features are similar for all three species when measured on Bruker spectrometer with the resolution of 0.25 cm^{-1} but there are different features that can be used for discrimination between species [60]. Spectral signature of samples deposited on PE, PTFE, or PC (Polycarbonate) substrates in most cases is practically the same as for powder samples although the intensities of spectral features are small. There are some differences in spectra of one material that depend on the sample preparation procedure. For example, samples from milled spore material are more absorbing and demonstrate the clearest structure. Dilution of spore material with silica does not change significantly specific features, since this is a featureless material in this range with low losses. Spectra of BG spores from different sources are very close. The transmission data are given in Fig. 11.7 for BG spores deposited on Teflon substrates.

Material was deposited as a rather thick layer in a small spot at the center of a card and air-dried. The transmission of a blank substrate used as a background. A direct scaling with spore mass is observed. Increasing the mass of the spores in the BG sample leads to a consistent decrease in the overall transmission while the fine resonant structures remain discernable.

At the same time, spectra of spores in the form of films were not always consistent since films were not uniform but included islands of several hundred micrometers size that is comparable with the wavelength of radiation. Transmission spectra in this case can depend on physical properties of films and demonstrate features similar to diffraction pattern originated from much bigger units than spore size. The artifacts are easily to recognize since they produce the cosine-like spectrum. In most cases however the vibrational frequencies from a film and a powder samples are reasonably reproduced.

The absence of spherical symmetry in the spore structure (they have the form of rods) explains the sensitivity of their transmission spectra to the direction of electric field of radiation. Turning the sample by 90° changes the spectroscopic pattern slightly. Different substrates may orient the spore material in different ways and changing the alignment might affect the observed spectral features as well.

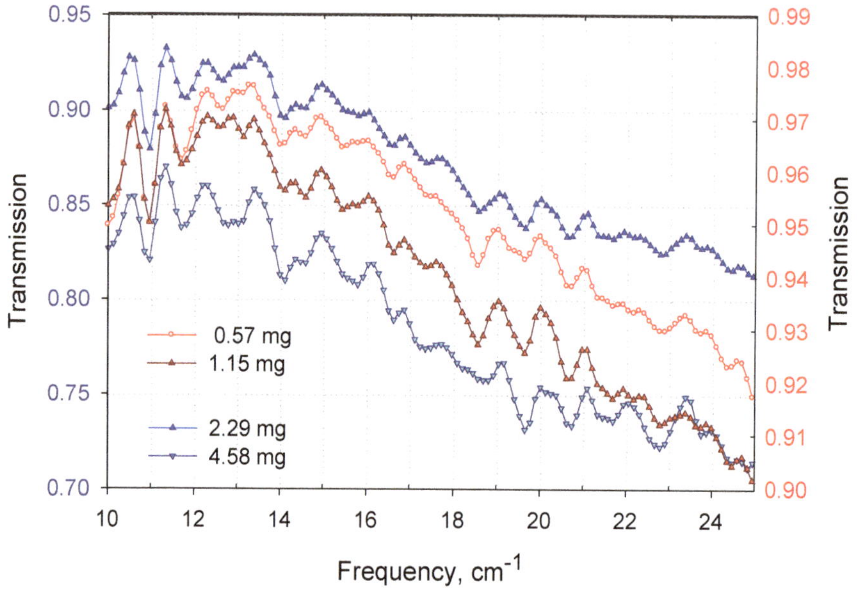

Fig. 11.7 Transmission spectra of *B. subtillis* spore dried gel samples (not orientated), relative to the PTFE substrate, made with different amounts of material. For the plots in *red* and *brown*, use the right axe

Characterization of BG spores was extended into sub-range less than 10 cm^{-1}using photomixing [42, 43] and FTIR technologies [58, 59]. In addition to this being the region where remote sensing is most viable, the lower frequency range of 60–300 GHz exhibits absorption bands with minimal overlap where the form of spectral lines can be characterized along with the location of the absorption-peak frequencies. This scientific information has to be invaluable for the design of future sensors to achieve effective levels of detection and identification [65]. BG powdery mix of spores with vegetative cell wall material diluted in PE and concentrated BG samples were measured between 200 and 1200 GHz using photomixing technology. Normalized transmission from concentrated BG revealed two wide signatures between 252 and 290 GHz and other centered around 425 GHz [43]. The same signatures of the dilute-BG sample detected at 250 and 415 GHz. are evident, but now centered around 260 and 420 GHz. In the same work, the transmission spectrum through the BG aerosol sample was obtained by FTS [43]. As expected, the broadband attenuation present in the powder samples is missing in the aerosol. The 250 and 415 GHz signatures of the dilute-BG sample are evident, but now centered around 260 and 420 GHz. In addition there are other, even stronger, signatures such as the one centered around 500 GHz. Unfortunately, these aerosol signatures have proven difficult to reproduce, in part because of difficulties in suspending the BG material and its sensitivity to electrostatic conditions. The 260 and 420 GHz signatures display a linewidth of about 14 GHz, about twice the instrumental limit (8 GHz).

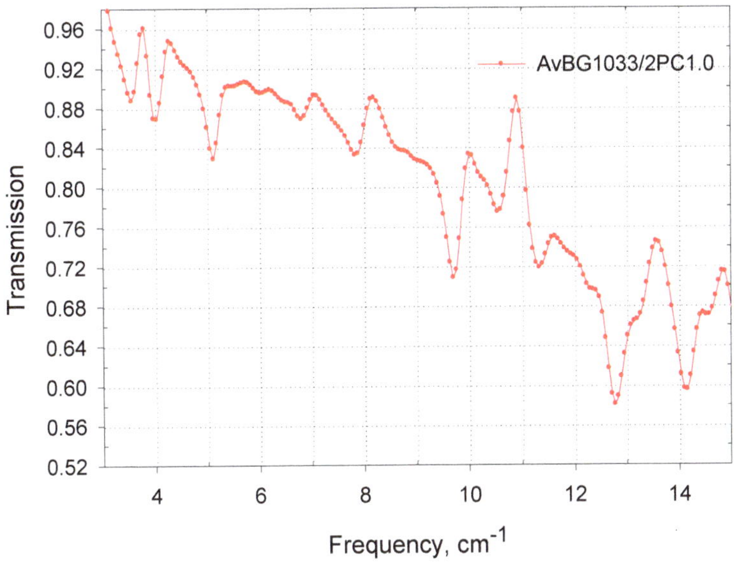

Fig. 11.8 Transmission spectrum from a BG spore air dried film sample on PC membrane with 0.4 mkm pores

In [58] transmission spectra were measured in the spectral range 3–25 cm^{-1} using modified commercial FTIR spectrometer. Bruker IFS-66v, with a Si-bolometer (1.7°K) was equipped with an additional source developed at the National Radio Astronomy Observatory, which consists of a microwave solid-state noise source followed by a microwave power amplifier chain, and a frequency multiplier. This source provides two orders of magnitude higher power over a multi-octave band (2–15 wave numbers, i.e., 60–450 GHz) range in comparison with a conventional Mercury arc lamp. Since the noise source uses only solid-state and waveguide components, it has excellent stability, repeatability and reliability. It is very compact, operable using the standard bench-top DC power supplies and can be switched on and off electronically. The source was attached through the spectrometer window used for Raman configuration. Figure 11.8 shows our results of transmission spectrum from a BG spore air-dried film sample on PC membrane with 0.4 μm pores.

The modes at 9.7 cm^{-1} (291 GHz) and at 14.1 cm^{-1} (423 GHz) are close to those detected by photomixing technology. There is additional strong mode at 12.7 cm^{-1} that is not detected in [42]. The mode at 14.1 cm^{-1} is also seen in spectra on Fig. 11.7 [60] from conventional FTIR characterization, although the sample preparation procedures, the substrates and the amount of material are very different.

Terahertz time-domain spectroscopy was used to measure absorption coefficient and refractive index spectra of BG spores thick film (0.5 mm) on PE substrate in the frequency range between 0.2 and 2 THz (6.7–67 cm^{-1}) [66]. Five broad bands were observed but four of these bands were assigned to water. The authors believe that the mode at 1.54 THz belongs to biological molecules present in the spores. This finding however was not confirmed by the independent measurements.

Reflection mode from BG spores and biological macromolecules was characterized in parallel with transmission to extract absorption coefficient and refractive index spectra using interference technique analysis in the range 10–25 cm^{-1} [52]. Reflection spectra from biological macromolecules and species reveal resonance features similar to those demonstrated earlier for transmission. Absorption coefficient grows with frequency from 20 $^{-1}$ to 70 cm^{-1}.

Progress in sample preparation techniques in the last several years significantly enhanced experimental spectroscopic features, thus resulting in more reliable signatures from natural macromolecules and bacterial cells/spores [64]. We have measured transmission spectra of lyophilized and in vitro cultured bacterial cells from two bacterial species of *E. coli* and *B. subtilis*, spores of *B. subtilis,* and DNA from *E. coli* cells using commercial FTIR infrared spectrometer, Bruker IFS66v equipped with a mercury lamp and with a Si-bolometer operating at 1.7 K for signal detection in the range of 10–25 cm^{-1} with a moderate spectral resolution of 0.25 cm^{-1}. A very simple sample preparation technique has been applied. *BG* spores and lyophilized *BG* cells were mixed with distilled water in 1:5 ratio and set aside for 24 h. Half of the mixture with cells was heated at 100 °C. Liquid and air-dried samples for the spectra measurements were prepared from the three suspensions (spores, cells, and thermally treated cells) by pipetting 5–30 μl aliquots onto a surface of polytetrafluoroethylene (PTFE) disposable IR cards (International Crystal Laboratory) or other substrates. The PTFE cards are characterized by a high transmission level (~98 %) and by an absence of strong peaks in the frequency range of 10–25 cm^{-1}. High temperature treatment (~ 100 °C) was used as a method of destruction of living cells and it was found to have a detectable effect on transmission/ absorption spectra by changing certain resonance frequencies. Effects that desiccation has on the transmission spectra of spores were studied as well. It has also been demonstrated that different substrates can modify the orientation of bacterial cells in the sample. The orientation effects are more pronounced in DNA spectra since molecules can be significant in length (compared to width) and therefore can be preferentially oriented during sample preparation. In addition, absorption spectra have been studied for quantitative characterization of samples with different amount of biomaterial. The most significant experimental result from the current study is demonstration of many similarities in spectroscopic features of the *E. coli* DNA and the entire cell—the experimental fact that confirms the contribution from molecular components into spectroscopic signatures of bioparticles.

The comparison of experimental absorption spectra of relatively small macromolecules with molecular dynamics (MD) simulation was used to confirm that observed spectroscopic features are caused by fundamental physical mechanism of interaction between the THz radiation and vibrational modes within biological molecules. Particularly, we demonstrated that the spectroscopic signatures of microorganisms originate from the combination of low frequency vibrational modes or group of modes at close frequencies (vibrational bands) within molecular components of bacterial cells/spores, with the significant contribution from the DNA.

The characterization usually started from material that still contains water. The spectroscopic features from wet samples are very strong and reproducible on the

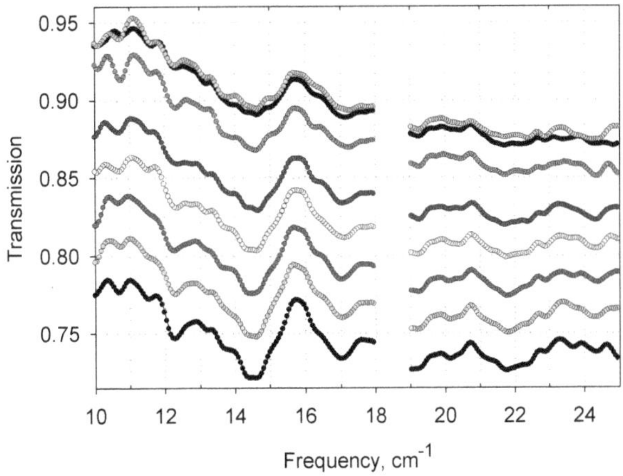

Fig. 11.9 Transmission spectra of BG spores, with varying water content

frequency scale as it is demonstrated in Fig. 11.9. The positions of the transmission minima (absorption peaks) on the frequency scale are fairly reproducible. The peak intensities vs. water content have maximum at approximately 20–30% and reduce significantly in dry samples (on the top).

We used very thin substrates (10–12 μm) to eliminate or at least reduce interference effects. Nevertheless, a broad interference pattern is observed in our most thick dried samples (~80 μm). It is overlapping with more narrow and therefore well distinguishable vibrational modes as it is demonstrated in Fig. 11.10. The first interference maximum occurs at zero frequency (not shown) and the second broad maximum is observed around 27 cm^{-1}. This graph also shows the reproducibility of vibrational features in spectra measured with two different beam-splitters. The standard deviation from a set of similar samples measured at the same conditions is usually less than 0.3–0.4% in the entire sub-range specified for a particular beam-splitter.

The absence of spherical symmetry in the cell and spore structure (they have the form of rods) explains the sensitivity of their transmission spectra to the direction of electric field of radiation. Turning the sample changes slightly the spectroscopic pattern. Different substrates may orient the spore material in different ways and changing the alignment might affect the observed spectral features as well. The samples in [64] were preferably oriented in specific direction, and orientation effects do present in measured spectra of spores and cells (see Fig. 11.11 for DNA). Transmission minima change intensity and positions on the frequency scale while the sample is measured at different angles relative to the polarization of the light. The effects are stronger in spectra of cells because they are much longer than spores.

To characterize bio-material rather than individual samples, transmission data were recalculated into spectra of specific absorption for 1 mg of dry material. Beer's law was not used because of multiple reflections at surfaces of thin samples and because sample thickness was not uniform in most cases. However measured transmission spectra can be used to retrieve the absorption coefficient spectra and refractive index of biological material by applying an interference spectroscopy technique to take into account the Fresnel losses at the interfaces [53, 67]

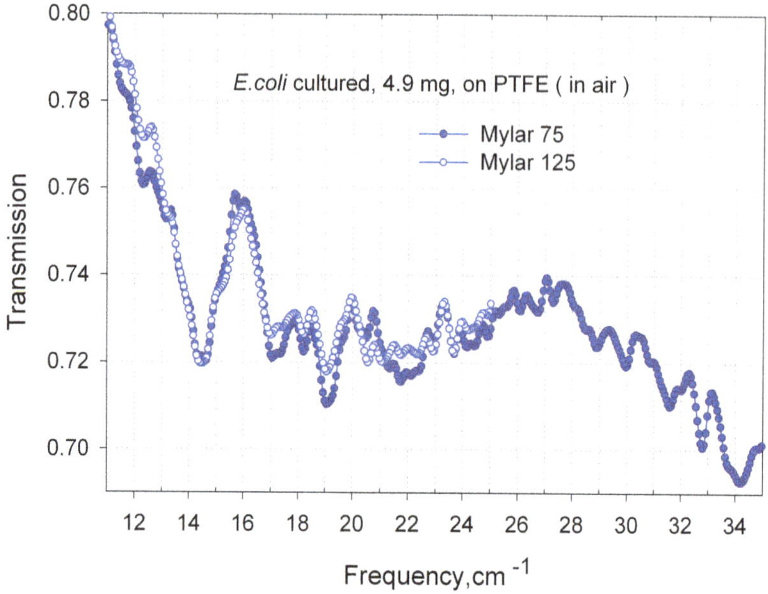

Fig. 11.10 Interference effects in transmission spectra from cultured *Ec* cells measured with two beam-splitters in the extended frequency range

Fig. 11.11 Orientation effects in transmission spectra of *E. coli* DNA on PTFE card

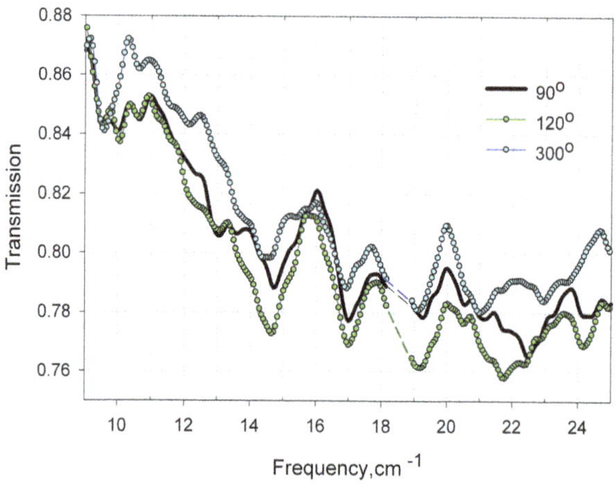

Figure 11.12 compares absorption spectra of BG and *E. coli* cells. The shape of the curves, peak positions and their intensities are rather close. Although some specificity could be found, particularly in the 9–13 cm^{-1} and 21–24 cm $^{-1}$ ranges, better spectral resolution is required for more certain discrimination result that will be discussed in the following section (Highly resolved sub-THz characterization of biological molecules and biological cells/spores).

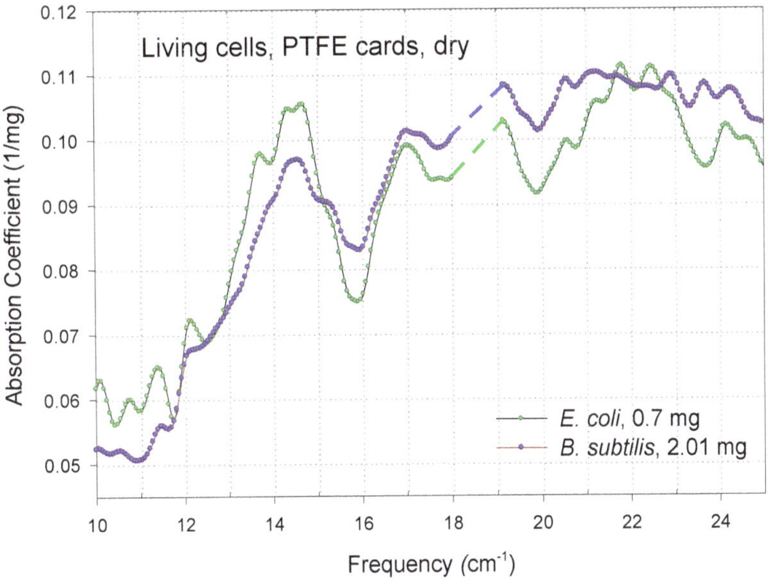

Fig. 11.12 Sub-THz absorption spectra of *B. subtilis* and *E. coli* living cells on PTFE substrate, dry [64]

THz Characterization of Bioparticles in Aerosols

We have characterized transmission spectra of BG cells in the form of aerosol using a gas cell equipped with a micro-fan inside a Bruker FTIR spectrometer. Transmission of a clean gas cell was recorded as a background. Very small portion of the total loaded amount of BG cells was dispersed by the gas flow, thus creating aerosol with low concentration, less than ~ 5.10^{-3} mg/ml. Transmission of bio-material was very high ~99%, with the intensity of spectral features less than 1%. Figure 11.13 compares absorption spectra of BG cells and spores in bulk and BG cells in aerosol. Although particles in aerosol are not oriented, here is certainly some correlation in absorption peaks for aerosol and bulk samples (partially oriented, thickness 20–80 μm). At the same time, spectral lines for aerosol sample are sharper compared to the features from bulk material. Significant broadening and damping of spectral features in bulk samples is observed which increases in thicker bulk samples having more material. This effect can probably explain the sensitivity of sub-THz spectroscopic characterization to specificities of sample preparation techniques.

Overall, the spectral response of both, relatively small and macro molecules, as well as complete cellular biological samples, revealed spectral features caused by interactions between radiation and the biological material. Multiple resonances due to low frequency vibrational modes within biological macromolecules, components of bacterial organisms, have been demonstrated experimentally in the sub-THz frequency range. The results indicate that detection of biological agents in aerosol can be potentially a very promising technique [12].

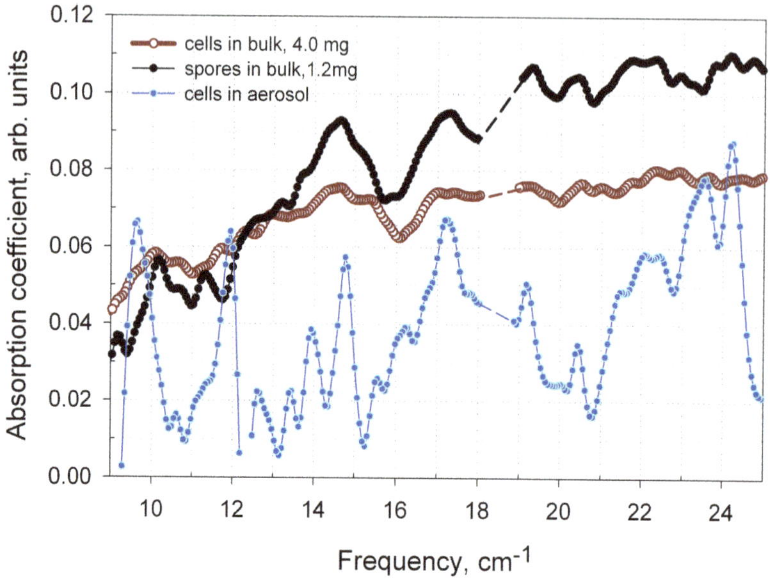

Fig. 11.13 Comparison between absorption spectra of BG cells and spores in bulk [63] and BG cells in aerosol (our group results presented in [12]). Spectral resolution 0.25 cm^{-1}

Highly Resolved Sub-THz Characterization of Biological Molecules and Biological Cells/Spores

Improving the sensitivity, the spatial and especially spectral resolutions for biological materials characterization remained important issues in THz resonance spectroscopy for biological materials. Vibratess LLc has recently developed a new spectroscopic sensor with imaging capability that demonstrated significant improvement in all three aspects [23, 24]. The new technique that is based on a local EM field enhancement of THz radiation in sub-wavelength regions and improved coupling of radiation to materials is described in [24, 68]. It utilizes THz radiation distribution and propagation through a sub-wavelength-slit conductive structure [69–71]. This transmission, frequency domain sub-THz spectroscopic bio-sensor (Fig. 11.14), integrated with a multi-channel chip for a sample material is operating at room temperature with high sensitivity, high spectral resolution, and a spatial resolution below the diffraction limit through the use of the discontinuity edge effect [68]. The system utilizes an electronically tunable sub-THz source of radiation (315–480 GHz) developed by the VDI Ltd with the average output power of ~0.2 mW and the current spectral resolution limit of ~40 MHz. A Schottky diode detector (VDI) for monitoring the THz signal is integrated with a replaceable beam lead microprobe. The system also provides a motorized three dimensional scanning capability of a detector probe relative to a sample holder with the precision of one micron. It already demonstrated better than 100 μm spatial resolution and a high sensitivity with signal/noise ratio greater than 10^3 at room temperature operation without requirements for evacuating or purging with nitrogen.

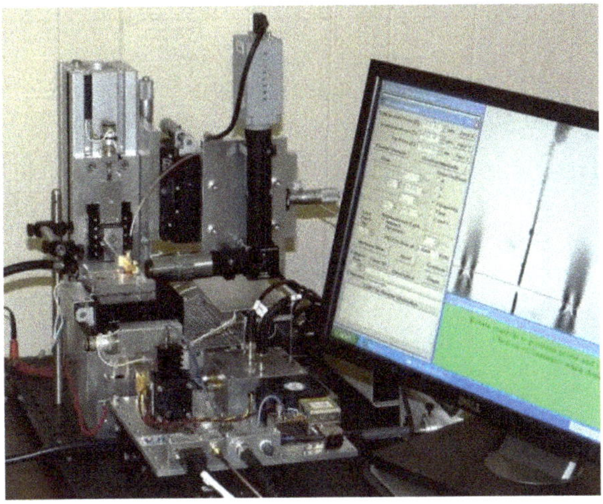

Fig. 11.14 Frequency domain sub-THz spectroscopic bio-sensor (Vibratess, LLc)

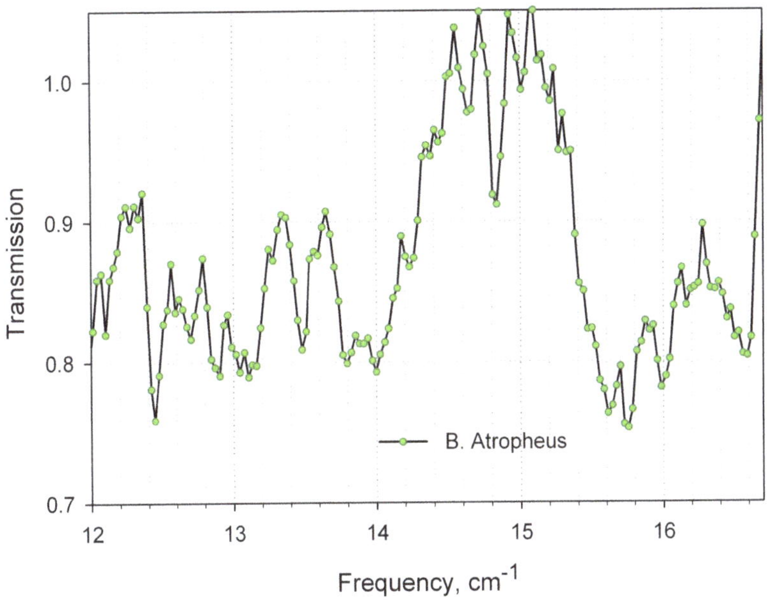

Fig. 11.15 Transmission spectra of *B. atropheus* spores (dried suspension in water) measured with Vibratess' frequency domain spectroscopic sensor

The test results demonstrated enlargement of the sensitivity by almost an order of magnitude compared to a Bruker FTIR spectrometer operating with the liquid helium cooled bolometer (1.7 K), with 20 times less amount of material under the test, and a required sample area less than 0.2×0.2 mm^2, which is currently limited by the size of a microdetector opening.

Figure 11.15 shows first results on transmission spectrum of *B. atropheus spores* as measured on Vibratess spectroscopic instrument. Spectral line widths are less

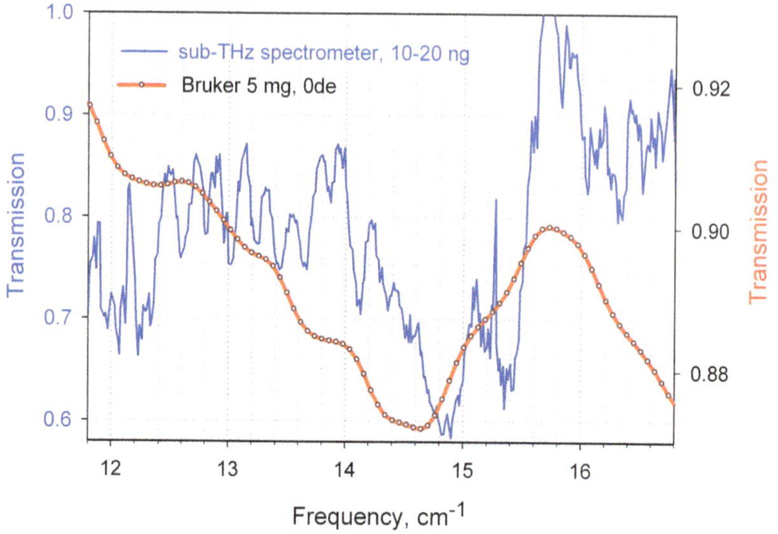

Fig. 11.16 Comparison transmission spectra of *E. coli* cells measured with two instruments: Bruker FTIR and THz Vibratess spectroscopic biosensor

than 0.2 cm^{-1}. Transmission spectra from *E. coli* cells obtained with a classical FTIR system and a new Vibratess bio-sensor are compared in Fig. 11.16.

More spectral features are resolved in THz spectrometer with much higher features intensities compared to the Bruker FT spectrometer. Almost each spectral feature with transmission minimum or shoulder observed in Bruker FT spectrometer is resolved by a new THz spectrometer in one or several more narrow lines and strongly enhanced in intensity (more than 10 times). Small shift on frequency can be due to a different orientation of cells relatively to the electric field of radiation in two instruments. These latest results promise significantly improve the discriminative capability of sub-THz vibrational spectroscopy.

Conclusions

The spectral response of both, relatively small and macro molecules, as well as complete cellular biological samples, revealed spectral features caused by interactions between THz radiation and the biological material. Multiple resonances due to low frequency vibrational modes within biological macromolecules, components of bacterial organisms, have been demonstrated experimentally in the sub-THz frequency range. Improving spectral resolution lead to higher discriminative capability of sub-THz vibrational spectroscopy. These results together with higher sensitivity in low concentration samples indicate that detection of biological agents in aerosol can be potentially a very promising technique.

Acknowledgments This work is supported by contracts from the ARO #W911NF-08-C-0049 & #W911NF-10-C-0076 and by the Defense Threat Reduction Agency, grant #HDTRA1-08-1-0038.

References

1. Van Zandt LL, Saxena SK (1992) Identifying and Interpreting Spectral Feaures of DNA in the Microwave-Submillimeter Frequency Range. In: Sarma RH, Sarma MH (eds) Structure and Function, vol 1: Nuceleic Acids. Adenine Press, Schenectady, N.Y., pp 237–243, and references therein
2. Duong TH, Zakrzewska K (1997) Calculation and analysis of low frequency normal modes for DNA. J Comput Chem 18 (6):796–811. doi:10.1002/(SICI)1096-987X(19970430)18:6<796::AID-JCC5>3.0.CO;2-N
3. Smye S, Chamberlain J, Fitzgerald A, Berry E (2001). The interaction between terahertz radiation and biological tissue. J. Phys. Med. Biol. 46: R101–112
4. Mie scattering. http://www.philiplaven.com/p2.html. Accessed 20 Januari 2014. Reyleigh scattering http://hyperphysics.phy-astr.gsu.edu/hbase/atmos/blusky.html. Accessed 20 Januari 2014
5. Globus T, Bykhovskaia M, Woolard D, Gelmont B (2003) Sub-millimetre wave absorption spectra of artificial RNA molecules. J Phys D: Appl Phys 36 (11):1314–1322. doi:10.1088/0022-3727/36/11/312
6. Globus T, Khromova T, Woolard D, Gelmont B (2004) Terahertz Fourier transform characterization of biological materials in solid and liquid phases. Proc SPIE 5268:10–18. doi:10.1117/12.519172
7. Korter TM, Plusquellic DF (2004) Continuous-wave terahertz spectroscopy of biotin: vibrational anharmonicity in the far-infrared. Chem Phys Lett 385 (1–2):45–51. doi:10.1016/j.cplett.2003.12.060
8. De Lucia FC THz spectroscopy – techniques and applications. In: IEEE Microwave Theory and Techniques Society International Microwave Symposium Digest, vol 3, Seattle, WA, USA, 2–7 June 2002. pp 1579–1582. doi:10.1109/MWSYM.2002.1012158
9. Payne JM (1989) Millimeter and submillimeter wavelength radioastronomy. Proc IEEE 77 (7):993–1017. doi:10.1109/5.30751
10. Phillips TG, Keene J (1992) Submillimeter astronomy [heterodyne spectroscopy]. Proc IEEE 80 (11):1662–1678. doi:10.1109/5.175248
11. Waters JW, Froidevaux L, Read WG, Manney GL, Elson LS, Flower DA, Jarnot RF, Harwood RS (1993) Stratospheric ClO and ozone from the Microwave Limb Sounder on the Upper Atmosphere Research Satellite. Nature 362 (6421):597–602. doi:10.1038/362597a0
12. Woolard DL, Brown ER, Pepper M, Kemp M (2005) Terahertz Frequency Sensing and Imaging: A Time of Reckoning Future Applications? Proc IEEE 93 (10):1722–1743. doi:10.1109/JPROC.2005.853539
13. Heilweil EJ, Plusquellic DF (2007) Terahertz Spectroscopy of Biomolecules. In: Dexheimer SL (ed) Terahertz Spectrocopy: Principles and Applications. Optical Science and Engineering. CRC Press, Boca Raton, pp 269–297
14. Johnson TJ, Valentine NB, Sharpe SW (2005) Mid-infrared versus far-infrared (THz) relative intensities of room-temperature *Bacillus* spores. Chem Phys Lett 403 (1–3):152–157. doi:10.1016/j.cplett.2004.12.095
15. Globus T, Woolard D, Bykhovskaia M, Gelmont B, Werbos L, Samuels A (2003) THz Spectroscopic Sensing of DNA and Related Biological Materials. Int J High Speed Electron Syst 13 (04):903–936. doi:10.1142/S0129156403002083
16. Bykhovskaia M, Gelmont B, Globus T, Woolard DL, Samuels AC, Duong TH, Zakrzewska K (2001) Prediction of DNA far-IR absorption spectra based on normal mode analysis. Theor Chem Acc 106 (1–2):22–27. doi:10.1007/s002140100259

17. Globus T, Bykhovskaia M, Gelmont B, Woolard DL (2002) Far-infrared phonon modes of selected RNA molecules. Proc SPIE 4574:119–128. doi:10.1117/12.455149
18. Li X, Globus T, Gelmont B, Salay LC, Bykhovski A (2008) Terahertz Absorption of DNA Decamer Duplex. J Phys Chem A 112 (47):12090–12096. doi:10.1021/jp806630w
19. Bykhovski A, Li X, Globus T, Khromova T, Gelmont B, Woolard D, Samuels AC, Jensen JO (2005) THz absorption signature detection of genetic material of E. coli and B. subtilis. Proc SPIE 5995:59950N.1–10. doi:10.1117/12.629959
20. Bykhovski A, Globus T, Khromova T, Gelmont B, Woolard D, Bykhovskaia M (2006) An analysis of the THz frequency signatures in the cellular components of biological agents. Proc SPIE 6212:62120H.1–10. doi:10.1117/12.665272 and Bykhovski A, Globus T, Khromova T, Gelmont B, Woolard D (2007) Analysis of the THz Frequency Signatures in the Cellular Components of Biological Agents. Int J High Speed Electron Syst 17 (02):225–237. doi:10.1142/S012915640700445X
21. Bykhovski A, Globus T, Khromova T, Gelmont B, Woolard D (2008) Resonant Terahertz Spectroscopy of Bacterial Thioredoxin in Water: Simulation and Experiment. In: Woolard D, Jensen J (eds) Spectral Sensing Research For Water Monitoring Applications And Frontier Science And Technology For Chemical, Biological And Radiological Defense. Selected Topics in Electronics and Systems, vol 48. World Scientific, Singapore, pp 367–375. doi:10.1142/9789812833242_0033
22. Alijabbari N, Chen Y, Sizov I, Globus T, Gelmont B (2012) Molecular dynamics modeling of the sub-THz vibrational absorption of thioredoxin from E. coli. J Mol Model 18 (5):2209–2218. doi:10.1007/s00894-011-1238-6
23. Globus T (2010) Low-Terahertz Resonance Spectroscopy for Fingerprinting of Biological and Organic Materials. Paper presented at the Chemical and Biological Defense Science and Technology Conference, Orlando, 15–19 November 2010
24. Globus T, Moyer AM, Gelmont B, Khromova T, Lvovska MI, Sizov I, Ferrance J (2013) Highly Resolved Sub-Terahertz Vibrational Spectroscopy of Biological Macromolecules and Cells. IEEE Sens J 13 (1):72–79. doi:10.1109/JSEN.2012.2224333
25. Globus T, Moyer A, Gelmont B, Sizov I, Khromova T (2013) Dissipation Time in Molecular Dynamics and Discriminative Capability of Sub-THz Spectroscopic Characterization of Biological Molecules and Cells. Paper presented at the Chemical and Biological Defense Science and Technology Conference, Las Vegas, 14–18 November 2011
26. Crowe TW, Globus T, Woolard DL, Hesler JL (2004) Terahertz sources and detectors and their application to biological sensing. Philos Trans R Soc London, Ser A 362 (1815):365–377. doi:10.1098/rsta.2003.1327
27. Beetz CP, Ascarelli G (1982) Far-infrared absorption of nucleotides and poly(I)·poly(C) RNA. Biopolym 21 (8):1569–1586. doi:10.1002/bip.360210808
28. Giordano R, Mallamace F, Micali N, Wanderlingh F, Baldini G, Doglia S (1983) Light scattering and structure in a deoxyribonucleic acid solution. Phys Rev A 28 (6):3581–3588. doi:10.1103/PhysRevA.28.358
29. Lindsay SM, Powell J (1983) Light scattering studies of the lattice vibrations of DNA. In: Clementi E, Sarma RH (eds) Structure and Dynamics: Nucleic Acids and Proteins. Adenine Press, New York, pp 241–259
30. Edwards GS, Davis CC, Saffer JD, Swicord ML (1984) Resonant Microwave Absorption of Selected DNA Molecules. Phys Rev Lett 53 (13):1284–1287. doi:10.1103/PhysRevLett.53.1284
31. Wittlin A, Genzel L, Kremer F, Häseler S, Poglitsch A, Rupprecht A (1986) Far-infrared spectroscopy on oriented films of dry and hydrated DNA. Phys Rev A 34 (1):493–500. doi:10.1103/PhysRevA.34.493
32. Powell JW, Edwards GS, Genzel L, Kremer F, Wittlin A, Kubasek W, Peticolas W (1987) Investigation of far-infrared vibrational modes in polynucleotides. Phys Rev A 35 (9):3929–3939. doi:10.1103/PhysRevA.35.3929
33. Weidlich T, Powell JW, Genzel L, Rupprecht A (1990) Counterion effects on the far-IR vibrational spectra of poly(rI) poly(rC). Biopolym 30 (3–4):477–480. doi:10.1002/bip.360300324

34. Powell JW, Peticolas WL, Genzel L (1991) Observation of the far-infrared spectrum of five oligonucleotides. J Mol Struct 247 (0):107–118. doi:10.1016/0022-2860(91)87067-R.
35. Gabriel C, Grant EH, Tata R, Brown PR, Gestblom B, Noreland E (1987) Microwave absorption in aqueous solutions of DNA. Nature 328 (6126):145–146. doi:10.1038/328145a0
36. Ferguson B, Zhang X-C (2002) Materials for terahertz science and technology. Nat Mater 1 (1):26–33. doi:10.1038/nmat708
37. Terahertz time-domain spectroscopy. http://en.wikipedia.org/wiki/Terahertz_time-domain_spectroscopy. Accessed 29 January 2014
38. Globus T, Woolard D, Crowe T, W., Khromova T, Gelmont B, Hesler J (2006) Terahertz Fourier transform characterization of biological materials in a liquid phase. J Phys D: Appl Phys 39 (15):3405–3413. doi:10.1088/0022-3727/39/15/028
39. Markelz AG, Roitberg A, Heilweil EJ (2000) Pulsed terahertz spectroscopy of DNA, bovine serum albumin and collagen between 0.1 and 2.0 THz. Chem Phys Lett 320 (1–2):42–48. doi:10.1016/S0009-2614(00)00227-X
40. Walther M, Fischer B, Schall M, H, Jepsen PU (2000) Far-infrared vibrational spectra of all-trans, 9-cis and 13-cis retinal measured by THz time-domain spectroscopy. Chem Phys Lett 332 (3–4):389–395. doi:10.1016/S0009-2614(00)01271-9
41. Han PY, Tani M, Usami M, Kono S, Kersting R, Zhang XC (2001) A direct comparison between terahertz time-domain spectroscopy and far-infrared Fourier transform spectroscopy. J Appl Phys 89 (4):2357–2359. doi:10.1063/1.1343522
42. Brown E, Siegel P, Samuels A, Woolard L (2003) High-Resolution Transmission Measurements of *Bacillus Subtilis* Between 300 and 500 GHz. Paper presented at the International Symposium on Spectral Sensing Research, Santa Barbara, CA, USA, 2–6 June 2003
43. Brown ER, Bjarnason JE, Chan TLJ, Lee AWM, Celis MA (2004) Optical attenuation signatures of *Bacillus subtillis* in the THz region. Appl Phys Lett 84 (18):3438–3440. doi:10.1063/1.1711167
44. Brown ER, Bjarnason J, Chan TLJ, Driscoll DC, Hanson M, Gossard AC (2004) Room temperature, THz photomixing sweep oscillator and its application to spectroscopic transmission through organic materials. Rev Sci Instrum 75 (12):5333–5342. doi:10.1063/1.1808912
45. Verghese S, McIntosh KA, Calawa S, DiNatale WF, Duerr EK, Molvar KA (1998) Generation and detection of coherent terahertz waves using two photomixers. Appl Phys Lett 73 (26):3824–3826. doi:10.1063/1.122906
46. Bjarnason JE, Brown ER (2005) Sensitivity measurement and analysis of an ErAs:GaAs coherent photomixing transceiver. Appl Phys Lett 87 (13):134105.1–3. doi:10.1063/1.2058205
47. Demers JR, Logan Jr RT, Bergeron NJ, Brown ER (2008) A coherent frequency-domain THz spectrometer with a signal-to-noise ratio of 60 dB at 1 THz. Proc SPIE 6949:694909.1–9
48. Brown ER, Bjarnason JE, Fedor AM, Korter TM (2007) On the strong and narrow absorption signature in lactose at 0.53 THz. Appl Phys Lett 90 (6):061908.1–3. doi:10.1063/1.2437107
49. Majewski A, Abreu R, Wraback M (2007) A high resolution terahertz spectrometer for chemical detection. Proc SPIE 6549:65490B.1–8. doi:10.1117/12.719485
50. Majevski A, Bansleben D, Wraback M (2008) A High Resolution Terahertz Spectrometer for Chemical Detection. Paper presented at the International Symposium on Spectral Sensing Research, Hoboken, NJ, USA, 23–27 June 2008
51. Egert S, Peri D, Abramovich A (2010) Spectroscopic Study of Containers and Their Content Using a High-Resolution THz System. IEEE Sens J 10 (3):379–383. doi:10.1109/JSEN.2009.2037291
52. Globus T, Parthasarathy R, Khromova T, Woolard DL, Swami N, Gatesman AJ, Waldman J (2004) Optical characteristics of biological molecules in the terahertz gap. Proc SPIE 5584:1–10. doi:10.1117/12.580838
53. Parthasarathy R, Globus T, Khromova T, Swami N, Woolard D (2005) Dielectric properties of biological molecules in the Terahertz gap. Appl Phys Lett 87 (11):113901.1–3. doi:10.1063/1.2046730
54. Rønne C, Thrane L, Åstrand P-O, Wallqvist A, Mikkelsen KV, Keiding SR (1997) Investigation of the temperature dependence of dielectric relaxation in liquid water by THz reflec-

tion spectroscopy and molecular dynamics simulation. J Chem Phys 107 (14):5319–5331. doi:10.1063/1.474242
55. Globus T, Khromova T, Gelmont B, Woolard D, Tamm LK (2006) Terahertz characterization of dilute solutions of DNA. Proc SPIE 6093:609308.1–12. doi:10.1117/12.646529
56. Rodriguez-Saona LE, Khambaty FM, Fry FS, Calvey EM (2001) Rapid Detection and Identification of Bacterial Strains by Fourier Transform Near-Infrared Spectroscopy. J Agric Food Chem 49 (2):574–579. doi:10.1021/jf000776j
57. Brown ER, Khromova TB, Globus T, Woolard DL, Jensen JO, Majewski A (2006) Terahertz-Regime Attenuation Signatures in *Bacillus subtilis* and a Model Based on Surface Polariton Effects. IEEE Sens J 6 (5):1076–1083. doi:10.1109/JSEN.2006.881354
58. Globus T, Woolard DL, Khromova T, Partasarathy R, Majewski A, Abreu R, Hesler JL, Pan S-K, Ediss G (2004) Terahertz signatures of biological-warfare-agent simulants. Proc SPIE 5411:25–32. doi:10.1117/12.549128
59. Majewski AJ, Miller P, Abreu R, Grotts J, Globus T, Brown E (2005) Terahertz signature characterization of bio-simulants. Proc SPIE 5790:74–84. doi:10.1117/12.603658
60. Globus T, Woolard D, Samuels A, Khromova T, Jensen J (2003) Sub-millimeter Wave Fourier Transform Characterisation of Bacterial Spores. Paper presented at the International Symposium on Spectral Sensing Research, Santa Barbara, CA, USA, 2–6 June 2003
61. Globus T, Theodorescu D, Frierson H, Khromova T, Woolard D (2005) Terahertz spectroscopic characterization of cancer cells. Proc SPIE 5692:233–240. doi:10.1117/12.594391
62. Globus T, Khromova T, Bykhovski A, Gelmont B, Woolard D (2007) Terahertz Sensing of Bio-Water Contaminents Using Vibrational Spectroscopy. Int J High Speed Electron Syst X (X):1–14
63. Dorofeeva T (2011) Characterization of Biosimulants Using Sub-THz Vibrational Spectroscopy. M. Sc., University of Virginia, Charlottesville
64. Globus T, Dorofeeva T, Sizov I, Gelmont B, Lvovska M, Khromova T, Chertihin O, Koryakina Y (2012) Sub-THz Vibrational Spectroscopy of Bacterial Cells and Molecular Components. Am J Biomed Eng 2 (4):143–154. doi:10.5923/j.ajbe.20120204.01
65. Woolard DL, Brown ER, Samuels AC, Jensen JO, Globus T, Gelmont B, Wolski M Terahertz-frequency remote-sensing of biological warfare agents. In: In: IEEE Microwave Theory and Techniques Society International Microwave Symposium Digest, vol 2, Philadelphia, PA, USA, 8–13 June 2003. pp 763–766. doi:10.1109/MWSYM.2003.1212483
66. Yu B, Alimova A, Katz A, Alfano RR (2005) THz absorption spectrum of *Bacillus subtilis* spores. Proc SPIE 5727:20–23. doi:10.1117/12.590951
67. Globus T, Ganguly G, Roca i Cabarrocas P (2000) Optical characterization of hydrogenated silicon thin films using interference technique. J Appl Phys 88 (4):1907–1915. doi:10.1063/1.1305855
68. Gelmont B, Globus T, Bykhovski A, Lichtenberger A, Swami N, Parthasarathy R, Weikle R (2012) Method of Local Electro-Magnetic Field Enhancement of Terahertz (THz) Radiation in Sub Wavelength Regions and Improved Coupling of Radiation to Materials through the Use of the Discontinuity Edge Effect, US Patent 8,309, 930, 26 November 2012
69. Parthasarathy R, Bykhovski A, Gelmont B, Globus T, Swami N, Woolard D (2007) Enhanced Coupling of Subterahertz Radiation with Semiconductor Periodic Slot Arrays. Phys Rev Lett 98 (15):153906.1–4. doi:10.1103/PhysRevLett.98.153906
70. Gelmont B, Parthasarathy R, Globus T, Bykhovski A, Swami N (2008) Terahertz (THz) Electromagnetic Field Enhancement in Periodic Subwavelength Structures. IEEE Sens J 8 (6):791–796. doi:10.1109/JSEN.2008.923222
71. Gelmont B, Globus T (2011) Edge Effect in Perfectly Conducting Periodic Subwavelength Structures. IEEE Trans Nanotechnol 10 (1):83–87. doi:10.1109/TNANO.2010.2064785

Part III
Standoff Sensor Systems for Bioaerosol Detection

Chapter 12
Introduction to Stand-Off Detection of Biological Warfare Agents

Per Jonsson and Göran Olofsson

Introduction to Stand-Off Detection of Biological Warfare Agents

In order to have the time to establish an effective defense against a biological attack, it is important to have an early warning. Furthermore, in mapping a cloud originating from biological attack, it is crucial to know what areas are affected. There is therefore a need for the development of methods to detect biological aerosols and discriminate between them and background aerosols, remotely or at a standoff distance. Preferably such a system should discriminate biological warfare agents from naturally occurring aerosols. This is, however, a very challenging task as the difference between non-harmful bacteria, other biological matter, and virulent microbes often is very subtle.

Suggested technologies to address these challenges include several standoff technologies covering a broad region of the electromagnetic spectrum. These technologies include spectrally resolved ultraviolet laser-induced fluorescence (UV-LIF) at several different excitation wavelengths, infrared depolarization, and long-wave infrared (LWIR) differential scattering (DISC)[1].

At present, there are ongoing development programs in Canada (IR elastic scattering for cloud mapping and UV-LIF for classification), France (similar approach), and USA (LWIR DISC). There are research programs in several other countries.

[1] Laser Based Stand-Off Detection of Biological Agents. Final Report of Task Group SET-098/RTG-55 (2010). RTO-TR-SET-098 AC/323(SET-098)TP/265. NATO Science and Technology Organization, Neuilly-sur-Seine Cedex, France. http://www.cso.nato.int/Pubs/rdp.asp?RDP=RTO-TR-SET-098. Accessed 27 Jan 2014.

P. Jonsson (✉)
Division of Sensor and EW Systems, FOI—Swedish Defence Research Agency,
Olaus Magnus väg 42, 581 11 Linköping, Sweden
e-mail: per.jonsson@foi.se

G. Olofsson
Division of CBRN Defence and Security, FOI—Swedish Defence Research Agency,
Cementvägen 20, 901 82 Umeå, Sweden
e-mail: goran.olofsson@foi.se

P. Jonsson et al. (eds.), *Bioaerosol Detection Technologies*, Integrated Analytical Systems,
DOI 10.1007/978-1-4419-5582-1_12, © Springer-Verlag New York 2014

Chapter 13
Spectrally Resolved Laser-Induced Fluorescence Lidar Based Standoff Biodetection System

Jean-Robert Simard, Sylvie Buteau and Pierre Lahaie

Introduction

With the increase in terrorist activities and the relatively easier access to biological warfare observed over the last 20 years, the need to monitor regularly, if not continuously, the atmosphere over areas that may extend to tens of square kilometers for its biological aerosol content has become more and more a critical issue. To achieve this task in a useful manner, such monitoring must survey the wide area of interest within a minute and report a threat within seconds of detection. Furthermore, because of the extent of the countermeasures to implement and the possible presence of people over the monitored area, such systems must provide a high probability of detection (>90 %) with a false alarm rate better than once a week, and be deployable over spaces crowded with people while exerting no harm of any kind (and ideally be imperceptible). These objectives, considering the small size (1–10 μm) and very low concentration (as low as a few agent-contained particles per liter of air) that represents an aerosolized biological threat are very difficult to achieve. Nevertheless, it is becoming more and more accepted by expert members of international working groups that this demanding task, within the present state of the technology and the ones in sight, will be best performed by the combination of two types of instruments: optical standoff monitoring systems (also referred as a standoff optical triggers) and point sensing systems capable of sampling air contaminants on request at locations linked to the information provided by the standoff optical trigger. The first has the advantage of having no consumable beside electrical energy and is able to detect rapidly and track an aerosol threat over a wide area 24 h a day, 7 days a week. The second confirms the presence of a biological threat and, ideally, the prophylaxis to implement in addition to providing forensic information. This chapter aims at reviewing a promising standoff sensing system based on lidar technology,

J.-R. Simard (✉) · S. Buteau · P. Lahaie
Defence Research and Development Canada: Valcartier Research Centre,
2459, de la Bravoure Road, Quebec, QC G3J 1X5, Canada
e-mail: Jean-Robert.Simard@drdc-rddc.gc.ca

P. Jonsson et al. (eds.), *Bioaerosol Detection Technologies,* Integrated Analytical Systems,
DOI 10.1007/978-1-4419-5582-1_13, © Springer-Verlag New York 2014

leaving the description of the challenges associated with point sensing/sampling systems to third parties.

Direct detection of bioaerosols from a standoff position using lidar technologies has been pursued at least since the 1980s. However, the main challenge faced during these early years, and still today, has always been to differentiate human threatening bioaerosols from all other aerosols present in the atmosphere. Classical elastic cloud mapping will detect equally well, for equivalent diffusing aerosol clouds, dust raised from the ground or a pollen puff issued from a blooming forest. Therefore, quite early through this effort it was recognized that additional sources of information must be exploited to achieve standoff bioaerosol threat detection with a satisfactory false alarm rate. First, the polarization state of the return scatters was investigated as a means to differentiate bioaerosols from inorganic dust and shown to have limited success [1]. Then, bulk laser-induced fluorescence (LIF) demonstrated a high sensitivity to detect aerosols of biological origin at ranges extending to more than 10 km with little contribution from inorganic aerosols [2]. This was an important advance in demonstrating sensitive standoff bioaerosol monitoring. Nevertheless, even with the bulk LIF approach, unacceptable false alarm rates from naturally occurring bioaerosols and other fluorescing aerosols still prohibited effective wide area biological threat monitoring. It is only when LIF collected while keeping the spectral information and observing that fluorescing aerosols from different origins (including bioaerosols) carry spectral differences measurable from kilometer ranges that standoff biodetection with a high probability of detection and acceptably low false alarm rate was at reach [3].

The fluorescence induced from biological materials originates from aromatic amino acid residues that are intrinsic constituents of nearly all proteins [4]. Only three of the 20 amino acids, tryptophan, tyrosine, and phenylalanine [5], have aromatic chemical structures (benzene-like ring) capable of emitting fluorescence with nanoseconds response time when excited with UV light [6]. This short response time allows the spatial localization of the source of the fluorescence with precision below a meter by using the "time of flight" technique associated with Lidar technology (see the section "Fundamentals" below). The types of amino acid residues as well as how they are bound within the proteins dictate the spectral quantum yield (the energetic efficiency of the compound to generate the fluorescence as a function of the excitation wavelength) as well as the spectral distribution of the induced fluorescence [7]. As an example, pure tryptophan has excitation and emission maxima around 280 and 350 nm, respectively [6]. However, when its aromatic residues are embedded in the reduced nicotinamide adenine dinucleotide (NADH), a coenzyme having an important role in the cellular energetic production mechanisms and present in all living cells[1], a second excitation peak appears around 340 nm [7] that induces a fluorescence maximum around 460 nm [8]. Even if the spectral structures of the auto-fluorescence of these aromatic residues show little spectral structures in

[1] In our work, we have chosen to target this specific fluorophore because of its important role in leaving organisms. Additional arguments for this choice are the good atmospheric transmission and the commercial availability of energetically efficient lasers (tripled YAG) in that spectral band.

comparison with IR spectroscopy, it is sufficient to provide a classification of the compound at the origin of that signal. This property is at the base of the capability associated with spectrally resolved LIF lidar standoff bioaerosol detection systems.

This chapter reviews in detail spectrometric LIF lidars that have demonstrated important capabilities in detecting bioaerosols from a standoff position. First, the section "Fundamentals" reviews the supporting model associated with spectrometric lidars. It revisits the basic approach and associated approximations that lead to the construction of a simple applied model with all the classes of contributing parameters identified. Then, the section "Robustness, Specificity and Sensitivity of Spectrometric LIF Lidars in Standoff Bioaerosol Detection" reviews the advances performed to assess the robustness, specificity, and sensitivity of the information obtained from fluorescing aerosols with spectrometric LIF lidars. Also, the sensitivities of this technology, measured at multiple occasions with the *Bacillus globigii* (BG) subgroup, are reported. The section "Spectral Classification of Bioaerosols" describes in details a statistical mathematical tool, the Mahalanobis distance, to exploit the information from spectral LIF lidars and allowing the effective standoff detections and classifications of bioaerosols. Finally, a conclusion highlights the capabilities in standoff biodetection reported in the present document and briefly identifies future avenues that may lead to improved standoff biodetection systems.

Fundamentals

In order to understand how lidar technology can provide standoff detection (or early warning for the presence) of biological threats in aerosol form, it is important to shortly revisit the fundamental model associated with this type of instrument. This review is in large part inspired by the work of Measures [9].

At the fundamental level, the optical signal collected by all lidars can be defined as a differential equation relating the element of light collected and sent to a detector and the element of volume from where this element of light originated (see Fig. 13.1). This equation can be expressed as

$$dP_\lambda(\lambda_0,\vec{r},t) = J_\lambda(\lambda_0,\vec{r},t)\, p_\lambda(\vec{r})\, dV, \tag{13.1}$$

where dP_λ is the spectral element of light power detected at a time t resulting from elastic or inelastic scattered laser light with an excitation wavelength λ_0 and within a volume element dV located at position $\vec{r}$, $J_\lambda(\lambda_0,\vec{r},t)$ is the spectral radiance generated by the laser light incident within that volume dV and responsible for the detected spectral element of light[2] and $p_\lambda(\vec{r})$, the geometric factor, is the fraction

[2] It is important to note the time delay that exists between the moments when the laser light is fired and reach the volume element dV where the spectral radiance is generated and the moment it is detected by the lidar. This delay depends on the time of flight between this volume element and the lidar instrument. This time variable will be introduced later in this document.

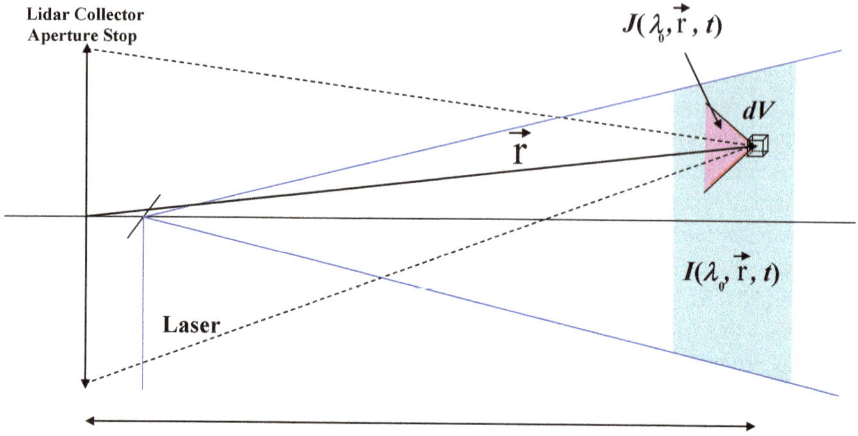

Fig. 13.1 Scheme illustrating some of the fundamental parameters defining the optical signals collected by lidars. The *blue area* represents the propagating laser pulse and the *pink area* represents the spectral radiance induced by the laser pulse on the fluorescing aerosols contained in the volume element *dV*

(or probability) of this element of induced spectral radiance in reaching the lidar detector.

The spectral radiance J_λ is detailed by introducing the laser irradiance I (W/m^2) incident at the volume dV as

$$J_\lambda(\lambda_0,\vec{r},t) = \beta_\lambda(\lambda_0,\vec{r})I(\lambda_0,\vec{r},t), \tag{13.2}$$

Where β_λ is defined as the volume spectral backscattering coefficient at position $\vec{r}$. This coefficient takes into account the physical characteristics of the scatterers and depends on the excitation wavelength. It can be defined, for a given scatterer, as

$$\beta_\lambda(\lambda_0,\vec{r}) = N(\vec{r})\frac{d^2\sigma_\lambda(\lambda_0)}{d\Omega d\lambda}, \tag{13.3}$$

where $N(\vec{r})$ is the volume density of the scatterers at position $\vec{r}$ and $d^2\sigma_\lambda(\lambda_0)/d\Omega d\lambda$ is the spectrally distributed backscattering cross-section of the scatterers per units of solid angle and wavelength for an excitation wavelength λ_0.[3]

Finally, the geometric factor $p_\lambda(\vec{r})$ introduced in Eq. (13.1) can be detailed further as

$$p_\lambda(\vec{r}) = \frac{A_{AS}}{|\vec{r}|^2} \times t_\lambda^a(\vec{r}) \times t_\lambda^c \times \xi(\vec{r}), \tag{13.4}$$

[3] The definition of β_λ in Eq. (13.3) can be generalized as a summation over the different types of scatterers present in the probed volume to model aerosol mixture lidar detection.

where A_{AS} is the total area of the aperture stop of the lidar collector including obscuring components, t_λ^a and t_λ^c are the atmospheric and lidar collector spectral transmissions, respectively, for a collected photon originating from a location $\vec{r}$. Finally, $\xi(\vec{r})$ is the overlap function describing the fraction of scattered photons emitted at location $\vec{r}$ and collected by the aperture stop that reach the detector. This last parameter is regularly referred to as the geometrical form factor or obscuration factor and is one of the main optimizing criteria for an adequate optical design of a lidar transmitter.

Eqs. (13.1)–(13.4) form the fundamental model of most lidar systems. To refine further this model for the specific case of standoff biosensors, Eq. (13.1) must be integrated over the spatially resolved (or probed) volume, over the resolved time interval associated with the electronics of detection and, in the particular case of the inelastic scattering, over the resolved spectral interval of the lidar collector. In order to achieve a manageable analytic expression after these integrations, a series of simplifications are required. These are:

1. The maximum radial distance $|\vec{r}_r|$ from the optical axis in the probed volume is much smaller than its range R ($|\vec{r}|^2 = |\vec{r}_r|^2 + R^2 \approx R^2$). This is expected in most lidar systems. With this approximation, R is the range variable along the lidar optical axis.[4]
2. The variability of the parameter $1/R^2$ over the probed range interval is negligible. This is valid as long as the range interval of the resolved volume is much smaller than the range to that volume.
3. The concentration of targeted scatterers is averaged within the probed volume.
4. The response time of the inelastic scatterers to return the signal following the laser irradiation is much smaller than the laser pulse duration or the detection electronic time resolution. This allows neglecting the convolution of the return signal with the response time of the inelastic scatterers.
5. The laser irradiation energy lying outside the central flat region of the field of view of the lidar collector is negligible. This is achieved by an optimal lidar optical design.
6. The variation of the atmospheric attenuation within a probed volume is considered negligible.

Simplified Model Derivation

With the simplifications inserted in Eqs. (13.1)–(13.4) and integrating over the probed volume V delimited by the collector field of view irradiated by the laser pulse for a given instant t, Eq. (13.1) is reduced to

$$P_\lambda(\lambda_0,t) = \int_V dP_\lambda(\lambda_0,\vec{r},t) = \int_V \beta_\lambda(\lambda_0,\vec{r}) I(\lambda_0,\vec{r},t) p_\lambda(\vec{r}) dV, \tag{13.5}$$

[4] For single optical axis lidar, this axis is defined by the laser path. When the lidar transmitter optical alignment is optimized, the collector field-of-view should be centered on the laser path axis.

$$= \frac{A_{AS}}{R_0^2} t^a_{\lambda_0}(R_0) t^a_{\lambda}(R_0) t^e_{\lambda_0} t^c_{\lambda} \xi(R_0) N(R_0) \frac{d^2\sigma_\lambda(\lambda_0)}{d\Omega d\lambda} \int_V I_L(\lambda_0, r, t) dV, \quad (13.6)$$

where R_0 is the distance between the aperture stop of the lidar collector and the center of the probed volume V, $\xi(R_0)$ is the overlap function of the collector at the range R_0 and I_L is the laser pulse irradiance spatial distribution at the instant t within the volume V with the attenuations created by the emitter optics $t^e_{\lambda_0}$ and the atmosphere $t^a_{\lambda_0}(R_0)$ between the probed volume and the lidar system taken into account and reported outside the integral. Then, the photonic energy collected by the lidar within the electronic resolved time Δt is given by

$$E_\lambda(\lambda_0, t, \Delta t) = \int_t^{t+\Delta t} P_\lambda(\lambda_0, t)\, dt$$
$$= \int_t^{t+\Delta t} \frac{A_{AS}}{R_0^2} t^a_{\lambda_0}(R_0)\, t^a_{\lambda}(R_0)\, t^e_{\lambda_0} t^c_{\lambda}\, \xi(R_0)\, N(R_0) \frac{d^2\sigma_\lambda(\lambda_0)}{d\Omega d\lambda} \int_V I_L(\lambda_0, r, t)\, dV\, dt \quad (13.7)$$

To further simplify Eq. (13.7), it is necessary to insert the relation between the time where a signal resulting from firing the laser is being detected and the range from where this detected signal is originating. Taking the instant that the laser pulse leaves the lidar transmitter as the time origin (t=0) and noting that the light pulse has to travel up to and back from the element of volume where the laser light has been scattered, it is straightforward to establish that

$$t = \frac{2R}{c} \text{ and } dt = \frac{2dR}{c}. \quad (13.8)$$

Furthermore, by introducing the definition of I_L which is the element of laser energy dE_L incident on an element of area dA perpendicular to the laser axis per time interval dt and substituting Eq. (13.8) in the volume integral of Eq. (13.7), this volume integral is reduced to

$$\int_V I_L(\lambda_0, r, t)\, dV = \int_V \frac{dE_L}{dt\, dA} dA dR = \frac{c}{2} \int_V dE_L = \frac{cE_L}{2}, \quad (13.9)$$

where E_L is the total laser pulse energy and c is the speed of light. Inserting the last result in Eq. (13.7) and integrating over time, the collected photonic energy is derived as

$$E_\lambda(\lambda_0, R_0, \Delta R) = E_L\, \xi(R_0) \frac{A_{AS}}{R_0^2} t^a_{\lambda_0}(R_0) t^a_{\lambda}(R_0)\, t^e_{\lambda_0} t^c_{\lambda}\, N(R_0) \Delta R \frac{d^2\sigma_\lambda(\lambda_0)}{d\Omega d\lambda}. \quad (13.10)$$

In this last result, ΔR reports formally half the distance travelled at the speed of light during the time integration Δt, not the range interval probed by the lidar. The range interval probed by the Lidar results from the spatial convolution of ΔR with the laser pulse length and the aerosol scattering decay life time reported in spatial dimension once multiplied by the speed of light. The resulting probed range interval is centered at R_0. Only when the laser pulse length and the aerosol scattering decay lifetime equivalent length are negligible in comparison with ΔR, then ΔR is associated with the lidar probed range interval.

Finally, by introducing the quantum yield Ψ_λ^{PC} describing the efficiency of the lidar detector to convert the collected photons into electrons, the electronic conversion factor κ defining the number of electronic counts produced by each detected photon[5] and the number of laser pulses n_p fired to produce the signal of interest $S_\lambda(\lambda_0, R_0, \Delta R)$, this signal $S_\lambda(\lambda_0, R_0, \Delta R)$ resulting solely from the laser irradiation (or corrected for non-laser-induced parasitic signal contributions) can be derived from Eq. (13.10) as[6]

$$S_\lambda(\lambda_0, R_0, \Delta R) = \kappa\, \Psi_\lambda^{PC}\, n_p\, E_\lambda(\lambda_0, R_0, \Delta R)$$

$$= \underbrace{\kappa \Psi_\lambda^{PC}}_{\substack{\text{electronic}\\ \text{conversion}}}\ \underbrace{n_p E_L}_{\substack{\text{laser}\\ \text{energy}}}\ \underbrace{\xi(R_0)\frac{A_{AS}}{R_0^2}}_{\text{geometry}}\ \underbrace{t_{\lambda_0}^a(R_0) t_\lambda^a(R_0)}_{\text{atmospherics}}\ \underbrace{t_{\lambda_0}^e t_\lambda^c}_{\text{optics}}\ \underbrace{N(R_0)\Delta R}_{\substack{\text{cloud}\\ \text{column}}}\ \underbrace{\frac{d^2\sigma_\lambda(\lambda_0)}{d\Omega d\lambda}}_{\text{scatterers}}. \tag{13.11}$$

Eq. (13.11) expresses the spectrally distributed signal detected by a lidar as a function of its electronic conversion efficiency, the total laser energy sent to the probed volume located at a range R_0, the geometric parameters, the atmospheric and optic transmissions, the targeted cloud column[7] and the spectral cross-section of the scatterers disseminated in the probed volume. This equation forms the basic model dictating the spectral magnitude of the detected signal as a function of key lidar parameters. Only the convolution with the lidar spectrometer spectral response has not been introduced in this model.

In the following sections, this model will be further developed and adapted as a function of the phenomenology being the object of the discussion.

---

[5] These electronic counts may be those produced by the readout electronics of a Photo Multiplier Tube (PMT) output, a CCD camera, or any other electronic transducer components. It should be derived from the sum of counts resulting from a detected photon distributed over several (spectral) detector elements if spectrometric detection is performed. It is also important to note that if the electronic transducer is performing photon counting detection, κ is equal to 1 by definition.

[6] In this equation, the spectral resolution limited by the dispersive element is not introduced for simplification. This effect can be visualized as a convolution of the collected spectra with the spectral impulse response of that dispersive element.

[7] The scatterer concentration in the cloud column is averaged either over the whole range interval ΔR defined by the lidar design or the cloud thickness itself. If the lidar range interval is greater than the region where the targeted cloud is located, the effective lidar range interval is determined by the cloud column, reflecting that a fixed number of scatterers contributes to the return signal.

Robustness, Specificity, and Sensitivity of Spectrometric LIF Lidars in Standoff Bioaerosol Detection

Before exploiting with success the spectral information of LIF from bioaerosols, properties such as robustness and specificity associated with this phenomenon must be assessed. These properties will dictate to which extend the intrinsic LIF can be useful for detecting and discriminating bioaerosols. First, the robustness of spectral LIF is investigated by evaluating the stability of spectral features, the impact of growth conditions and preparation and, dissemination techniques and external environmental conditions. Then, the spectral specificities associated with fluorescence induced from different types of aerosols are reported. Finally, the challenges associated with the sensitivity evaluation of spectrometric LIF lidars will be briefly depicted. Additionally, since the choice of the excitation wavelength impacts on these properties, it will be referred all along this section. Note that most experimental results reported in this section were produced with the standoff integrated bioaerosol active hyperspectral detection (SINBAHD) system. Readers interested to learn more about this system are invited to consult Simard et al. [3].

Robustness of Fluorescence Spectral Features

The spectrally distributed LIF signal detected with lidar technology as modeled in the section "Fundamentals" includes intrinsic information about the probed aerosols, opening the path to discrimination among bioaerosols from a standoff position. However, to exploit efficiently this classification capability, it is paramount that spectral information carries a minimum of robustness. So the question is what level of robustness can be expected from these bioaerosols' fluorescence-induced spectra. The vast majority of reported fluorescence spectra in the literature appear to be repeatable, therefore robust, for a given type of material [10, 11]. An exception was reported by Manninen et al. [12]; they observed fluorescence spectra significantly different from two separate particles of dry male-specific bacteriophage 2 (MS2) which unexpected behavior was attributed to irregular drying of the aerosolized phages. They did not observe this spectral induced fluorescence spectra from the corresponding MS2 suspension. It is worth underlining the fact that measurements coming from single particle (point detection-based techniques) are highly dependent of the sample homogeneity in opposition to lidar measurements, which are a result of the averaging of all particles present in the probed volume, see Eq. (13.11).

Although fluorescence spectra appear to be robust for a given material, it is known that several factors affect its spectral features to various degrees. These factors include growth conditions [13, 14], bio-material preparation [15] (washed vs. unwashed, additives such as stabilizers, anti-caking, etc.), material state [12, 16] (dry vs. wet), physiological age [17] and, to a lesser extent, dissemination techniques and external environmental conditions. Heaton [14] has reported significant spectral variability within the *Bacillus globigii* (BG, also refereed as *Bacillus subtilis var niger*) and *B. anthracis* (Ba) sub-groups. They have excited, at 355 nm, four BG samples having different growth and preparation methodologies including deactivation by irradiation. For the Ba samples, they had two different strains; one was grown as a reference and the others were prepared with three different procedures. Hill [10] have observed repeatable fluorescence spectra from uniform single particles with strong or weak variation between washed and unwashed samples depending on the excitation wavelength (266 versus 351 nm). Using a lab-sized lidar aerosol chamber and using a 355 nm laser, Laflamme et al. [15] have shown that culture media and washing procedure did influence to various degrees the spectral signatures of three bio-simulants (BG, *B. thuringiensis* (Bt) , *Erwinia herbicola* (EH)). However, he also observed that the spectral maxima appeared to be invariant. This is in agreement with Christesen and Ong's results [18] who performed measurements on 12 different samples of killed Ba. They observed that fluorescence emission peak location was independent of the Ba strain, growth stage, and the medium in which it was grown. Various groups have observed that materials exhibit stronger fluorescence in their dry state compared to wet [12, 16]. Moreover, Hill et al. [16] observed a spectral shift to the longer wavelength emitted by dry material in comparison with one in water suspension. It must be noted that these results were obtained from bulk samples rather than airborne particles. Steinvall et al. [19] have reported stable BG fluorescence spectra excited at 355 nm and measured at three concentrations. Buteau et al. [20] observed a significant degree of spectral robustness for both BG and EH while excited at 351 nm from samples of different origins, disseminated in open air with various methods based on wet slurries and dry powder and at different time period (2001 and 2005). These results were especially surprising when considering the low signal-to-noise ratio obtained during the measurements. However, additional measurements performed later in 2006 by the same team showed noticeable variability with the two previously acquired data set in the case of BG simulant [21]. These results acquired with the Canadian system SINBAHD [3] are reported in Fig. 13.2. Each spectrum in this figure results from the average of fluorescence spectra induced with 1000 laser pulses fired at a BG cloud disseminated during several minutes. Each collected spectrum was first corrected from background contributions. The averaged spectrum was normalized by the area under the curve (or equivalently, the total fluorescent energy) but was not corrected by the sensor spectral response or the spectral transmission of the atmosphere. Table 13.1 presents the different reported parameters of the BG samples studied at these occasions.

The source of the discrepancy between the BG 2006 results with the ones from the previous trials was not identified by the authors [21] but could be related to the climatic conditions or the fluorescing background aerosol populations since signals were obtained from open-air releases. Although results from lidar measurements are representing an averaged response of numerous aerosol particles, the ambient conditions might have a significant impact since they will act upon the entire cloud column, see Eq. (13.11). In particular, fluorescing background aerosol population and its spatiotemporal evolution will inevitably have an effect on any type of biosensors including lidar measurements [22].

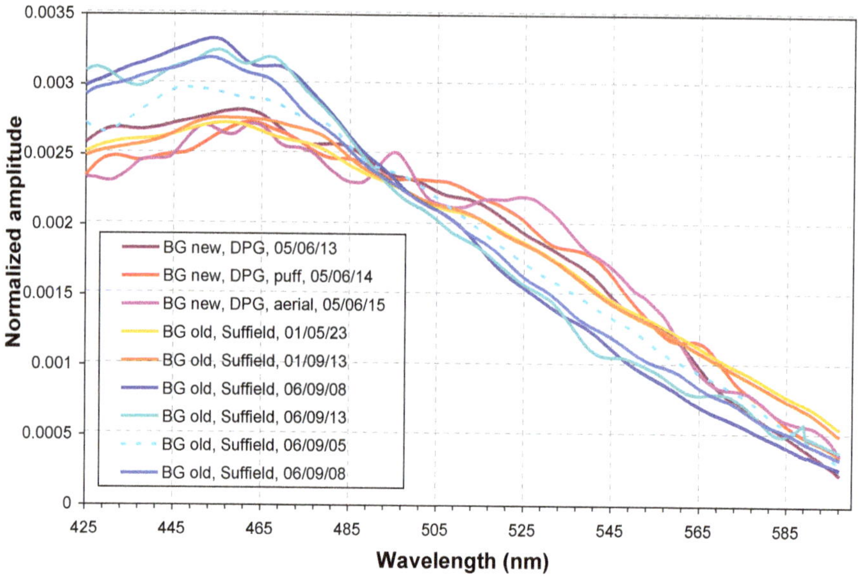

Fig. 13.2 Normalized spectral signatures obtained with the SINBAHD system for different types of BG disseminated with various methods during trials held at Suffield in 2001 and 2006, and during the JBSDS Increment Demo II trial performed in 2005

Table 13.1 Specific information on BG samples studied during trials held at Suffield, Canada in 2001 and 2006, and the joint biological standoff detection system (JBSDS) Demo II field trial in 2005, DPG, Utah, USA

Testing	Material name	Origin	Preparation	State	Dissemination
Suffield (2001)	Old BG	US DPG	Liquid suspension containing 1% BG slurry	Wet	Micronair agricultural sprayer
DPG (2005)	New BG	Denmark	Milled at the same size as the old BG	Dry	Skil Blower-point + puff and aerial
Suffield (2006)	Old BG	US DPG	2 g in 20 L H_2O (concentration 0.01%)	Wet	Micronair agricultural sprayer

Signature Specificity, Excitation Wavelengths, and Spectral Resolution

The fluorescence spectral characteristics of biological materials depend on various parameters but the predominant one, besides the material itself (including its state, preparation, age, etc.) is the excitation wavelength of the probing sensor. A

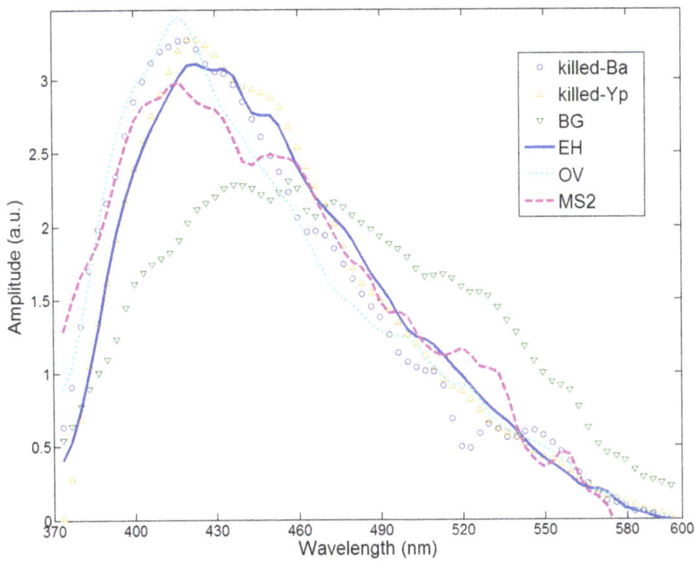

Fig. 13.3 Background subtracted fluorescence spectra of different materials excited at 351 nm gathered with the SINBAHD instrument at the JBSDS Increment Demo II trial [26]

useful way to visualize the material fluorescence specificity and its dependence with wavelength is the fluorescence excitation–emission matrix (EEM), which reports the amount of auto-fluorescence (Emission) once excited at a given wavelength (Excitation) [16, 18]. The ARL team [16] have measured fluorescence EEM of several types of biological materials and also inorganic particles and oil. It must be noted that these measurements were done on bulk materials (dry) or suspensions (wet). Although their results are not directly applicable to airborne particles as stated in their work [16], their results are useful to have an appreciation of the specificity of material's spectra and its dependence with the excitation wavelength.

To date, the most commonly use wavelength regions for lidar-based biological standoff detection [23–25] are the 266 and 355 nm, two harmonics of the energetic efficient YAG solid-state laser. Depending on the excitation wavelength and materials, discrepancies in the LIF spectra can be observed or not [10]. Figure 13.3 presents an example of fluorescence spectra of various materials excited at 351 nm during the joint biological standoff detection system (JBSDS) increment II Tech Demo III trial [26]. Each of the spectra are corrected for background signals and averaged over 1000 acquisitions, each resulting from single laser pulse, made by the SINBAHD system during a release of a particular material, and are not corrected for the instrument or the atmospheric spectral transmissions [26].

The spectra of Fig. 13.3 show high spectral specificity for numerous fluorescing materials but modest specificity for others. Indeed, the laser-induced fluorescent spectrum of EH is quite similar to the one of irradiated *Yerstinia pestis* (Yp), both being vegetative cells. The similarities show the challenge associated with the discrimination of different biological materials based on spectral LIF measurements. Cheng et al. [27] observed very similar fluorescence spectra from four different bacteria including *B. subtilis var niger* (BG, also refereed as *Bacillus globigii*) and Bt, all excited at 275 nm. Furthermore, with their induced fluorescence spectrometric detection system capable of producing variable excitation wavelengths from 242 to 362 nm, they demonstrate that this similar spectral characteristic is observable at all excitation wavelengths between 265 and 290 nm. They came to the conclusion that fluorescence signals with very high signal-to-noise ratios would be required to have any possibility of distinguishing spectra of these four specific bacteria when excited in this spectral range. The main outcome from most work performed on the spectral LIF characteristics of biological materials is that significant signal is available with the current state of detection technologies. However, discrimination between many biological materials is challenging and depends largely on the excitation wavelength. The choice of the best excitation wavelength depends on the discrimination objectives, varying from recognizing biological from non-biological aerosols up to differentiating biological materials [16], and the targeted range of operation since atmospheric absorption decays rapidly below 300 nm. Discrimination would certainly beneficiate using multiple excitation wavelengths. In fact, this has been demonstrated for single particle probing (point detector technologies) with excitation around 266 and 355 nm [28–30].

Another important characteristic of spectral LIF lidar techniques is the spectral resolution of the acquired signal. Currently, most spectral lidar detection systems are either based on an array of photomultiplier tubes (PMT) [25, 31] or intensified charge-coupled device (ICCD) camera [3, 24]. The first provides a 2D spectral and range resolved dataset while the second gives a linear wide dynamic range spectrum that may be designed with high spectral resolution over large spectral intervals. Aside from the difference in detection dynamic, a key detection issue is the spectral resolution needed to capture the finest spectral features (if any) of the fluorescence induced from materials of interest targeted for classification. Using an ICCD-based lidar system at 355 nm, Farsund et al. [24] have analyzed experimental data at different spectral resolutions. They came to the conclusion that 10–20 bands in the 350–700 nm spectral intervals are sufficient to discriminate between ovalbumin (OV), BG, Bt, and five interfering pollens. Heaton [14] has also presented results where the spectral resolution of the collected fluorescence was degraded from 1 to 34 nm without significant impact on the classification capabilities. Nonetheless, it is worth emphasizing the fact that finer resolution than the targeted spectral structure could present advantages, such as better photon counting capability and better resolved atmospheric Raman lines (N_2, O_2 and H_2O), which are useful references to estimate the transmission of the atmosphere.

Before concluding this section, it must be emphasized once more that the conditions of measurement affect the collected spectral information. Although limited work has been published on this matter, one thing is certain, background aerosols (aerosols naturally occurring in the atmosphere) should be well characterized to effectively detect and classify the arrival of new bioaerosol materials in minute quantity [22]. Finally, to exploit adequately the spectral signatures gathered with different spectrometric LIF instruments and compare or combined their respective spectral libraries, a proper spectral calibration methodology should be used to correct the acquired spectra for the spectral radiometric response of the sensors, and to a lesser extent, the spectral transmission of the atmosphere during the acquisitions.

Standoff Spectral LIF Sensor Sensitivity Limit

Standoff spectral sensor sensitivity limit can be defined for the detection and the classification of remote aerosol clouds. For detection, this limit is associated with the magnitude of the scalar signal (as bulk induced-fluorescence or elastic scattering from classical Lidar cloud mapper) returned by aerosols. For classification, this limit is associated with the ability to differentiate various fluorescing materials based on the spectral specificity of the detected signal. This later property is best exploited using multivariate analysis. Multiple multivariate techniques can be found in the literature, some being reviewed in the section "Spectral Classification of Bioaerosols" below, and most of them are powerful tools to perform spectral classification even under very weak signals. Regardless of the multivariate analysis chosen, the sensitivity of a sensor system, usually characterized as a minimal detectable concentration, depends on the magnitude of the detected signals. However, since this sensitivity is related to the probability of detection, it is highly desirable to assess this characteristic through a receiver operating characteristic (ROC) [32]. However, producing these ROC curves requires large amounts of measurements and they are therefore rarely available. The sensitivity is quite often presented as four times the standard deviation of the signal resulting from the processed background clutter and is referred as the "4σ sensitivity." For successful spectral classifications, the required minimum signal-to-noise ratio of the spectral fluorescence signal processed by a multivariate technique must be greater than this reference quantity (e.g., 4σ) for a given end member or a combination of end members part of a library[8]. Essentially, the goal is to have a reference level to assess the sensitivities of a standoff detection system as a function of the concentration of the probed materials. In the following discussion, the term sensitivity refers to the limit of detection (LoD) associated with the multivariate processed spectral LIF signal acquired during the dissemination of a given type of aerosols versus the level of that signal before or after that dissemination. As a result, such sensitivity evaluation will depend on the concentrations and the types of fluorescing background aerosols at that location during this measurement. The dependency of this sensitivity definition with the environmental conditions is considered acceptable when compared to the

[8] In addition to the end members of a library, an extra concurrent end member may be associated with the measured LIF spectrum of background aerosols. This is particularly advantageous in a stable and homogeneous atmosphere and facilitates the detection of the arrival of new aerosols that may be associated with end members of the library. This is further discussed in section "Spectral classification of bioaerosols".

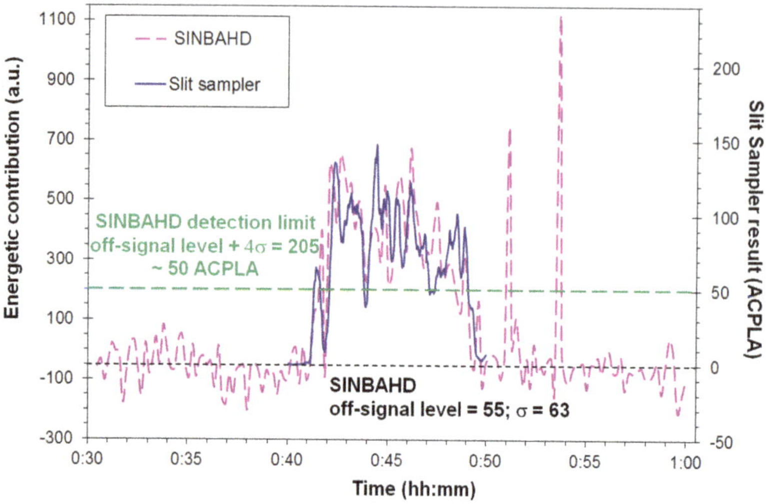

Fig. 13.4 Sensitivity evaluation based on temporal correlation of the Canadian lidar standoff system, SINBAHD, and Slit Sampler referee data for a 20-m cloud of BG at a range of 990 m during an open-air release performed at DRDC Suffield, Canada, in 2007 [33]

extensive amount of work required to produce a full ROC curve for the given environmental aerosol conditions.

Based on Eq. (13.11), for a fixed range, range interval, and well-defined type of scatterers, it is possible to linearly relate the signal detected by the sensor, once corrected for the atmospheric transmission, with the averaged scatterer concentration in the probed volume [3]. This makes possible to express the sensor sensitivity limit as a scatterer concentration. Figure 13.4 presents an example of a sensitivity evaluation after multivariate process achieved by the SINBAHD system for a 20-m long BG cloud released in open-air at a range of 990 m [33]. From this example, the 4σ sensitivity of the lidar system is derived as 50 agent containing particles per liter of air (ACPLA).

In the previous paragraph, we defined the sensitivity as a function of the concentration of a given fluorescing scatterers. The next issue is how such concentration should be derived. Should it be the total number of particles per liter (ppl) of air, only the ones having a diameter between 1 and 10 μm referred as the respirable range, only those viable derived as ACPLA, colony forming unit (CFU) for bacteria, or plaque-forming unit (PFU) for viruses. The best choice for the referee sensor in determining sensitivity is still under debate [34, 35]. Currently, the most used referee sensors are the aerodynamic particle sizer (APS) providing particle counts as a function of their size distribution and viability sensors such as slit and impingers samplers. The outputs from these two families of referee sensors are related but this relation depends on numerous factors such as the type of material [33] and the dissemination method [36].

The evaluation of standoff sensor sensitivity expressed in the referee sensor unit is highly dependent on the ability to confidently characterize the challenging

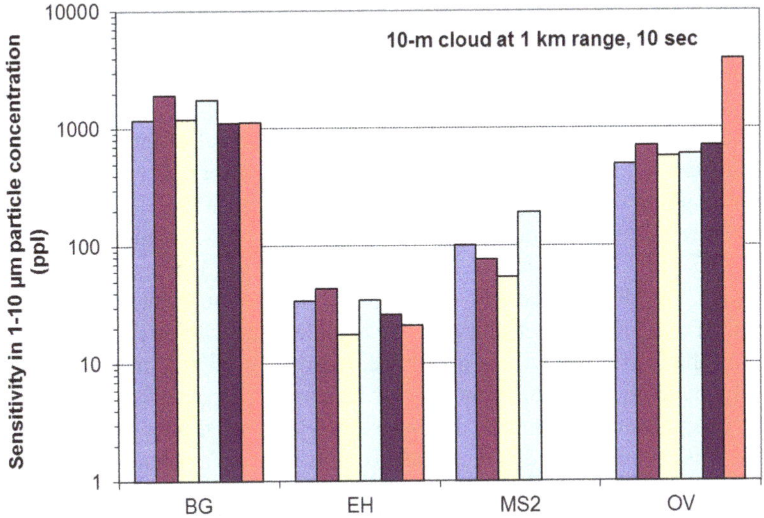

Fig. 13.5 4σ sensitivity extracted from open-air measurements performed by SINBAHD at Suffield, Canada in 2008: 10 s integration of a 10-m cloud at 1 km [36]. The different colored bars report sensitivities obtained from different releases that may have been done several days apart

biocloud. This characterization will depend on the selected referee sensor as mentioned previously and also on the achievable spatio-temporal correlation between the air volume characterized by the point referee sensors and the one probed by the lidar. During open-air releases, as in the example shown in Fig. 13.4, an important difficulty is to collocate in space the volume sampled by the point referee sensor and the one by the lidar. This collocation is always an approximation because of the different geometry of the sampled volumes (point: more or less cubic; lidar: a long cylinder). This approximate collocation will generate nonnegligible correlation variability between the two signals observed as linear amplitude discrepancies over the acquisition time period. Nevertheless, these discrepancies are expected to be partially reduced due to the time distribution of the analyzed collected data. Figure 13.5 presents SINBAHD sensor 4σ sensitivities for several open-air releases of four types of biological simulants made at Suffield, Canada in 2008 [36]. The sensitivities are reported in terms of concentration of aerosols having a diameter between 1 and 10 µm in a 10-m long cloud located at a range of 1 km and averaged over a period of 10 seconds.

To overcome the spatial correlation challenge when assessing sensitivity, the approach pursued by several countries is to build dedicated lidar adapted bioaerosol chambers that have the capability to generate well-controlled and well-confined bioaerosol clouds [24, 36, 37]. These chamber facilities allow a better characterization and correlation between standoff and referee systems but do not provide representative background aerosol conditions. However, a solution has been proposed [38] to disseminate representative background aerosols simultaneously with the challenging simulants of biological agent to overcome this limitation with confining chambers. Currently, biological detection systems are normally both challenged with chamber and open-air releases to best capture their capabilities in assessing sensitivity as well as their probability of detection versus the false alarm rate in representative environment conditions.

The optical transmission and the ambient lighting level of the surrounding atmosphere can affect the sensitivity of standoff biosensors. The atmosphere has four main optical windows offering multi-kilometric transmissions: the Ultraviolet/Visible/Near-InfraRed window (300–1000 nm), the short-wavelength infrared window (1–2.5 μm), the mid-wavelength infrared window (3–5 μm) and the Long-Wavelength InfraRed window (8–14 μm). If either the emission or the collection channels of a standoff biodetectors operate outside these windows[9], the radiometric quality of the instrument will be greatly degraded as a function of the range, most probably reducing in an important manner its standoff biodetection capability. On the other hand, sun light can also degrade severely the sensitivity since the spectral LIF signal is mainly located in the visible band where sun's radiance is at its maximum. This degradation occurs during day, with transitions at sunrise and sunset, when parasitic day lights are collected simultaneously with the lidar signal scattered by the aerosols of interest. There are fundamentally four approaches to mitigate sensitivity degradation due to sun light contribution. First, spectral filters are inserted to block most light outside the lidar wavelength interval of interest. This is the classic method for quasi single wavelength elastic lidar but cannot be applied to spectrometric lidars where wide collecting spectral intervals are needed. A second method is to design very narrow lidar collecting FOVs. Lidars having FOVs of few hundreds of microradians is common and some are even as narrow as several tens of micro-radians. However, with very narrow lidar FOVs, the challenges are to find a multi-watts laser source with a high quality M^2 and to maintain the optical alignment between the lidar's emitter and collector with a few micro-radians accuracy. A third approach is to minimize the laser pulse repetition frequency (PRF) while keeping the same laser average power. This is equivalent to maximize the peak power of each laser pulses, or equivalently, to minimize the number of fired laser pulses to deliver a given laser energy while minimizing simultaneously the day light contribution responsible of the lidar sensitivity degradation. Finally, the last method is to use the shortest time interval resolved by the lidar detector while maximizing the lidar signal return by the cloud of interest. For classical elastic cloud mapper, this minimum time period corresponds to the range interval resolved by the lidar. This is usually a few meters (or about 7 ns).[10] For spectrometric lidars, the maximum time interval containing the signal of interest is defined by the cloud thickness (see Eq. (13.11)), which is usually less than a few tens of meters. Therefore, during

[9] Note that the wavelength intervals defining the atmospheric windows are representatives. There are severe attenuation bands contained within the NIR, SWIR and MWIR windows and must be taken into account in all Lidar designs, especially spectrometric Lidars.

[10] It is also common to 'switch-on' the gain of the Lidar detector only for the time taken by the light to reach the maximum range and back. This has the advantages to avoid common detection electronic saturations at the moment the laser pulse leaves the Lidar emitter.

day time operation, it is important to limit the day light signal contribution to the time interval where laser induced fluorescence is expected from a probed aerosol cloud.

Spectral Classification of Bioaerosols

To avoid lexical ambiguity with the vocabulary common in microbiology, the term classification (instead of identification as in artificial intelligence or statistical signal processing) of bioaerosols is used in spectrometric standoff biodetection. This refers to the spectral process by which the class of bioaerosols forming part or the totality of a bio-cloud is identified. It is therefore a process of identification using the spectral specificities of the bioaerosol that are attributed to this class. Mathematically, identification operators developed by artificial intelligence and statistical signal processing are numerous. Depending on the data and the operating conditions, one of them can be chosen. If the data are Gaussian or if their distribution is elliptically shaped, the Mahalanobis distance [39] is a good choice for many reasons: it has a simple form; it requires only the mean and covariance of the signal of interest and background as parameters; if the data are Gaussian, it follows a Gamma distribution and thresholds are easy to compute; the parameters (mean and covariance) are easy to compute; the parameters associated with the signatures of the materials of interest (mean and covariance) can be acquired externally and stored in a database. There is no proof however that spectrally resolved LIF from aerosols obeys Gaussian or even elliptical statistics. But, this assumption can be invoked because the noise of such signal follows very often such model. For ICCD detectors, the detector noise contribution originates mainly from photon shot noise and electronic noise. In this context, assuming Gaussian noise is reasonable. An important difficulty with the fluorescent signatures of biological agents, simulants or interferents are that they are not orthogonal in the sense that they may overlap each other in the multivariate space. Therefore, discriminate between them can be challenging.

In addition of the Mahalanobis distance, many mathematical operation or methods can be used to perform the discrimination between the spectra of various fluorescing materials. Among them, there are the spectral angle, the matched filter [40], adaptive matched filters as the GLRT [41] and the Rx Algorithm [42], neural networks or support vector machine. It is not the intent here to provide an exhaustive description of each of these methods and their possible implementation. Instead, we chose to detail the Mahalanobis distance method because it accounts for both the variability of the backgrounds and of the signatures. It is also simpler than neural networks and support vector machines.. It has also the advantage of not requiring a large training set. However, the evaluation of the performance of these methods is required and will be made in future studies. For that reason, only the performance of the Mahalanobis distance method will be analyzed in the following paragraphs.

One approach in implementing the Mahalanobis distance is to determine first if a measurement is different from the background signals that have been previously acquired at a given location. If it is close to the previously acquired background, it can be used to update the background statistics (mean and covariance). If the difference exceeds a given threshold, an attempt to classify this signal is performed. The measurements will most of the time come from a location where a background probing has not been performed previously due to the wide area under surveillance. In that case, background will not be updated. However, the background statistics will have to be adjusted to the different location to introduce the adequate parameters in the Mahalanobis distance (mean and covariance). These parameters vary mostly as a function of the distance to the sensor. Potential variations in the statistical parameters of the background in azimuth and low height elevation are assumed negligible unless a global change in the aerosol state of the atmosphere is in progress (weather front) or if some of these directions are associated with localized artifacts of the scene under surveillance (blooming trees, chimneys etc.). The Mahalanobis distance, as a detection operator, has been selected because of its simplicity, its robustness and the success it obtained as the operator in the Rx algorithm. It is used to perform the discrimination between biological agents and interferants. The spectra of the materials display many similarities and, usually, do not have strong spectral features and are smooth shaped. Nevertheless, these weak spectral variations can be discriminated when the signal is strong enough. Statistical signal processing operators such as the Mahalanobis distance can be very sensitive to discriminate a material from the others even if the differences are small. However, the mean and covariance matrix of the materials and of the backgrounds are both needed to be exploited effectively with this method.

In the following, we develop a model for the statistical parameters, the mean and the covariance matrix related to the lidar return. These results are then used to develop a procedure to condition the measured signal to serve as an input to the Mahalanobis distance estimation as well as input to the computation of this estimator. We begin with a signal description, followed by the development of a model for the statistical parameters and we finish with the details of the computation procedure involved in the classification.

Signal Description

In the section "Fundamentals", a general lidar equation (Eq. (13.11)) has been derived to describe the signal returned from fluorescing aerosols. This result forms the basis for the construction of the statistical model and the subsequent signal processing procedure described in the following sections. In Eq. (13.11), some of the parameters can be modeled using statistical distributions. To obtain this model, bioaerosols are assumed naturally variable. Their behaviour can be described by a probability distribution detailing their fluorescent characteristic variability. It is common in life science to model random variable using a Gaussian distribution. Out of the usability advantages of the Gaussian model, photon shot noise is modeled by a Poisson distribution which, for a large number of photons can be approximated as a Gaussian distribution. The transmittance is a parameter that can be described using a random variable because of the error on its estimation.[11] Many parameters of the lidar equation can be considered random variables but most will be neglected in the following development. Examples of such parameters are the electronic conversion factor and the number of photons emitted in the laser pulse. The quantum yield of the photocathode represents the probability that an electron will be emitted when a photon is absorbed. Other random variables that are not considered are the position of particles along the laser beam and the radial position of the particle relative to the lidar optical axis. The photon counting mode of detection is assumed. This will yield simpler equations while keeping all the important concepts. One must consider the difference between the value of a statistical parameter and its estimation by experimental sampling, the last being itself a random variable while the first being an evaluation from a model.

In the following development, it is assumed that the laser energy in photons, the field of view and the illuminated field is the same for each fired pulses. When an irradiated aerosol absorbs a photon, there is a probability that an induced fluorescing photon will be collected and detected by the lidar. Based on Simard et al. [3], this probability can be derived from the fluorescence cross-section defined as the product of the fluorescence quantum yield Ψ_{λ_0} of the aerosol, its geometric projected area A and normalized spectral signature ζ_λ. In this definition, the normalized signature can be interpreted as the probability density for a photon to be emitted by the aerosol in a given spectral band knowing that a photon has been emitted, and the quantum yield, the ratio of the emitted fluorescence energy to the received inducing laser energy, can be interpreted as the probability that a photon be emitted knowing a UV photon has been absorbed. Another random variable is the number of fluorescent particles irradiated by the laser beam that can be seen by the lidar spectrometric collector. Based on these considerations, the expectation value in number of photons $N_{pf}(\lambda)$ for any spectral acquisition can be derived from Eq. (13.11) as

$$N_{pf}(\lambda) = \frac{Q'_\lambda}{R_0^2}\tau_\lambda(R_0)N(R_0)\Delta R\frac{d^2\sigma_\lambda(\lambda_0)}{d\Omega d\lambda}, \tag{13.12}$$

where $\tau_\lambda(R_0)$ combines $\xi(R_0)$, $t^a_{\lambda_0}(R_0)$, and $t^a_\lambda(R_0)$. Q'_λ combines all the other terms in Eq. (13.11) except κ. Eq. (13.12) expresses the expected value in photons before being read out by the detector electronics. All the system parameters have been grouped together to facilitate their computation from measured data. A very important issue is to take into account the fluorescence induced on specific types of bioaerosols of interest mixed with the one produced by the background aerosols. Finally, another important aspect is the estimation of the statistics of the spectral parameters. These issues will be addressed in the following sections.

---

[11] It is important to note that the extent of the validity of this choice for the distribution and associated error theory is still to be demonstrated with experimental results. However, it is anticipated that this initial modeling choice will provide valuable tools to assess the capability of LIF based standoff biodetection technology.

Signal Mean and Covariance Matrix Estimation

This section aims to present the formalism to obtain the mean and the covariance matrix from basic principles. Basically, the mean shall look like Eq. (13.12). The covariance will be built using a very similar formalism. In Eq. (13.12), the parameters related to random variables are $N(R_0)$, the number of particles within the probed volume, the transmittance $\tau_\lambda(R_0)$ and the spectral fluorescence cross-section $d^2\sigma_\lambda(\lambda_0)/d\Omega d\lambda$. The covariance matrix for the fluorescence can be expressed in terms of these parameters and, for now, will be derived solely from the variability of that cross-section upon the aerosols of the probe volume. To produce this derivation, all aerosols will be considered statistically independent and receive the same laser irradiance. With these assumptions, the average number of photons returned from a single aerosol during an acquisition can be considered to be a random variable because it depends on the fluorescence cross-section, itself a random variable. As a result, the lidar return from a single aerosol *i* can be obtained by considering the lidar equation for a single aerosol and assuming a pencil beam for which the power incident on the aerosol decreases with the inverse of the square of the distance and the return from the single aerosol decreases also with the inverse of the square of the distance since it is unresolved by the optical system. This allows defining the number of detected photon n_{pfi} for the *i*-th aerosol in the illuminated volume as

$$n_{pfi}(\lambda) = \frac{Q_\lambda}{R_o^4}\tau_\lambda(R_o)\frac{d^2\sigma_\lambda^i(\lambda_0)}{d\Omega d\lambda}, \tag{13.13}$$

where $d^2\sigma_\lambda^i(\lambda_0)/d\Omega d\lambda$ is the spectrally distributed backscattering fluorescence cross-section of the aerosol *i*. Q_λ reports the effect of the optical system on the fluorescence return without the geometric aspect of the probed volume. However, it does not include the parameters influencing the size of the laser beam. In the following, we will consider that all aerosols are located at the same range R_0. It is also interesting to note that the $1/R_0^4$ range dependency of the signal is in accordance with the problem of a nonresolved target, a case well known in active sensing. In addition, it is important to observe that the total number of aerosols in the probed volume is a random variable and the concentration $N(R_0)$ in Eq. (13.12) is a mean derived from this random variable. To keep the information associated with this random characteristic, a Poisson distribution is associated with this quantity and is combined with Eq. (13.13) to produce a probabilistic model for the number of particles and their spectral characteristics. It permits also to express the average of the total number of photons to be detected by the lidar $\langle Npf(\lambda)\rangle$ as

$$\langle N_{pf}(\lambda)\rangle = Q_\lambda \tau_\lambda(R_0)\sum_{N_a=0}^{\infty} p(N_a)\sum_{i=1}^{N_a}\frac{1}{R_0^4}\left\langle\frac{d^2\sigma_\lambda^i(\lambda_0)}{d\Omega d\lambda}\right\rangle, \tag{13.14}$$

where $p(N_a)$ is the probability that N_a aerosols be present in the probed volume irradiated by the laser. Equation (13.14) is a probabilistic representation of the integration over all the aerosols. The transition from Eq. (13.12) to Eq. (13.14) can be described as the addition of the contribution from each aerosol in the probed volume while conserving the probability of occurrence that N_a aerosols be present in that volume. The Poisson distribution is discrete and is defined as

$$p(j) = \frac{\beta^j \exp(-\beta)}{j!}, \tag{13.15}$$

where β is the mean and variance of the distribution and corresponds to the average of the number of particles contained inside the probed volume as

$$\beta = N(R_o)\Delta R \theta_L^2 R_0^2, \tag{13.16}$$

where θ_L is the angular divergence of the laser beam. With these definitions, Eq. (13.14) can be rearranged to yield Eq. (13.17). The summation in Eq. (13.14) is the expression for the average of the Poisson distribution. Introducing the average leads to Eq. (13.18).

$$\left\langle N_{pf}(\lambda)\right\rangle = \frac{Q_\lambda \tau_\lambda(R_0)}{R_0^4}\left\langle \frac{d^2\sigma_\lambda(\lambda_0)}{d\Omega d\lambda}\right\rangle \sum_{N_a=0}^{\infty} N_a p(N_a), \tag{13.17}$$

$$\left\langle N_{pf}(\lambda)\right\rangle = \frac{Q_\lambda \tau_\lambda(R_0)}{R_0^2}\left\langle \frac{d^2\sigma_\lambda(\lambda_0)}{d\Omega d\lambda}\right\rangle N(R_o)\Delta R\, \theta_L^2 \tag{13.18}$$

The laser beam full angular width θ_L is explicit in Eq. (13.18) contrary to Eq. (13.12) where it is embedded in Q_λ. Equation (13.18) is similar to Eq. (13.12) except that emphasis was placed on the fact that the statistical parameters are the mean of their respective random variables. This formalism to express the mean equation is also used to build the expression of the covariance. The covariance, however, is a second-order moment. The lidar return covariance is derived by the evaluation of the following expression

$$C(\lambda_1, \lambda_2) = \left\langle N_{pf}(\lambda_1) N_{pf}(\lambda_2)\right\rangle - \left\langle N_{pf}(\lambda_1)\right\rangle \left\langle N_{pf}(\lambda_2)\right\rangle. \tag{13.19}$$

Before inserting Eq. (13.13) in Eq. (13.19), the effect of the statistical independence between particles must be taken into account. This assumption of statistical independence is used because it is assumed that there is no influence from a given particle on any other particle. Having information about the spectral response of a given particle does not provide information on any other particle's spectral response. Its impact is given by Eq. (13.20) and translates the statistical independence property. This property is very important since it will permit a very large simplification of the covariance derivation since it provides a justification for the elimination of all the cross-terms involving different particles and enables the expression of the covariance of a group of particles as the sum of the covariance of the individual particles. Eq. (13.20) is valid for $i \neq j$.

$$\langle S_i(\lambda_1)S_j(\lambda_2)\rangle = \langle S_i(\lambda_1)\rangle\langle S_j(\lambda_2)\rangle \tag{13.20}$$

In this equation, S_n is a variable representing any spectral parameter for a given aerosol, n being the identifier of a particle. Using this simplification, it can be said that for a given set of aerosols totalizing N_a particles, the covariance is the addition of the particles' individual covariance matrices. This can be expressed as

$$C_{N_a}(\lambda_1,\lambda_2) = N_a\left(\frac{1}{R_0^4}\right)^2 Q_{\lambda_1}Q_{\lambda_2}\tau_{\lambda_1}(R_0)\tau_{\lambda_2}(R_0)$$
$$\left(\left\langle \frac{d^2\sigma^i_{\lambda_1}(\lambda_0)}{d\Omega d\lambda}\frac{d^2\sigma^i_{\lambda_2}(\lambda_0)}{d\Omega d\lambda}\right\rangle - \left\langle \frac{d^2\sigma^i_{\lambda_1}(\lambda_0)}{d\Omega d\lambda}\right\rangle\left\langle \frac{d^2\sigma^i_{\lambda_2}(\lambda_0)}{d\Omega d\lambda}\right\rangle\right), \tag{13.21}$$

where C_{N_a} is the covariance matrix for the N_a particles. From this expression, we can introduce the definition of the spectral covariance of the cross-section of the aerosol i, $C_{\sigma,i}(\lambda_1,\lambda_2)$, as

$$C_{\sigma,i}(\lambda_1,\lambda_2) = \left\langle \frac{d^2\sigma^i_{\lambda_1}(\lambda_0)}{d\Omega d\lambda}\frac{d^2\sigma^i_{\lambda_2}(\lambda_0)}{d\Omega d\lambda}\right\rangle - \left\langle \frac{d^2\sigma^i_{\lambda_1}(\lambda_0)}{d\Omega d\lambda}\right\rangle\left\langle \frac{d^2\sigma^i_{\lambda_2}(\lambda_0)}{d\Omega d\lambda}\right\rangle, \tag{13.22}$$

Upon introducing this last expression, Eq. (13.21) is rewritten as

$$C_{N_a}(\lambda_1,\lambda_2) = N_a\left(\frac{1}{R_0^4}\right)^2 Q_{\lambda_1}Q_{\lambda_2}\tau_{\lambda_1}(R_0)\tau_{\lambda_2}(R_0)C_{\sigma,N_a}(\lambda_1,\lambda_2) \tag{13.23}$$

Eq. (13.23) is the expression for the case where there is N_a particles in the considered volume. Once each cases of particle number are added together using the weighting probability, the final covariance is derived as

$$C(\lambda_1,\lambda_2) = \left(\frac{1}{R_0^4}\right)^2 Q_{\lambda_1}Q_{\lambda_2}\tau_{\lambda_1}(R_0)\tau_{\lambda_2}(R_0)C_{\sigma,N_a}(\lambda_1,\lambda_2)\sum_{N_a=0}^{\infty} N_a p(N_a). \tag{13.24}$$

Again, the term inside the summation is the mean of the Poisson distribution for the number of particles in the probed volume provided by Eq. (13.16). Introducing this expression for the total number of particles in Eq. (13.24) provides the following result:

$$C(\lambda_1,\lambda_2)=\frac{1}{R_0^6}Q_{\lambda_1}Q_{\lambda_2}\tau_{\lambda_1}(R_0)\tau_{\lambda_2}(R_0)C_{\sigma,N_a}(\lambda_1,\lambda_2)N(R_o)\Delta R\theta_L^2. \tag{13.25}$$

From this last equation, we can conclude that the spectral covariance of the Lidar return is linear with the number of aerosols in the monitored volume. However, Eq. (13.25) also predicts a variation of the spectral covariance of the collected signal with the inverse sixth power of the distance between the probed volume and the sensor. At first, this last result may appear surprising. But it can be deduced from the previous result that every aerosol is statistically independent, each aerosol return is proportional to the inverse fourth power and the spectral covariance is a second-order moment of the number of aerosol. Also from Eq. (13.25), the spectral covariance of the lidar return is expected to be proportional to the divergence θ_L^2 of the laser beam, the range interval of the probed volume ΔR, the square of the delivered laser energy E_L and of the combined atmospheric transmittance $t_{\lambda_0}^a(R_0)t_{\lambda}^a(R_0)$, and, the specific covariance of the aerosol $C_{\sigma,i}(\lambda_1,\lambda_2)$.

Fluorescence from Diverse Atmospheric Aerosols

As expected, the atmosphere is not clean; it contains dust of various origins depending on the monitored area. Many sources, human or natural, can generate dust or aerosols of all kind. Most of the time, if the particles are of a mineral origin, they do not fluoresce (or very little). However, aerosols made of organic materials are a predominant source of induced fluorescence from the atmosphere. Examples of sources from human origins are associated with industrial, transport, or domestic activities. They can be seen as aerosol pollution sources. Natural sources of fluorescing aerosols are associated with dust from drying biological materials, pollens, animals and many others. The origins of these aerosols and their associated concentrations can change with time. Brutal changes, like those shown in Fig. 13.6, can be observed in the atmospheric LIF background signals and must be accounted in the design of an efficient LIF lidar classification procedure.

In Fig. 13.6, left graphic, the color lines designate different events during this atmospheric monitoring. The black lines are anomalies containing only a single measurement. The pink lines are near-anomalous signals. Light and dark green signals are anomalies which came in burst and exceed clearly the threshold set to 60 for the Mahalanobis distance, a statistical spectrometric parameters associated with the measured LIF states of the atmospheric fluorescing aerosols. The Mahalanobis distance D_M is defined mathematically as

$$D_M=(x-x_m)^t C^{-1}(x-x_m). \tag{13.26}$$

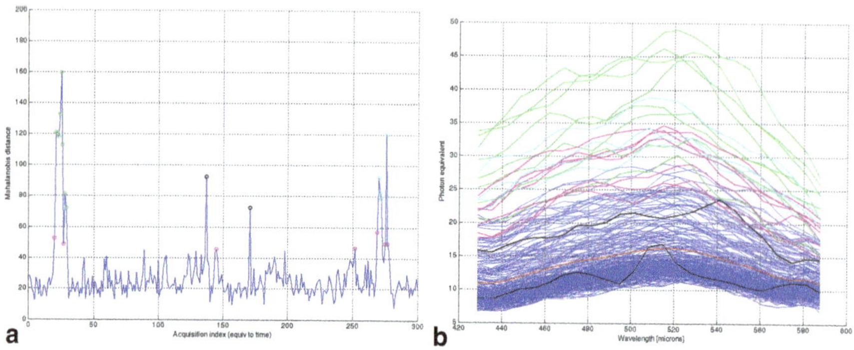

Fig. 13.6 Example of fluorescence from the background

As discussed earlier, the input spectral vector x is a single acquisition with a lidar system resulting from a single or multiple binned laser pulses and x_m is a spectral mean vector. Depending on its application, x_m and C contained in Eq. (13.26) can have different sources. In the case of background monitoring, for example, the mean is the result of the computation of the average of many single contiguous acquisition obtained in the near past. The Covariance C is obtained in a similar way for background monitoring. In the case of classification, the values are obtained quite differently. In this case, x is an input vector obtained from the lidar system. x_m and C are parameters that are scaled and constrained to respect the conditions in which the signal have been acquired. They originate from a database and are the characteristics of the aerosol of interest. x_m may be associated with a single type of fluorescing aerosol (as in Eq. (13.18)) or a quasi-stable atmospheric background aerosol state as it is discussed here. In parallel with x_m, the covariance matrix may correspond to a single type of fluorescing aerosol (as in Eq. (13.25)) or a quasi-stable atmospheric background aerosol state as discussed here. The Mahalanobis distance is a comparison operator from the field of statistical signal processing which is optimal for multivariate data having a normal distribution. The method to compute the Mahalanobis distance and how it can be exploited will be reviewed in the following section addressing fluorescing aerosol classification.

For standoff bioaerosol monitoring, the background fluorescence will be embedded with the signal from the specific aerosolized materials of interest. From a signal processing point of view, this combination of spectral signals must be taken into account. This implies that the spectral statistical characteristics of the background fluorescing signal must be acquired before being mixed with fluorescing aerosols of interest. Since this background signal evolves with time, the signal processing procedure must monitor this evolution continuously. A complete description of this continuous monitoring of the background signal evolution is not detailed in the present document. Nevertheless, this background monitoring procedure will involve comparing each new spectral LIF acquisition with the previous one, searching for variation justifying the initiation of a classification procedure of a suspect event. If there is no variation exceeding a defined Mahalanobis distance threshold between two successive acquisitions, the new acquisition will replace the previous one and will be use in the subsequent comparison.

Classification Procedure

The classification procedure is the process by which an acquired LIF spectrum x is analyzed and compare with a database of fluorescing materials. The classification procedure described in this subsection is based on the one developed for the Canadian BioSense system. This system is based on two Lidar technologies: an eye-safe classical Lidar cloud mapper monitoring an area of interest for suspect aerosol clouds and a UV LIF spectrometric Lidar probing specific air volumes for the presence of bioaerosol threats. During the monitoring of an area of interest, the controlling algorithm may initiate a classification procedure following three distinct scenarios. First, the system operator may use a manual override and asks for the specific classification of a return signal from a selected air volume of interest. Second, it can be directly requested by the cloud mapper after detecting a suspect aerosol cloud. Third, it can be initiated because an anomaly was detected while monitoring the fluorescence background. When one of these three situations occurs, the system changes its state to classification and tries to compare the spectral LIF acquisition with a database of material's signatures. This database contains a number of materials, threatening or not, for which minimal spectral noise laboratory measurements have been obtained.

The classification is performed using a comparison of a measurement with the content of the database after being modified to take into account the specific conditions of that measurement. These modifications are made using the monitored background statistics, the distance to the probed volume and the signal level. Lidar technology provides measurements that are proportional to the concentration of aerosols, the range interval and inversely proportional to the square of the range of the probed air volume as shown in Eq. (13.18). These parameters also affect the covariance following Eq. (13.25). Two other parameters, the laser power and the overall transmittance, will also affect the mean and covariance of the collected spectra. Finally, the last parameter of importance in Eqs. (13.18) and (13.25) is directly related to the aerosols properties, the fluorescing cross-section. This cross-section can be defined as the product of the quantum yield, the geometric projected area and the normalized spectral signature of the fluorescing aerosols [3]. The statistical characteristics of these three parameters should ideally be included in the reference database. However, from a practical point of view, it may be sufficient to get their means and covariances only for the normalized spectral signature. The Mahalanobis distance D_M introduced in the previous section is the mathematical operator that has been selected to compare an acquire spectrum with the reference database and perform the classification.

To compute D_M when a classification is requested, the mean x_m and the covariance C taken from the database are adjusted to the parameters associated with the acquired spectral vector (range, laser power, and global optical transmission). Then, the acquired spectral vector x and the covariance taken from the database are corrected from the contribution of the background induced parasitic fluorescence. To achieve these background corrections, the spectral means from the two background monitoring zones obtained during the surveillance phase are interpolated at the range at which the spectral vector to classify has been taken.[12] This interpolation and associated corrections are performed in two steps. First, the background mean is interpolated while taking into account the inverse square law (Eq. (13.18)) before being subtracted from the acquired spectral vector.[13] Second, the covariance associated with the background parasitic contribution is also interpolated but this time taking into account the inverse sixth power square law (Eq. (13.25)). The resulting interpolated background covariance is added to the corresponding one in the covariance database used for the classification. In the following section, it is assumed that the acquired spectral vector corrected from parasitic contribution will have a positive spectral integral. This is equivalent to assuming that the acquired spectral vector being classified has substantially more signal from a biocloud of interest than from the parasitic background fluorescence.

Computation of the Mean and Covariance from the Database

To compute the mean and covariance of the parameter to be inserted in Eq. (13.26), we have to compare the structure of Eq. (13.18) for the mean and Eq. (13.25) for the covariance. In the database, the mean and the covariance materials to be identified are stored in such a way that at a standard distance, the same scaling factor can be used because they both are linear with respect to the number of particles. To compute the mean signal x_m from the reference database, the fluorescence cross-section in Eq. (13.18) is further detailed [3] as,

$$\frac{d^2\sigma_\lambda(\lambda_0)}{d\Omega d\lambda} = \frac{\Psi_{\lambda_0} A}{4\pi}\zeta_\lambda^{\lambda_0}, \tag{13.27}$$

where, as introduced earlier, Ψ_{λ_0} is the aerosol effective quantum yield, A is the projected geometric area of the inelastic scatterer, and $\zeta_\lambda^{\lambda_0}$ is the normalized spectral cross section for an excitation wavelength λ_0. These parameters describing the spectral cross section of specific fluorescing aerosols are stored in the database in order to compute the covariance of the mean signals. To populate the reference data base, specific acquisitions from clouds made of known aerosols are performed. The spectra acquired at these occasions are corrected from the parasitic background, summed and used to weight $\zeta_\lambda^{\lambda_0}$. The results are the mean signal x_m and are stored in the reference database. In parallel, by inserting Eqs. (13.27), (13.18) and (13.22) in Eq. (13.25), the covariance can be re-written as

$$C(\lambda_1,\lambda_2) = \frac{Q_{\lambda_1}Q_{\lambda_2}\tau_{\lambda_1}(R_0)\tau_{\lambda_2}(R_0)}{R_0^6}\left(\frac{\Psi_{\lambda_0}A}{4\pi}\right)^2 N(R_o)\Delta R\theta_L^2 C_\zeta(\lambda_1,\lambda_2), \tag{13.28}$$

[12] This background sampling is performed regularly at different ranges during the surveillance procedure.

[13] In this approach, the changes in laser power and in the transmission for different ranges between the two background samplings are neglected in the interpolation of the spectral mean and covariance.

where C_ζ is the covariance matrix of the normalized spectral cross section $\zeta_\lambda^{\lambda_0}$ defined as $\left\langle \zeta_{\lambda_1}^{\lambda_0} \zeta_{\lambda_2}^{\lambda_0} \right\rangle - \left\langle \zeta_{\lambda_1}^{\lambda_0} \right\rangle \left\langle \zeta_{\lambda_2}^{\lambda_0} \right\rangle$. In order to facilitate the comparison of the measurement with the database entries, the scaling parameter Q_M is introduced. It is computed from the measurement and will be used to estimate the value of both the mean and the covariance from the database. This scaling parameter is defined as

$$Q_M = \int_{\lambda_o}^{\lambda_1} \frac{x - \mu_\lambda}{Q_\lambda \tau_\lambda} d\lambda = \frac{\Psi_{\lambda_0} A}{4\pi R_0^2} N(R_o) \Delta R \theta_L^2 \int_{\lambda_o}^{\lambda_1} \zeta_\lambda^{\lambda_0} d\lambda, \tag{13.29}$$

where x is the measurement used for comparison, μ_λ and τ_λ are the value of the background fluorescence and of the transmittance at the measurement location, respectively. The second right-hand equality member of this equation is the analytical counterpart of the measurement. Equation (13.28) defines almost completely the spectral covariance for an entry of the reference database. This expression has the advantage of confining all spectral parameters within the element-to-element product of the overall transmittance, the parameter expected to have the highest impact on the error susceptible to be made during a classification procedure with the covariance matrix of the normalized spectral cross-section. To make this expression of a covariance entry usable in the classification procedure, an error consideration term must be included in its estimation. These error terms will inevitably generate more clutter for a class entry of the database associated with a given type of fluorescing aerosol.

Using the database entry, that are the signatures' mean ${}^D\zeta_\lambda^{\lambda_0}$ and covariance ${}^D C_\zeta(\lambda_1, \lambda_2)$, with Q_M and the database quantum yield ${}^D\Psi_{\lambda_0}$ and aerosol's surface ${}^D A$, the mean and covariance to be inserted in Eq. (13.26) for computation of the Mahalanobis distance are

$$x_m = Q_\lambda \tau_\lambda(R_0) Q_M {}^D\zeta_\lambda^{\lambda_0} \text{ and} \tag{13.30}$$

$$C(\lambda_1, \lambda_2) = \frac{Q_{\lambda_1} Q_{\lambda_2} \tau_{\lambda_1}(R_0) \tau_{\lambda_2}(R_0)}{R_0^4} \left(\frac{\Psi_{\lambda_0} A}{4\pi} \right) Q_M {}^D C_\zeta(\lambda_1, \lambda_2). \tag{13.31}$$

To be complete, Eq. (13.31) should contain considerations for the error on the transmittance, the transmittance being an estimate according to a model or from any available atmospheric data. In the case of Eq. (13.30), the error in transmittance is not necessary since it is considered to have a zero mean. For the covariance, it is another matter. However, the development of the effect of the error of the transmittance on the covariance will result in a much more complex equation than Eq. (13.31) and is outside the scope of the present document. In the next subsection, the method to estimate the transmittance is discussed.

The Transmittance

The transmittance (optics and atmospherics) and its associated measurement error have been identified as key parameters in the classification procedure. A promising approach to obtain this information is to use the Raman signal returned from the atmospheric nitrogen collected simultaneously with the induced fluorescence. This approach has the advantage of providing directly the overall transmittance at the Raman wavelength once introducing the Raman cross-section of nitrogen and its atmospheric concentration derived from the meteorological conditions (essentially the local air temperature and pressure). With the spectral transmittance of the lidar system and an atmospheric transmission model like MODTRAN, the overall transmittance can be extended from the nitrogen Raman wavelength to the rest of the spectral interval where induced fluorescence is collected. This is particularly straightforward to perform in the visible atmospheric window where the main attenuation, originating from molecular and aerosol scattering, are generally smooth functions of the wavelength.

On the other hand, the error in the overall transmittance must be estimated from experimental validation of computation model.

Database of Materials

To perform the classification procedure, a reference database of fluorescing aerosols must be constructed. This database should include detailed characteristics on each fluorescing aerosol entry as the spectral mean, the covariance matrix statistics, the average size and the quantum yield. This level of details will make possible the estimation of the concentration of a given detected fluorescing aerosol cloud. An important challenge in obtaining these parameters is to make the spectrometric LIF lidar acquisitions from reproducible and well-characterized disseminated clouds of bioaerosols of interest. To achieve this task, lidar-adapted and lab-sized bioaerosol chambers are being developed.

Conclusions

Standoff biodetection has been recognized as an important capability necessary to monitor effectively aerosolized biological threatening event that may occur over a wide area where critical infrastructures and personnel are located. Such challenge involves continuous monitoring of areas that may cover tens of square kilometers with a refresh rate of the order of a minute. Furthermore, this capability must detect with a high probability minute quantity of bioaerosols with a false alarm rate approaching once a week or better. Presently, the technology that shows the best potential to meet these requirements is light detection and ranging (lidar). In this chapter, several characteristics of a specific lidar applied to the standoff biodetection are detailed. It is the spectrally resolved LIF lidar.

In order to assess the capabilities of spectrometric LIF lidars in standoff biodetection, a fundamental model developed specifically for this type of lidar is presented. This model has the advantages of describing the expected signal as a function of seven groups of key parameters: the aerosol spectral cross-section, the aerosol cloud column, the transmission of the lidar optics, the transmission characteristics of the atmosphere, the geometry of the detection, the laser energy and the optoelectronic conversion module.

Then, the experimental results obtained with a spectrometric LIF lidar prototype at the occasions of multiple trials are reported. These results were analyzed with the objectives of appraising the robustness and the specificity of the spectral properties of the fluorescence induced on different types of bioaerosols as well as the sensitivity, measured in ACPLA, achievable for a given range. Based on these results, good robustness of the spectral characteristics has been observed over several years with simulants of biological agents as long as the growth recipes do not change in an important manner. However, even if the recipe changes are important, the modification on the spectral signature is still sufficiently moderate to be associated with a specific type of bioaerosols that may represent a threat while being quite distinct from benign fluorescing aerosols found naturally in the atmosphere. Also from the analysis of the results obtained at these trials performed over multiple years, sensitivity of a few tens of ACPLA has been obtained for clouds of the BG subgroup 10-m thick located at a range of 1 km.

For an optimum exploitation of the information provided by spectrometric LIF lidars, an effective multivariate tool is described. This tool has the advantages of taking into account the random variability of some of the parameters identified in the fundamental model. From this development, a statistical model for the mean and the covariance of the spectral acquisition is obtained. Then, a description of how this statistical model is used to monitor wide area of interest for a possible threatening biological cloud is detailed.

The results presented here support important capabilities associated with lidar technologies to effectively perform the standoff detection of very small concentrations of biological agents in aerosol forms at multi-kilometer ranges and within seconds. With a scanning process, this technology has the capability to monitor multiple tens of square kilometers with minute refresh rate with only electrical power as consumable. With lessons learned from bulk-induced fluorescence lidars that have shown false alarm rate approaching once a day, spectrally resolved induced fluorescence lidars are expected to extend the false alarm rate to once a week. The lidar instrument and the range provided by this capability may be scaled down. There are ongoing works to develop miniature spectrometric LIF lidars having an effective range of about 100 m. This smaller device, once networked, will provide an important and completely new capability in the prevention, surveillance, and alert management over critical indoor, semi-enclosed and outdoor venues of various geometric complexities. These new systems will improve in an important manner evacuation planning and crowd control, as well as medical casualty management under an aerosolized biological event. In either case, short- and long-range LIF lidar standoff biodetection system have shown important capabilities that make this technology a valuable contender for the 24 h/7 days optical trigger needed in a complete wide area bioaerosol defense suite, such suite including point bioaerosol detection technologies required to confirm a bioaerosol event detected by the optical trigger and identify precisely medical countermeasures to implement.

References

1. Evans BTN, Roy G, and Ho J (1990) The detection and mapping of biological simulants V: statistical investigation of LCM limitation. Report DREV-4616/91 (Defence Research Establishment, Valcartier, Canada)
2. Tiee JJ, Hof DE, Karl RR, Martinez RJ, Quick CR, Cooper DI, Eichinger WE, Holtkamp DB (1992) Remote detection of biological particles and chemical plumes using UV fluorescence lidar. Paper presented at the 16th International Laser Radar Conference, Langley Research Center (NASA), USA, Part 1, p.189-192 (SEE N92-29228 20–35)
3. Simard J-R, Roy G, Mathieu P, Larochelle V, McFee J, Ho J (2004) Standoff sensing of bioaerosols using intensified range-gated spectral analysis of laser-induced fluorescence. In: IEEE Trans. on Geoscience and Remote Sensing 42(4): 865–874
4. Li J-K, Asali EC, Humphrey AE, Horvath JJ (1991) Monitoring cell concentration and activity by multiple excitation fluorometry. Biotechnology Progress 7(1): 21–27
5. Chen RF (1990) Fluorescence of proteins and peptides. In: G. G. Guilbault (ed) Practical Fluorescence, 3rd ed. Marcel Dekker, New York
6. Gabor M (1999) Intrinsic Fluorescence of Proteins and Peptides. In: The Use of Fluorescence in Research into Protein Structures. Available on the Public Web. http://dwb.unl.edu/Teacher/NSF/C08/C08Links/pps99.cryst.bbk.ac.uk/projects/gmocz/fluor.htm). Accessed 7 Aug 2012
7. Dawson RB (1985) Data for biochemical research (3rd ed.). Clarendon Press, Oxford, 122
8. Lakowicz JR, Szmacinski H, Nowaczyk K, Johnson ML (1992) Fluorescence lifetime imaging of free and protein-bound NADH. Proc. Natl. Acad. Sci. U.S.A. 89 (4): 1271–1275
9. Measures RM (1984) Laser-Remote-Sensor Equations. In: Laser Remote Sensing: Fundamentals and Applications. John Wiley & Sons, New York
10. Hill SC, Pinnick RG, Niles S, Pan YL, Holler S, Chang RK, Bottinger J, Chen BT, Orr CS, Feather G (1999) Real-time measurement of fluorescence spectra from single airborne biological particles. Field Analytical Chemistry and Technology, 3:221–239
11. Buteau S, Simard J-R, Rowsell S, Roy G (2010) Bioaerosol standoff detection and correlation assessment with concentration and viability point sensors. SPIE Europe Security and Defense: Optics and Photonics for Counterterrorism and Crime Fighting, SPIE 7838: 78380J1–78380J12
12. Manninen A, Putkiranta M, Saarela J, Rostedt A, Sorvajärvi T, Toivonen J, Marjamäki M, Keskinen J, Hernberg R (2009) Fluorescence cross sections of bioaerosols and suspended biological agents. Applied Optics, 48(22): 4320–4328
13. Bronk BV, Reinisch L (1993) Variability of Steady-State Bacterial Fluorescence with Respect to Growth Conditions. Appl. Spectroscopy, 47:436–440
14. Heaton H I (2005) Principal-components analysis of fluorescence cross-section spectra from pathogenic and stimulant bacteria. Applied Optics, 44(30):6486–6495
15. Laflamme C, Simard J-R, Buteau S, Lahaie P, Nadeau P, Déry B, Houle O, Mathieu P, Roy G, Ho J, Duchaine C (2011) Effect of growth media and washing on the spectral signatures of aerosolized biological simulants. Applied optics, 50(6):788–796
16. Hill SC, Mayo MW, Chang RK (2009) Fluorescence of Bacteria, Pollens, and Naturally Occurring Airborne Particles: Excitation/Emission Spectra. ARL-TR–4722
17. Agranovski V, Ristovski Z, Hargreaves M, Blackall PJ, Morawska L (2003) Performance evaluation of the UV APS: Influence of physiological age of airborne bacteria and bacterial stress. Journal of Aerosol Science, 34:1711–1727
18. Christesen DS, Ong KK (1998) Fluorescence spectroscopy of biological agents - 1. *Bacillusa Anthracis*. Edgewood Research Development and Engineering Center report, ERDEC-TR–466

19. Steinvall O, Jonsson P, Kullander F (2007) Performance for a standoff biological warfare agent detection lidar. Proc. SPIE 6739: 673912.1-673912.14
20. Buteau S, Simard J-R, Déry B, Roy G, Lahaie P, Mathieu P, Ho J, McFee J (2006) Bioaerosols Laser-Induced Fluorescence provides specific robust signatures for standoff detection. Proc. SPIE 6378: 637813.1-637813.12
21. Buteau S, Stadnyk L, Rowsell S, Simard J-R, Ho J, Déry B (2007) Spectrally resolved Laser-Induced Fluorescence for bioaerosols standoff detection. Proc SPIE 6756: 08.1-08.10
22. Jensen GJ (2007) Effect of atmospheric background aerosols on biological agent detectors. Final technical report, Science Application International Corporation, VA, USA
23. Buteau S, Lahaie P, Rowsell S, Rustad G, Baxter K, Castle M, Foot V, Vanderbeek R, Warren R, Marquardt J, Baynard T (2008) Final Report for Task Group (RTG-55) on Laser based stand-off detection of biological agents. NATO RTO SET098 TG55 final report
24. Farsund O, Rustad G, Kasen I, Haavardsholm TV (2010) Required spectral resolution for bioaerosol detection algorithms using standoff laser induced fluorescence measurements. IEEE Sensor Journal, 10(3):355–661
25. Hopkins RJ, Barrington SJ, Castle MJ, Baxter KL, Felton NV, Jones J, Griffiths C, Foot V, Risbey K (2007) UV-LIF lidar for standoff BW aerosol detection. Proc. SPIE 7484:748409-748409-11
26. Buteau S, Ho J, Simard J-R, Lahaie P, McFee J, Roy G, Mathieu P, Déry B (2008) Bioaerosol standoff monitoring using intensified range-gated laser-induced fluorescence spectroscopy. In: Kim, Young J., Platt, Ulrich (Eds.) Advanced Environmental Monitoring, Chap. 16, Springer-Verlag, p. 203–216.
27. Cheng YS, Barr EB, Fan BJ, Hargis PJ, Rader DJ, O'Hern TJ, Torczynski JR, Tisone GC, Preppernau BL, Young SA and Radlioff RJ (1999) Detection of bioaerosol using multiwavelength UV fluorescence spectroscopy. Aerosol Science and Technology 30:186–201
28. Sivaprakasam V, Huston A, Scotto C, Eversole J (2004) Multiple UV wavelength excitation of bioaerosols. Opt. Express 12(19):4457
29. Huang HC, Pan YL, Hill SC, Pinnick RG, Chang RK (2008) Real-time measurement of dual-wavelength laser-induced fluorescence spectra of individual aerosol particles. Opt. Express 16(21):16523–16528
30. Pan YL, Hill SC, Pinnick RG, Huang H, Bottiger JR, Chang RK (2010) Fluorescence spectra of atmospheric aerosol particles measured using one or two excitation wavelengths: Comparison of classification schemes employing different emission and scattering results. Opt. Express 18:12436–12457
31. Mierczyk Z, Kopczyński K, Zygmunt M, Wojtanowski J, Młynczak J, Gawlikowski A, Młodzianko A, Piotrowski W, Gietka A, Knysak P, Drozd T, Muzal M, Kaszczuk M, Ostrowski R, Jakubaszek M (2011) Fluorescence/depolarization lidar for mid-range stand-off detection of biological agents. Proc. SPIE 8037:80371J.1–9
32. Carrano J, Jeys T, Cousins D, Eversole J, Gillespie J, Heally D, Licata N, Loerop W, O'Keefe M, Samuels A, Schultz J, Walter M, Wong N, Billote B, Munley M, Reich E, Roots J (2005) Chemical and Biological Sensor Standards Study. Defense Advanced Research Projects Agency, Department of Defense, DTIC, Fort Belvoir, VA
33. Buteau S, Simard J-R, Rowsell S (2009) Bioaerosol standoff detection simultaneously refereed with particle concentration (ppl) and viability units (ACPLA). Proc. SPIE 7484:748408.1–12
34. National Research Council of the national academies (2008) A framework for assessing the health hazard posed by bioaerosols. The National Academies Press, Washington, D.C.
35. Valdes JJ, Mohr J, Mackay R, Tunia E, Mara A (2010) Total Agent per Liter of Air With Particle Size Distribution (TALAp): A New Unit of Measure for the Test and Evaluation of Biodetectors. International Test and Evaluation Association Journal, 31:417–425
36. Buteau S, Simard J-R, Rowsell S, Roy G (2010) Bioaerosol standoff detection and correlation assessment with concentration and viability point sensors. Proc. SPIE 7838:78380J.1–12
37. Baker J (2011) US Army Dugway Proving Ground: DPG as the Chem/Bio MRTFB Activity. http://www.dtic.mil/ndia/2011test/11626WednesdayBaker.pdf, Accessed 20 Sep 2012
38. Ratnesar-Shumate S, Wagner ML, Kerechanin C, House G, Brinkley KM, Bare C, Baker NA, Quizon R, Quizon J, Proescher A, Van Gieson E, Santarpia JL (2011) Improved Method for the Evaluation of Real-Time Biological Aerosol Detection Technologies. Aerosol Science and Technology, 45:635–644

39. Theiler J, Wohlberg B (2012) Detection of spectrally sparse anomalies in hyperspectral imagery. Paper presented at IEEE Southwest symposium on Image Analysis and Interpretation (SSIAI), Santa Fe, NM, USA, Apr 2012
40. Poor HV (1994) An introduction to signal detection and estimation. 2nd edn. Springer-Verlag, New-York
41. Kelly EJ (1986) An adaptive detection algorithm. In: IEEE Transactions on aerospace and electronic systems, AES-22(2):115–127
42. Reed IS, Yu X (1990) Adaptive Multiple band CFAR detection of an optical pattern with unknown spectral distribution. In: IEEE Transactions on acoustics, Speech and Signal processing, 38(10): 1760–1770

Chapter 14
Standoff Aerosol Size Determination based on Multiple-Field-Of-View of Elastic Scattering

Gilles Roy and Nathalie Roy

Introduction

The extinction coefficients of particles with size significantly smaller than the probing wavelength varies in an important manner as a function of their size; multiple wavelength lidars have been used with success to characterize background aerosol sizes [1–2]. For particles greater than the probing wavelengths, the extinction coefficients show little sensitivity over the wavelength and this technique becomes not applicable. However, a technique based on multiple scattered lights can be used.

It is now established that the use of Multiple-Field-Of-View (MFOV) lidar to measure multiple scattering produced in optically dense clouds is a source of retrievable information on particle size [3–4]. These innovative results and recent developments with gated-ICCD (intensified charged coupled device) cameras have led to the standoff measurement of bioaerosols size and concentration [5–7]. The difficulty with bioaerosol clouds is that they are sub-visible and therefore their low concentrations do not produce strong multiple scattering signals. In addition, they are mixed with background aerosols. However, it is possible to obtain size information and to retrieve optical depth of low concentration bioaerosols by using the MFOV lidar measurements when the background aerosol contribution is subtracted. The concept relies on the measurement of the forward scattered light as a function of FOV using the background aerosols at a given distance at the back of a sub-visible cloud.

G. Roy (✉) · N. Roy
Defence Research and Development Canada—Valcartier, 2459,
route de la bravoure, Québec G3J 1X5, Canada
e-mail: Gilles.roy@drdc-rddc.gc.ca

N. Roy
e-mail: nathalie.roy@drdc-rddc.gc.ca

P. Jonsson et al. (eds.), *Bioaerosol Detection Technologies*, Integrated Analytical Systems, DOI 10.1007/978-1-4419-5582-1_14, © Springer-Verlag New York 2014

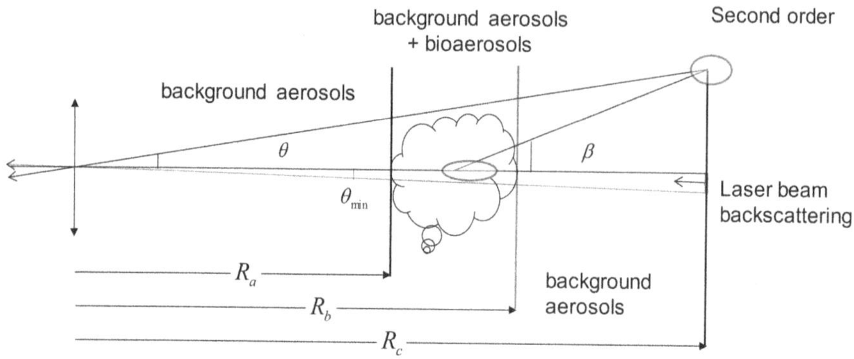

Fig. 14.1 Second-order scattering detected at an angle θ by the MFOV lidar; *ellipses* represent the phase functions

Theory

Figure 14.1 illustrates a second order scattering event: a forward scattering occurs at an angle β and is followed by a backscattering that is collected at an angle θ. For angles $\theta > \theta_{min}$, the scattered radiation comes from multiple scattering events. In addition of detecting multiply scattered radiations, segmenting the FOV is required to resolve the angular geometry for the proposed retrieval method. We thus speak of a MFOV receiver and MFOV lidar returns. The retrieval of information on the sub-visible particles cloud is based on some assumptions:

- The laser beam divergence is small;
- The backscatter power originating from single scattering events is concentrated inside the smallest field of view θ_{Min} of the detector;
- The extinction and the particle size distribution of the atmospheric aerosols are spatially homogeneous;
- The double scattered light is measured from the background aerosol at a given distance at the back of the bioaerosol cloud;
- The path length difference of the multiple scattered photons versus the single scattered photons is: $\Delta R = R_c\theta^2/2$. At 1 km distance and for a FOV of 5 mrad, the difference in path is 1.2 cm. This is negligible.

Detail discussion on the limit and feasibility of the technique will be discussed in the section named: Limits of the method.

The bioaerosol cloud has an extinction coefficient α_b and an albedo ω_{0b} while the background aerosol has an extinction coefficient α_a and an albedo ω_{0a}.

The single scattering lidar equation provides the backscatter light power, $P_s(R_c)$, coming from the probing distance R_c:

$$P_s(R_c) = P_0 \frac{c\tau}{2} \frac{A}{R_c^2} \omega_{0a} \alpha_a(R_c) p_a(r, R_c, \beta_{back} = \pi)$$
$$\exp\left[-2\int_0^{R_C} (\alpha_a(R) + \alpha_b(R) dR\right], \qquad (14.1)$$

where as usually, P_0 is the laser pulse power, c the speed of light, τ the pulse duration, A the area of the collecting optic, R_c the sounding distance, $p_a(r, R_c, \beta_{back} = \pi)$ the value of the phase function for a scattering angle β of 180°. Note the backscatter light is proportional to the background aerosol extinction coefficient at position R_c. The bioaerosol and the background aerosol attenuate the backscatter light over their respective value and spatial distribution. For compactness, the total extinction coefficient will be written $\alpha(R) = \alpha_a(R) + \alpha_b(R)$ and the scattering extinction coefficient $\alpha_s(R) = \omega_{0a}\alpha_a(R) + \omega_{0b}\alpha_b(R)$.

The essence of the MFOV lidar technique is the measurement of the scattered power as a function of the receiver field of view θ. The scattered power contains information on scatters size. In Fig. 14.1, the angle θ corresponds to one half of the total lidar FOV. According to previous assumptions, for angles $\theta > \theta_{min}$ the scattered radiation must come from multiple scattering events. The scattered power in the FOV interval $\Delta\theta_i = \theta_{i+1} - \theta_i$ coming from second scattering order can be calculated using:

$$P_D(R_c, \Delta\theta_i) = P_0 \exp\left[-2\int_0^{R_c} \alpha(R) dR\right] \frac{c\tau}{2} \frac{A}{R_c^2}$$
$$2\int_0^{R_c} [\alpha_s(R) \Delta L(r, R, \beta_{i+1} - \beta_i)] \alpha_s(R_c) p(r, R_c, \beta_{back}) \mathrm{d}R, \quad (14.2)$$

where $\Delta L(r, R, \beta_{i+1} - \beta_i)$ is the fraction of the energy contained in the ring delimited by $(\beta_{i+1} - \beta_i)$:

$$\Delta L(r, R, \beta_{i+1} - \beta_i) \equiv \int_{\beta_i}^{\beta_{i+1}} \int_0^{2\pi} p(r, R, \beta) \sin\beta d\beta d\phi, \qquad (14.3)$$

$p(r, R, \beta)$ and $p(r, R_c, \beta_{back})$ are the values of the phase function for the forward (β) and backward scattering angles ($\beta_{back} = \pi - \beta + \theta$) for a particle of radius r, R_a and R_b are the distances delimiting the bioaerosol cloud, R_c is the distance where the scattered radiation is measured, the quantity $[\alpha_s(R)p(r, R, \beta)]$ represents the forward scattering coefficient while $[\alpha_s(R_c)p(r, R_c, \beta_{back})]$ represents the backscattering coefficient, and φ is the azimuthal angle ranging from 0 to 2π. The FOV θ is easily related to the scattering angle β in a simple geometric relation: $\tan\beta = \frac{R_c \tan\theta}{R_c - R}$.

Now, we consider that $p(r, R_c, \beta_{back})$ is uniform over the backscattering angles considered and we set its value equal to $p(r, R_c, \beta_{back} = \pi)$ and then combining Eq. (14.1) to (14.3) we obtain:

$$P_D(\mathrm{R_c}, \theta_{i+1} - \theta_i) = 2P_s(R_c)\int_0^{R_c} \alpha_s \Delta \mathrm{L}(\mathrm{r}, \mathrm{R}, \beta_{i+1} - \beta_i)\mathrm{dz}. \quad (14.4)$$

The integral over R is broken in 3 intervals: $[0, R_a]$, $[R_a, R_b]$ and $[R_b, R_c]$. In absence of bioaerosol, we have for the three intervals, if Raleigh scattering is considered and background aerosol and Raleigh scattering are homogeneous (no range dependence):

$$\alpha_s(R)p(r, R, \beta) = \alpha_a \omega_{0a} p_a(r, \beta) + \alpha_R p_R(r, \beta), \quad (14.5a)$$

and in presence of bioaerosols in the interval $[R_a, R_b]$ we have:

$$\alpha_s(R)p(r, R, \beta) = \alpha_a \omega_{0a} p_a(r, \beta) + \alpha_b(R)\omega_{0b} p_b(r, \beta) + \alpha_R p_R(r, \beta), \quad (14.5b)$$

where α_R, $p_R(r, \beta)$ are the Rayleigh scattering coefficient and phase function;

The recovery of the effective size parameter is achieved by the application of a simple mathematic formula which only requires knowledge of basic lidar information such as the cloud distance and the sounding depth as well as two measurements realized with a MFOV imaging lidar. The first lidar measurement must be done in presence of the sub-visible cloud under studied (α_b>0), while the second, measured in absence of the bioaerosol cloud (where α_b=0), is used to quantify background aerosols. Thereafter, we apply the following equation to experimental data on a ring defined by:

$$\Delta P_{Norm}(\theta_{i+1} - \theta_i) = \frac{P_D(R_c, \theta_{i+1} - \theta_i, \alpha_b > 0)}{P_s(R_c, \theta_s, \alpha_b > 0)} - \frac{P_D(R_c, \theta_{i+1} - \theta_i, \alpha_b = 0)}{P_s(R_c, \theta_s, \alpha_b = 0)}, \quad (14.6)$$

$$= 2\varpi_{0b} \int_{z_a}^{z_b} \alpha_b \Delta L(r, R, \beta_{i+1} - \beta_i)dz. \quad (14.7)$$

Based on Fig. 14.1, the FOV θ is easily related to the scattering angle β with the simple geometric relation $\tan\beta = R_c \tan\theta/(R_c - \bar{R})$ where $\bar{R} = 0.5\,(R_a + R_b)$ and, R_a and R_b correspond to the limits of the small optical depth cloud under studied. The measurement of $\Delta P_{Norm}(\theta_{i+1} - \theta_i)$ can provide, in principle, information on the particles size and on the sub-visible cloud concentration. For that, we parameterize the phase function with a Gaussian fit using the definition of the effective diameter (d_{eff}) suggested by Bissonnette [8]:

$$p(\beta) = \frac{A_2}{\pi \omega_{0b}} y^2 e^{-A_1^2 y^2 \beta^2}, \tag{14.8}$$

where $y = (\pi\, d_{eff})/\lambda_0$, $A_1 = 0.544$, $A_2 = 0.139$ and $d_{eff} = 2<r^3>/<r^2>$. In the definition of d_{eff}, the <> denotes the average of the quantity between brackets among all aerosols of the cloud of interest. By combining Eqs. (14.4)–(14.8) and by using the relation $\beta \simeq R_c\theta/(R_c - \bar{R})$, we obtain

$$\Delta P_{Norm}(\theta_{i+1} - \theta_i) = 2\tau_b \frac{A_2}{A_1^2} \left\{ \exp\left[-A_1^2 y^2 \left(\frac{R_c}{R_c - \bar{R}} \right)^2 \theta_i^2 \right] - \exp\left[-A_1^2 y^2 \left(\frac{R_c}{R_c - \bar{R}} \right)^2 \theta_{i+1}{}^2 \right] \right\}, \tag{14.9}$$

where τ_b is the optical depth of the bio-aerosol cloud. The measurement of $\Delta P_{Norm}(\theta_{i+1} - \theta_i)$ provides the key information that relates, via Eq. (14.9), the normalized double-scattering measurement to the bioaerosols cloud optical depth and the encircled energy. Figure 14.2a and b shows the value of $\Delta P_{Norm}(\theta_{i+1} - \theta_i)$ as a function θ_{i+1} for particles having diameters of 10, 20 and 35 µm. The symbols are obtained by applying Eq. (14.6) and by using the Mie theory to calculate the bioaerosol phase function while the continuous curves are obtained by applying Eq. (14.9). The cloud extends from 120 to 130 m and the MFOV signals are measured at distances of 140 (Fig. 14.2a) and 178 m (Fig. 14.2b) respectively. The agreement between simulated results and predictions obtained via Eq. (14.9) is quite good for the positions of the maximums. However, for FOVs greater than those associated with the peak position of $\Delta P_{Norm}(\theta_{i+1} - \theta_i)$, the Gaussian model shows lower values since it takes into account only the diffraction component of the phase function neglecting its geometric counterpart, which is independent of the particle size. Because our interest is in the position of the diffraction peak, the discrepancy associated with the scattering geometric component has been deliberately ignored.

The signals displayed in Fig. 14.2 show that the position of the maximum encircled energy is dependent on the cloud distance relative to the lidar system, the sounding distances as well as it is highly related to the bioaerosols size. In [6], it is demonstrated, when the angular increments defining the MFOV probed rings are log-spaced, that the particle effective diameter d_{eff} can be derived from $\theta_{max,\,Log}$ defining the peak position of $\Delta P_{Norm}(\theta_{i+1} - \theta_i)$ as

$$d_{eff} \cong 0.58 \frac{\lambda_0}{\theta_{\max,Log}} \frac{R_c - \bar{R}}{R_c}. \tag{14.10}$$

The effective diameter is a function of the position of the cloud ($\bar{R}$), the distance at which the backscatter light is measured (R_c), the probing wavelength and the position of the maximum ($\theta_{\max,Log}$). Figure 14.3, shows the value of the $\theta_{\max,Log}$ as a

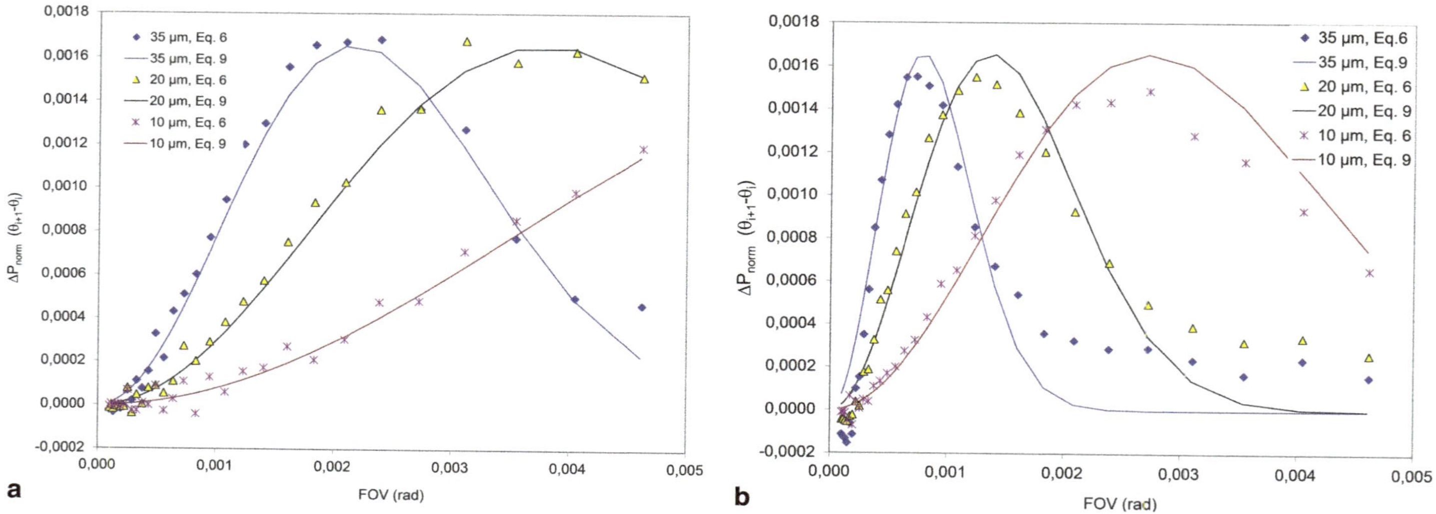

Fig. 14.2 $\Delta P_{Norm}(z_c, \theta_{i+1} - \theta_i)$ as a function of $0.5(\theta_{i+1} + \theta_i)$ for particle diameter of 10, 20 and 35 μm and for sounding distances of **(a)** 140 m and **(b)** 178 m respectively. From [6] with authorization from Applied Optics

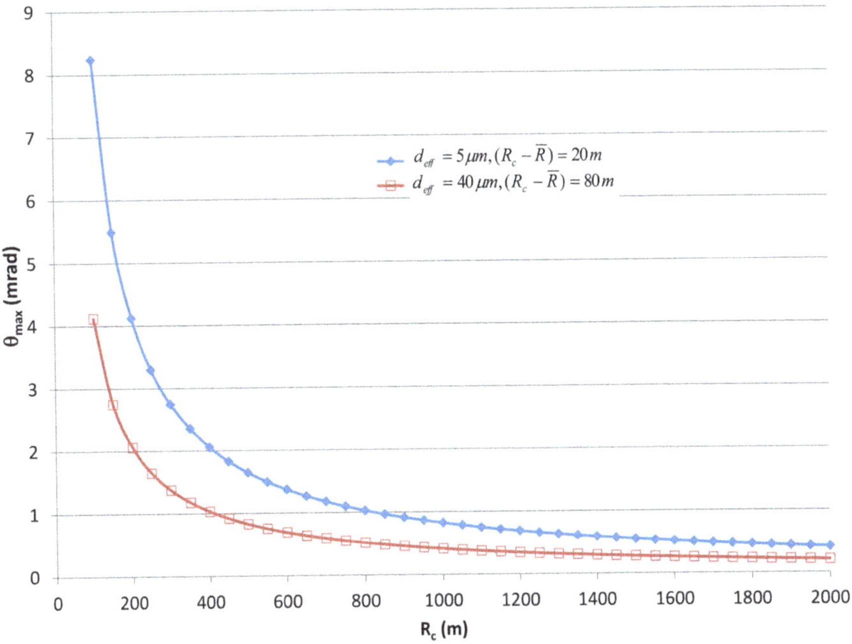

Fig. 14.3 $\theta_{max,Log}$ as a function of the probing distance, R_c for effective diameter, d_{eff} of 5 and 40 μm; the probing distance are 20 and 80 m respectively at the back of the cloud average distance from sensor

function of the probing distance, R_c for effective diameter, d_{eff} of 5 and 40 μm; the probing distance is 20 and 80 m at the back of the cloud average distance respectively. At close range, FOVs up to 8 mrad are required to resolve the 5 μm particles. At far range, FOVs around 0.2–0.5 mrad are required.

Limits of the Method

The recovering of particle parameters based on the measurement of multiple scattering will be successful as long as the size of particle of interest is compatible with the sounding wavelength of the laser, the field-of-views used, the width of the cloud and the sounding distances. Using Eq. (14.10) we can write the following constraint:

$$0.58 \frac{\lambda_0}{\theta_{FOV\max}} \frac{R_c - \bar{R}}{R_c} < d_{eff} < 0.58 \frac{\lambda_0}{\theta_{FOV\min}} \frac{R_c - \bar{R}}{R_c}, \tag{14.11}$$

where $\theta_{FOV\,min}$ and $\theta_{FOV\,max}$ correspond respectively to the smallest FOV (half angle) containing almost all laser energy and to the largest FOV covered by the detector.

With reasonably good optic at the emission and collection a $\theta_{FOV\,min}$ = 0.15 mrad and $\theta_{FOV\,max}$ = 8 mrad can be achieved.

Figure 14.4a shows the maximal and minimal effective diameters that could possibly be measured as a function of R_c for probing distance $R_c - \bar{R}$ equal to 20 and 80 m. Figure 14.4b also shows the maximal and minimal effective diameters that could possibly be measured as a function of $R_c - \bar{R}$ for a 20 m depth clouds with averages position $\bar{R}$ equal to 180, 380 and 780 m. The calculations have been performed using Eq. (14.11) at a wavelength of 355 nm. As an example, for a cloud ranging from 370 to 390 m and by performing the measurement 10 m at the back of the cloud, particles with effective diameter between 1 and 70 μm could be resolved in size. For the exact same cloud, if measurements are performed 200 m at the back of the clouds, particles with effective diameter within 4 and 236 μm could be resolved.

The dependence of the retrieved effective diameter over the probing distance $(R_c - \bar{R})$ has a strong impact on the success and on the limitations of the inversion algorithm developed. Considering the analysis of a cloud at a mean altitude $\bar{R}$ with a given lidar system (λ_0, $\theta_{FOV\,\min}$ and $\theta_{FOV\,\max}$ are fixed) combining two measurements with different probing distances can easily increase significantly the range of possible effective diameters than can be retrieved, as illustrated on Fig. 14.4a.

The proposed measurement method could be applied on clouds made of particles smaller than the wavelength. However, the equation that relates the effective diameter to the ratio of the wavelength over the maximum energy position ($\theta_{\max,Log}$), Eq. (14.10)), is no longer valid because the mathematical expression used to parameterize the phase function is based on diffraction effect, [6]. So, it is valid only for particle size greater than the wavelength. With proper parameterization of the phase function for particles smaller than the wavelength it should be possible the retrieve the effective diameter of particles smaller than the wavelength. However, further studies are required to parameterize the phase function of particles smaller than the probing wavelength into a suitable form. Under these conditions, the value of the refractive index should be taken into account.

Experimental Validation

Figure 14.5 illustrates the principle of operation of an imaging MFOV lidar. The position of the image plane is a function of the focal length and the object position, and it is necessary to adjust the image plane position in accordance with the cloud distance. A fused silica window is used to reflect part of the backscatter light on a conventional lidar module while the transmitted light through the quartz window is re-imaged on the ICCD. The conventional lidar channel is used to determine the bioarosol cloud position, an essential parameter to calculate the time delay to be applied to the ICCD camera. For the azimuthally resolved MFOV measurements, an intensified CCD (ICCD) camera is used. The camera is cooled to reduce dark signal and its associated shot noise and to range-resolve the recorded image, the

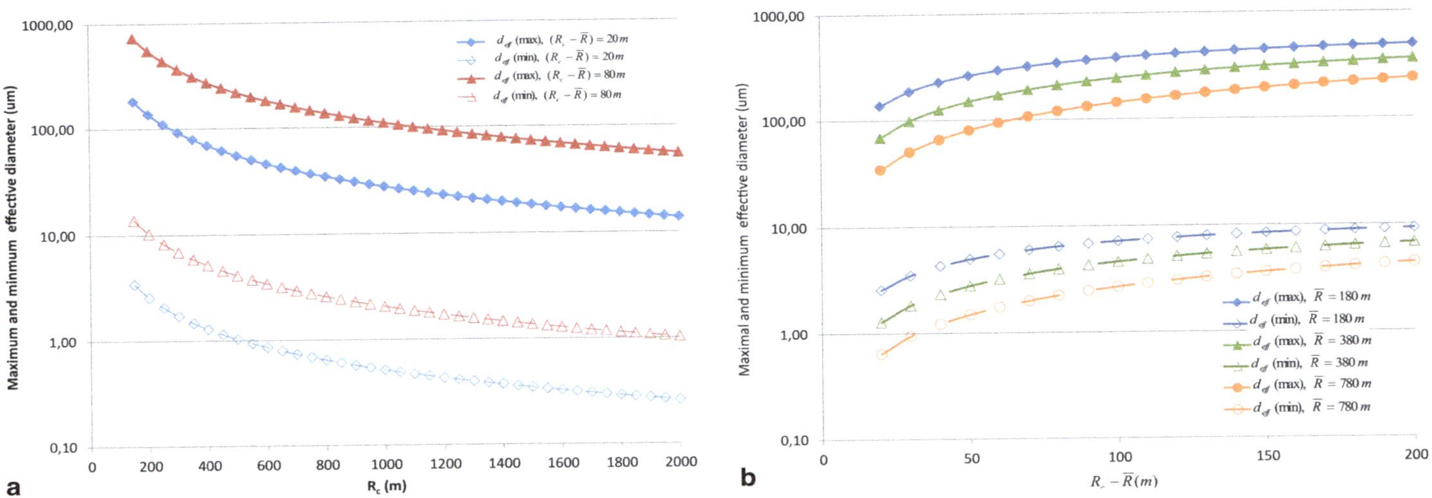

Fig. 14.4 **a** Maximum and minimum effective diameter as function of the probing distance, the probing distance of 20 and 80 m, respectively, at the back of the cloud average, the probing wavelength is 355 nm and the FOVs range is 0.15–8.0 mrad. **b** Maximum and minimum effective diameter as function of the probing distance at the back of the clouds, $R_c - \bar{R}$. Curves correspond to clouds average distance ($\bar{R}$) of 180, 380 and 780 m, the probing wavelength is 355 nm and the FOVs range is [0.15–8.0 mrad]

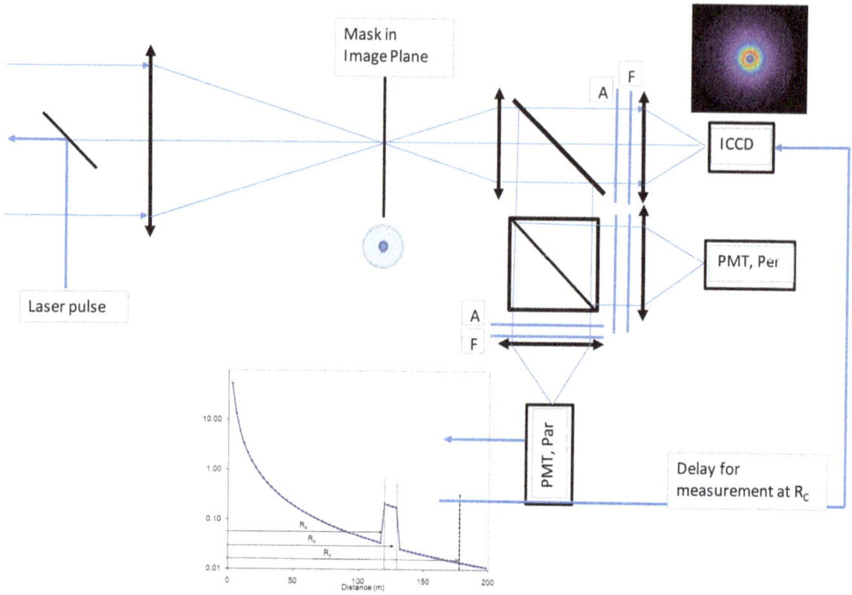

Fig. 14.5 A lidar with MFOV capability using an ICCD camera. A PMT is used to locate the cloud position and a mask is used to enhance the multiple scattering signal by attenuation of the central single scattering signal. *A* Attenuator, *F* laser line filter

micro-channel plate intensifier is gated with a pulse generator activated by the Q-switch of the lidar's laser.

The conventional lidar module illustrated on Fig. 14.5 has not a standard configuration since it is composed of a polarization detection module allowing discriminating of water clouds from bioaerosol clouds. The principle of operation is based on the simple facts that small homogenous liquid droplets suspended in air are spherical, and that spherical homogenous scatters do not depolarize the radiation at exactly 180°, whereas irregular-shaped solid particles give rise to significant depolarization. The polarization measurement module is nice to have but is not essential for the effective diameter recovery discussed in here.

The bioaerosol size characterization technique requires two lidar measurements: one in the presence of the sub-visible cloud under studied ($\alpha_b > 0$), and a second measured away from the biocloud (where $\alpha_b = 0$) to quantify background aerosols. Figure 14.6 resumes the schema of the data acquisition and treatment. For each measurement event, the number of pulses needed to sum on the camera chip before reading the CCD is determined to optimise the camera acquisition speed and reduce the noise level, especially the readout noise. The light background is then measured between laser pulses just prior to acquiring the lidar return by opening the camera gate the same number of times as for the lidar measurements. Finally, the background image is subtracted from the image of the lidar return, and the difference in pixel sensibility is corrected for the radial response of the whole system.

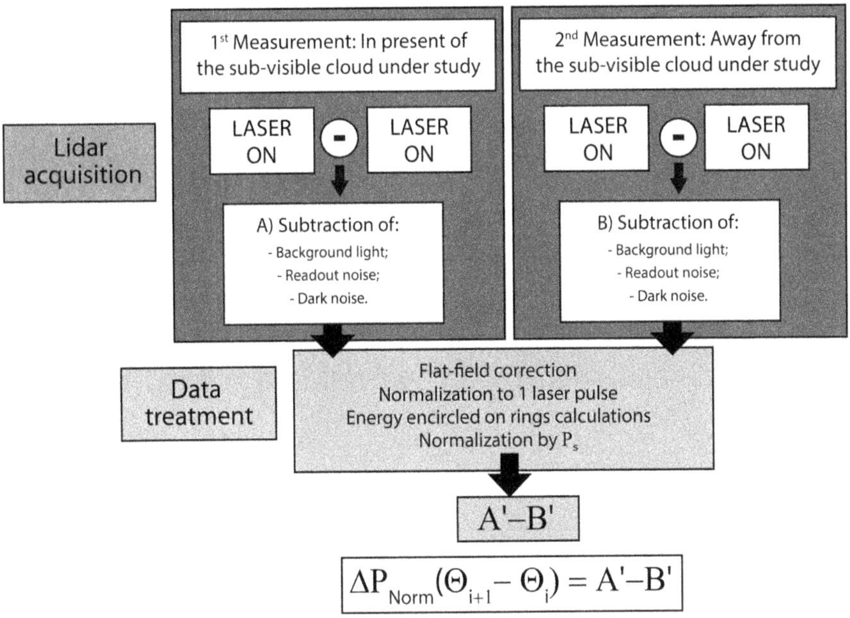

Fig. 14.6 Data acquisition and treatment schema. From [7] with authorisation from Applied Optics

Figure 14.7 illustrates the effective diameter retrieval method steps with the help of a typical dataset. Once the effective diameter is retrieved from Eq. (14.10), the optical depth can be easily obtained by using a simple mathematical formula that requires the same basic knowledge as for the size parameter retrieval. Since experimentally it is necessary to use a mask with a transmission T_{mask} to attenuate the first backscattering order in order to obtain a good dynamic range of the intensified CCD (ICCD) camera, the experimental optical depth is retrieved after rearrangement of Eq. (14.9) by using:

$$\tau_b = \frac{-\Delta P_{Norm}\left(\theta_{i+1} - \theta_i\right) T_{mask}}{\frac{A_2}{A_1^2}\left[e^{\left[-A_1^2 y^2 \left(\frac{R_c}{R_c - \bar{R}}\right)^2 \theta_i^2\right]} - e^{\left[-A_1^2 y^2 \left(\frac{R_c}{R_c - \bar{R}}\right)^2 \theta_{i+1}^2\right]}\right]}, \qquad (14.12)$$

The technique has been validated with diffractive "target" plates consisting of etched metallic disks of known size randomly distributed over a circular area of 15 cm in diameter [7]. The number of particles has been established so that the ratio of the area of the etched particles over the total surface is equal to 2%. Target plates were made with 5, 20 and 50 μm diameter etched circles. Table 14.1 shows a comparison of the mean diameter and standard deviation of the pollens as measured with the help of a microscope and the effective diameters retrieved with the MFOV

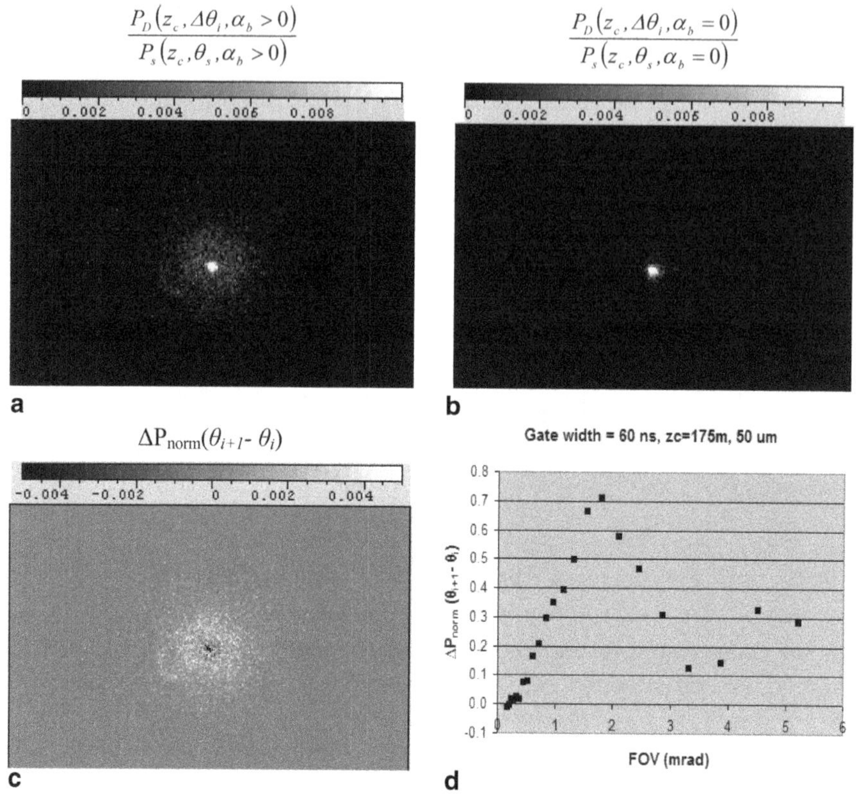

Fig. 14.7 (**a**) The normalized lidar returned on the background aerosols obtained in the presence of diffractive target with etched particles; (**b**) the same measurement in presence of reference target; Fig. 14.7c = Fig. 14.7a, b and d illustrated the image analysis in term of encircled energy in logarithmic spaced rings delimited by θ_i and θ_{i+1}. From [7] with authorization from Applied Optics

Table 14.1 Comparison of the mean diameter measured with a microscope and the MFOV lidar

Pollen type	Elm		Timothy	
Measurement technique	Photograph	Lidar	Photograph	Lidar
Mean diameter (μm)	26.8	29.1	34.3	35.5
Standard deviation (μm)	2.4	1.5	1.5	3.2

lidar measurements. The agreement between the two measurement techniques is very good. Experimental results have clearly showed that it is possible to characterize adequately target plates simulating aerosol clouds with fixed temporal characteristics, in some cases, even during daytime, with precisions on the size and optical depth better than 10% for optical depths as low as 0.005 and 0.016 during night and day, respectively. Similar experimental validations were also performed on clouds made of elm and timothy pollens with similar precisions [6].

The sensitivity of the MFOV technique at night is essentially limited by the shot noise related to the laser scattered signal itself while the aerosol background fluctuations in size distribution and in quantity are not significant error factors. In addition to the increased shot noise factor, the daytime measurements are also considerably affected by the background light fluctuations. So, daytime measurements require a proper subtraction of background light and an averaging of the signal over a longer period of time. Finally, the laser pulse-to-pulse energy fluctuations have little effect on the precision of the retrieval technique because of the normalization of the quantities by the single scattering return in Eq. (14.6).

Discussion and Conclusions

Evaluation of the size and concentration of fluorescing aerosols from a standoff position can be important information contributing to the reduction of the false alarm rate in detecting aerosolized biological threat. MFOV lidar can provide this information under some condition. In order to estimate such potential, a simple geometric model is reported and validated with a MFOV lidar prototype challenged with diffractive etched metallic disks simulating a cloud of aerosols having 5, 20 and 50 μm in size. From measurements done at ranges of a few hundreds of meters, precisions better than 10% on the diameter of the simulated aerosol cloud were obtained for optical depth as low as 0.005 and 0.016 during night and day, respectively. Similar experimental validations were also obtained with challenges made of clouds of pollens and were reported in open-literature.

The experimental proof of concept has been performed at night when the wind was mild. Successful measurements were also conducted during the day, but it becomes quickly apparent that under important background light intensity fluctuations and under large fluctuations of the amount of background aerosols (windy conditions) the subtraction of the background contribution was not always satisfactory. Daylight measurements successes are highly dependent on the lidar system optical components such as the interference filter bandwidth, the FOV values and the laser features such as its pulse width and its divergence. In this paper we have focused on the proof of concept. Further studies will be required to study the particularities of daylight measurement.

The present model aimed the standoff measurement of the effective diameter of a very small amount of bioaerosols. It is limited to second order scattering and it is reliable for small optical depths. Contributions from scattering order higher than 2 will start to be noticeable for optical depth higher than 0.3 [9]. Application of the present model to cloud with optical depth higher than 0.3 will require corrections/adjustments. Application of the measurement technique on clouds with optical depths ranging from 0.1 to 3 is required to study the problematic. Monte Carlo simulations could be a suited tool for that type of investigation.

Finally the experimental data have been obtained with an ICCD camera. It is not necessary to use an ICCD camera to do the measurements [4]. Whatever technique used, it is necessary to determine the FOV dependence of the sensor and to perform subtraction of the background radiation. It is also required that the measurement of the scatter light is done in the image plane, not at the focal point. The difference between the focal point and image plane is particularly important when sounding distances get short. In that case, it could produce notable alterations of the shape of the signal measured in function of the receiver FOV that will affect the retrieved position of the $\Delta P_{Norm}(\theta_{i+1} - \theta_i)$ peak and so the success of the method we developed. ICCD cameras present characteristics that make the measurements relatively easy. For example, they can be gated to match with the laser pulse width; they are programmable; the number of FOVs and their values can be defined prior to the analysis of data and the number of accumulations on the ICCD chips can be adjusted for the full dynamic range utilization of the camera.

References

1. D. Muller, U. Wandinger and A. Ansmann, “Microphysical particle parameters from extinction and backscatter lidar data by inversion with regularization: theory,” Appl. Opt. 38, 2346–2357 (1999).
2. I. Veselovskii, A. Kolgotin, D. Muller, D. Whiteman, “Information content of multiwavelength lidar data with respect to microphysical particle properties derived from eigenvalue analysis, ”Appl. Opt. 44, 5292–5302 (2005).
3. I. Veselovskii, Michail Korenskii, Vadim Griaznov, David N. Whiteman, Matthew McGill, Gilles Roy, Luc Bissonnette, “Information contain of data measured with multiple-field-of-view lidar,” Appl. Opt. 45, 6839–6848 (2006).
4. Luc R. Bissonnette, Lidar: Range-Resolved Optical Remote Sensing of the Atmosphere, chapter 3: Lidar and Multiple Scattering, Editor Claus Weikamp (Springer Series in Optical Sciences, 2005), p. 43–103.
5. N. Roy, G. Roy, L. R. Bissonnette, and J.-R. Simard, “Measurement of the azimuthal dependence of cross-polarized lidar returns and its relation to optical depth,” Appl. Opt. 43, 2777–2785 (2004).
6. G. Roy and N. Roy, “Standoff determination of particles size and concentration of small optical depth clouds based on double scattering measurements; concept and experimental validation with bioaerosols,” Appl. Opt. 47 no 9, 1336–1349 (2008).
7. N. Roy and G. Roy, “Standoff determination of particles size and concentration of small optical depth clouds based on double scattering measurements; validation with calibrated target plates and limitations for daytime and nighttime measurements,” Appl. Opt. 47 no 23, 4235–4252 (2008).
8. L. Bissonnette, “Multiple-scattering lidar equation,” Appl. Opt. 33, 6449–6465 (1995).
9. G. Roy, L. R. Bissonnette, C. Bastille, and G. Vallée, “Retrieval of droplet-size density distribution from multiple field-of-view cross-polarized lidar signals,” Appl. Opt. 38, 5202–5211 (1999).

Part IV
Outlook and Challenges

Chapter 15
Trends in Biological Detection

Per Jonsson and Torbjörn Tjärnhage

The "modern" phase of biodetector development was in principle initiated with the Gulf wars in the 1990s. The FLAPS development presented in Chap. 2 is an example of this. Another important incident was the Aum Shinrikyo sect that prior to its realization of the sarin attack in the Tokyo metro station in 1995, also was trying to produce and use both Anthrax and Botulinum toxin [1]. In the decade following after the Anthrax letter incident in the USA in 2001, quite massive bio-defense R&D programs both in military and homeland security area bloomed with a peak in 2005 [2]. One example of these programs, closely related to technical development in biodetection are for instance the US "Semiconductor UV Optical Sources" (SUVOS) program by Defense Advanced Research Projects Agency (DARPA). The program was basically development of novel UV light sources, but many of the applications brought into the different projects were aiming at improving biodetection devices. Good light sources are an essential part of spectroscopy based biodetection methods. Earlier (military) biodetector system was rather complex and exclusive. Costly UV lasers and detector components for instance, as well as relatively low serial production volumes, made the devices rather expensive. The exploitation of the technical land winnings from the SUVOS program led to smaller and cheaper biodetectors but with similar performances. Specific examples of the outcome from this effort was for instance the "low-cost bio-aerosol sensor" (LBAS) by ICx and the "tactical biological detector" (TAC-BIO) developed by Edgewood Chemical Biological Center (ECBC) later produced by Chemring Detection System (CDS) [3, 4].

P. Jonsson (✉)
Division of Sensor and EW Systems, FOI—Swedish Defence Research Agency,
Olaus Magnus väg 42, 581 11 Linköping, Sweden
e-mail: per.jonsson@foi.se

T. Tjärnhage
Division of CBRN Defence and Security, FOI—Swedish Defence Research Agency,
Cementvägen 20, 901 82 Umeå, Sweden
e-mail: torbjorn.tjarnhage@foi.se

P. Jonsson et al. (eds.), *Bioaerosol Detection Technologies*, Integrated Analytical Systems,
DOI 10.1007/978-1-4419-5582-1_15, © Springer-Verlag New York 2014

In addition to the previous examples, there was also another parallel trend at the same period of time to develop more advanced spectroscopic detection techniques. In this case that meant to use better laser sources with higher power, different and possible multiple excitation wavelengths. Higher spectral or temporal resolution and collection of more optical parameters from the same single particle were used. Examples of this direction are the fluorescence-based systems from Yale University [5]. The idea with this approach is to obtain better classification of the bioaerosol origin, often by multivariate data analysis methods applied to the numerous parameters measured.

In addition to US and EU programs, also several national programs as well as industrial efforts to develop new detection devices were active during this period. The list of biodetector devices is now quite populated, many of which were based on fluorescence and several of those are described in the fluorescence chapter in this book.

Apart from the more established systems like fluorescence-based systems and flame emission photometry, mass spectroscopy and LIBS-based systems appear reasonable well developed for bioaerosol detection purposes. It is still rather complex systems but they show promising results. Especially mass spectrometry should have a potential to provide a reasonable degree of identification capability. On the other end, Raman- and THz-based system will probably require more studies, development and an increase in sensitivity before aerosol detection systems will be mature.

Especially the price and robustness is a limiting factor for realizing new bioaerosol detection systems on the market. As many of the technologies described in this book are relying on high power, high quality radiation with specific wavelengths, the development of different laser sources is essential for the future development of biodetectors. The same is also valid for the development of high sensitivity light detectors and spectrometers. In several systems traditional photomultiplier tubes (PMT) have still better characteristics in many aspects but semiconductor photosensitive devices are approaching in performance and they have the potential to be smaller, cheaper and have more spectral channels. Technical advances will improve the performance of the mature technologies. It will also open up for systems with the less mature technologies as well as totally new advanced spectroscopic techniques not yet applied to bioaerosol detection.

As the specificity of the current available biodetector system has been an issue, it has become more common to combine two or even more biodetection technologies and use the collective result of the detectors. Ideally, the different detection technologies are chosen so that they are evaluating different properties of the bioaerosol and provide independent or *orthogonal* information into the subsequent data analysis step. This approach has the advantage that the response time will still be short and the false alarm rates have great potential to be reduced as more information is collected. Some early examples of combinations were fluorescence and LIBS by MIT Lincoln Laboratory [6] and fluorescence and mass spectroscopy by TNO and Delft University of Technology [7, 8]. However, the final systems tend to be rather complex and expensive.

More practically, this orthogonal solution can be done with a combination of two or more detectors in a parallel configuration and to use data fusion tools to provide a weighed detector response. The EU financed security project "*Two stage rapid biological surveillance and alarm system for airborne threats*"—TWOBIAS is an example of such concept [9]. In the project a technical demonstrator uses a biodetection system consisting of three different detection principles, in this case based on: fluorescence, flame emission spectroscopy and laser-induced breakdown spectroscopy. The fusion of the separate device responses can provide a good classification of the aerosol cloud as a population and is shown to improve the detection quality [10]. A sampling unit and a connected identification step is initiated when the detector stage and the data fusion concludes there is an alarm.

In the later years, the situation can be described with a more realistic view of what capability physical biodetectors can provide. Problems with false alarm rates make decision based on biodetectors only to be of low regret actions (i.e. initiate sampling, change HVAC system etc.) Complementary sampling and analysis is essential before a high regret action (evacuating a metro station or close down an airport for example) will be taken. Generally speaking, the different biodetector systems are now finding their (optimal) role in different applications. This can for instance be that the smaller and simpler systems are more suitable for sensor network arrays that combined provide a good coverage over a larger area. In other places, perhaps a more advanced stationary detector can monitor the air in a metro station. In some other applications, aerosol detectors are not really establishing. In a recent report made for the Department of Homeland Security—DHS [11], most listed devices suitable for first responders are indeed not bioaerosol detectors, but rather handheld or mobile identification systems.

The same situation can be seen for the standoff development. After the first boom during the first decade of 2000s, many programs was stopped or delayed when their complexity and limitation of performance were evident. Now several old and new programs are directed to more realistic concepts where the standoff advantages of large area coverage and overview are utilized. Although detection, classification and false alarm rate are not ideal for a standalone standoff system, some of the problems can be compensated for example, by using confirmation from strategically placed point detectors. In any case information about the extension and propagation of an aerosol cloud is important in most defense and security application.

One important trend during the last decade is test and evaluation of biodetectors. When numerous new detectors were developed as a result of the massive funding in the military and homeland security area, the need to compare and test the detectors became a crucial issue. The first major standardization effort was made by DARPA together with Defense Threat Reduction Agency (DTRA) in the USA in 2004 [12]. Both sensor metrics and testing were discussed. The study propose that the sensors should be evaluated with Receiver Operating Characteristic (ROC) curves that explicitly capture the performance trade-off between *sensitivity*, *probability of detection*, and *false positive rate*; and, it implicitly captures the performance trade-off with regard to *response time*. Beside these key sensor metrics the report suggests to include additional attributes such as unit cost, operation cost, maintenance, reliability, size, weight and power consumption which they presented in spider charts. A number of different scenarios useful for evaluation are also presented and were further developed in a later report [13].

The DARPA and DTRA reports highlighted that the way to test and characterize the detectors has varied between different labs and nations. Different methods and facilities have been used and the results of tests have varied accordingly. This is a reason to why there is prioritized activity within both NATO and EDA to start the process of harmonizing the test protocols and methods [14]. The ultimate goal is to have common test protocols and validated methods that will make it easier to compare test results between different test labs. It is beneficial for producers of detection equipment to have standard protocols that are used to extract the characteristic parameters of the detector. Equally important potential users and buyers of detection instrument can use testing standards in their product evaluation process.

A parallel trend to the testing and evaluation is the investigation of the complexity of the natural background. Since bioaerosol detection is finding a needle in the haystack, the knowledge of the "haystack" is as important as the knowing the characteristics of the "needle". The need for a better understanding of the naturally occurring atmospheric interfering constituents, both under different conditions and at different locations, has led to measurement campaigns both over time and at different locations, for example, [15].

The bioaerosol content in the atmosphere is not only of interest for biodefense applications. Lately there has been a growing interest from the area of environmental research. Several environmental studies have been performed [16–19] with systems that originally have been developed for biodefense purpose. Also new systems are under development dedicated to environmental studies using the knowhow from the biodefense area [20].

Another example of this passage is the mass spectrometer mentioned earlier that initially were developed by TNO and Delft University and financed by the Dutch ministry of defense. The system is now re-packaged with the fluorescence stage removed and transferred to the company BiosparQ BV as a general microbiological analytical tool and the intended applications are more general than for military use [21].

The last couple of decades have given us knowledge about the capability of detectors based on physical principles. However, important questions still remain to be answered; How much information that can be extracted from a single particle using physical methods? Will it be practically possible to build a system that measures all possible parameters on the same particle? Likely we will expect to see examples of research system that exploits the physical characterization to its maximum. These kinds of systems will definitively help addressing the above questions and if useful information can be extracted, technical and engineering development will be guided in right direction to improve current detection techniques.

To conclude the current and near future status: Physical biodetectors is certainly capable to serve their purpose and provide timely and good data especially in monitoring and large area surveillance. However, to follow the concentration of a specific type of bioaerosol under high and complex background levels, problems will certainly be noticeable.

In addition to the development of physical biodetectors to become more specific, there are tremendous advances in especially genetic analysis methods and eventually their response time will be short enough to compete with the response time of physical detectors. The major challenge here is to transfer and prepare the aerosol to the analysis stage in a fast and automatic manner.

Jim Ho has quite recently prepared a DRDC technical report [22] in which he summarizes the general challenges in the biodetection area but also propose possible strategies and ideas how incorporate more traditional biochemical methods in real time aerosol detection applications. That includes methods targeting the infectivity and virulence rather than the species of the bioaerosol content.

References

1. Wheelis M, Rózsa L, Dando M (eds) (2006) Deadly cultures: biological weapons since 1945. Harvard University Press, Cambridge, MA
2. Greenwood D (2007) Meeting the Chem–Bio Defense Challenge. Linc Lab J 17 (1):20–27
3. Cabalo J, DeLucia M, Goad A, Lacis J, Narayanan F, Sickenberger D (2008) Overview of the TAC-BIO detector. Proc SPIE 7116:71160D.1-11. doi:10.1117/12.799843
4. DeFreez R (2009) LIF bio-aerosol threat triggers: then and now. Proc SPIE 7484:74840H.1-15. doi:10.1117/12.835088
5. Huang HC, Pan Y-L, Hill SC, Pinnick RG (2010) Fluorescence-Based Classification with Selective Collection and Identification of Individual Airborne Bioaerosol Particles. In: Serpengüzel A, Poon AW (eds) Optical Processes In Microparticles And Nanostructures, A Festschrift dedicated to Richard Kounai Chang on his Retirement from Yale University. Advanced Series in Applied Physics, vol 6. World Scientific, Singapore, pp 153–167. doi:10.1142/9789814295789_0009
6. Hybl JD, Tysk SM, Berry SR, Jordan MP (2006) Laser-induced fluorescence-cued, laser-induced breakdown spectroscopy biological-agent detection. Appl Opt 45 (34):8806–8814. doi:10.1364/AO.45.008806
7. Van Wuijckhuijse AL, Stowers MA, Kleefsman WA, Van Baar BLM, Kientz CE, Marijnissen JCM (2005) Matrix-assisted laser desorption/ionisation aerosol time-of-flight mass spectrometry for the analysis of bioaerosols: Development of a fast detector for airborne biological pathogens. J Aerosol Sci 36 (5–6):677–687. doi:10.1016/j.jaerosci.2004.11.003
8. Stowers MA, van Wuijckhuijse AL, Marijnissen JCM, Kientz CE, Ciach T (2006) Fluorescence preselection of bioaerosol for single-particle mass spectrometry. Appl Opt 45 (33):8531–8536. doi:10.1364/AO.45.008531
9. TWOBIAS. http://www.twobias.com/. Accessed 24 February 2014
10. Pichon F, Aligne F, Feugnet G, Blatny JM (2013) A classifier fusion-based approach to improve biological threat detection. Paper presented at the 11th International Symposium on Protection against Chemical and Biological Warfare Agents, Stockholm, Sweden, 3-5 June 2013
11. Baird C, Colburn H, Seiner D, Straub T, Ozanich R, Bruckner-Lea C, Bartholomew R (2012) Biodetection Technologies for First Responders. PNNL-21713. Pacific Northwest National Laboratory, Richland, Washington. Available from: http://www.pnnl.gov/nationalsecurity/technical/chemical_biological/Biodetection_Technologies_for_First_Responders.pdf
12. Carrano JC, Jeys T, Cousins D, Eversole J, Gillespie J, Healy D, Licata N, Loerop W, O'Keefe M, Samuels A, Schultz J, Walter M, Wong N, Billotte W, Munley M, Reich E, Roos J (2004) Chemical and biological sensor standards study. Defense Advanced Research Projects Agency, Arlington VA. Available from: http://www.dtic.mil/cgi-bin/GetTRDoc?Location=U2&doc=GetTRDoc.pdf&AD=ADA458370

13. Carrano J, Jeys T, Eversole J, Gillespie J, Licata N, Loerop W, Munley M, O'Keefe M, Roos J, Samuels A, Schultz J, Shatz M, Wong N, D'Amico F, Casale AM, Holster SE, McGrath JF, Metrovich A, Murphy C, Nelson-Patel K, Reich E, Riisager T (2010) Chemical and Biological Sensor Standards Study II. Advanced Research Projects Agency and Defense Threat Reduction Agency, Arlington VA. Available from: http://www.dtra.mil/docs/system-documents/Chem_Bio_Sensor_Standards_Study_Vol_2_Oct_2010.pdf
14. EDA—Test & Evaluation of Biological Sampling, Identification and Detection Equipment http://www.eda.europa.eu/projects/projects-search/exchange-of-information-and-scoping-study-concerning-test-evaluation-of-biological-sampling-identification-and-detection-equipment-phase-1-(t-e-bio-dim-phase-1). Accessed 24 February 2014
15. Pan Y-L, Pinnick RG, Hill SC, Rosen JM, Chang RK (2007) Single-particle laser-induced-fluorescence spectra of biological and other organic-carbon aerosols in the atmosphere: Measurements at New Haven, Connecticut, and Las Cruces, New Mexico. J Geophys Res 112 (D24):D24S19.1-15. doi:10.1029/2007jd008741
16. Huffman JA, Treutlein B, Pöschl U (2010) Fluorescent biological aerosol particle concentrations and size distributions measured with an Ultraviolet Aerodynamic Particle Sizer (UV-APS) in Central Europe. Atmos Chem Phys 10 (7):3215–3233. doi:10.5194/acp-10-3215-2010
17. Gabey AM, Gallagher MW, Whitehead J, Dorsey JR, Kaye PH, Stanley WR (2010) Measurements and comparison of primary biological aerosol above and below a tropical forest canopy using a dual channel fluorescence spectrometer. Atmos Chem Phys 10 (10):4453–4466. doi:10.5194/acp-10-4453-2010
18. Gabey AM, Stanley WR, Gallagher MW, Kaye PH (2011) The fluorescence properties of aerosol larger than 0.8 μm in urban and tropical rainforest locations. Atmos Chem Phys 11 (11):5491–5504. doi:10.5194/acp-11-5491-2011
19. Pan Y-L, Hill SC, Pinnick RG, House JM, Flagan RC, Chang RK (2011) Dual-excitation-wavelength fluorescence spectra and elastic scattering for differentiation of single airborne pollen and fungal particles. Atmos Environ 45 (8):1555–1563. doi:10.1016/j.atmosenv.2010.12.042
20. Kiselev D, Bonacina L, Wolf J-P (2013) A flash-lamp based device for fluorescence detection and identification of individual pollen grains. Rev Sci Instrum 84:033302.1-7. doi:10.1063/1.4793792
21. BiosparQ. http://www.biosparq.nl. Accessed 24 February 2014
22. Ho J (2011) The future of biological detection: A 2011 update. DRDC Suffield TM 2011-135. Defence R&D Canada, Suffield. Available from: http://www.dvdtinc.com/publications

Index

Symbols
2,5-Dihydroxybenzoic acid (DHB), 179, 190

A
Absorption, 111–114, 116, 117, 127, 130, 160, 205, 211, 216, 223, 226, 241–245, 247, 251, 252, 254–257
Adenovirus, 226
Aerodynamic diameter, 19, 36, 66, 88, 191
Aerodynamic particle sizer (APS), 18–20, 23, 282
Aeromonas hydrophila, 217
Aerosol
  concentration, 63
  density, 19
  deposition, 33, 38, 40, 42
  jet, 79, 82
  particles, 6, 9, 12, 13, 19, 33, 40, 45, 64, 79, 102, 129, 148, 159, 179, 211
  sampling, 15, 82, 132, 170, 226
  settling velocity, 41, 58, 59, 66, 73, 74
  size distribution, 63
  size range, 4, 13, 25, 33, 210
  transport, 82, 132
Aerosol sodium detector (ASD), 149
Aerosol time-of-flight mass spectrometry system (ATOF-MS), 185
Agent containing particle per liter of air (ACPLA), 5, 122, 129, 155, 282, 297
Air viscosity, 65
Alditol acetate derivatization, 194, 195
Allergens, 24, 33, 37, 38, 43
Alternaria spp., 37
Amino acid, 24, 25, 39, 45, 114, 116, 117, 119, 127, 188, 190, 215, 270
Analogue to digital convertors (ADC), 23
Anemophilous, 37
Anthrax, 4, 11, 12, 116, 222, 317
AP2C, 146, 147, 149, 150
Arabitol, 38, 195, 196
Archaea, 36
Army Research Laboratory (ARL), 117, 121, 124, 126, 131, 279
Ascospores, 37
Aspergillosis, 43
Aspergillus flavus, 44
Aspiration efficiency, 69–71
Asymmetry factor, 14, 97
Atmospheric transmission, 282, 296
Atomic emission spectroscopy (AES), 112, 163
Auto-fluorescence, 270, 279
Avalanche photo-diodes (APD), 131

B
Bacillus anthracis, 40, 221, 226, 234, 277
Bacillus atrophaeus, 117, 119, 152, 158
Bacillus globigii, 12, 117, 226, 271, 277, 297
Bacillus megaterium, 214, 217
Bacillus subtilis, 117, 171, 214, 217, 220, 221, 226, 244, 280
Bacillus thuringiensis, 185, 214, 218, 226, 231, 232, 234, 251, 277, 280
Background
  image, 310
  particles, 79, 129, 155, 192
Backscatter power, 302
Backscattering, 45, 272, 288, 302–304, 311
Bacteria, 12–14, 24, 34–41, 97, 119, 120, 128, 144, 147, 152, 155, 159, 163, 170, 192–197, 214, 217, 234, 242, 251, 254, 280
Bacterial
  cells, 13, 14, 21, 33, 226, 234, 245, 246, 254
  shape analysis, 15

P. Jonsson et al. (eds.), *Bioaerosol Detection Technologies*, Integrated Analytical Systems, DOI 10.1007/978-1-4419-5582-1, © Springer-Verlag New York 2014

Basidiospores, 37
BAST, 123, 125, 131
Bioaerosol
  aerodynamic diameter, 19, 36, 66, 88, 191
  aging, 41
  ambient concentration, 34
  coagulation, 73
  non-viable, 43
  particles, 33, 34, 39, 40, 42, 45, 68, 79, 114, 116, 119, 123, 126, 127, 130, 131, 133, 169–171, 178, 179, 197, 210, 213
  size characterization technique, 310
  sources, 40
Bioaerosol Mass Spectrometry (BAMS), 192, 197, 198
Biofilm, 34, 39, 41
BioLert, 123
Biological
  identifier, 144
  macromolecules, 243, 245, 246, 250, 256, 257, 260
Biological Aerosol-Capture-Enrichment (BioACE), 124
Biological Agent Real Time Sensor (BARTS), 123, 128
Biological Agent Warning Sensor (BAWS), 122, 123, 125, 128, 131
Moniteur d'Alarme Biologique (Biological Alarm Monitor) (MAB), 145, 150–155, 163
Biological and Toxin Weapons Convention (BTWC), 3, 4
Biological warfare (BW), 4, 40
Biological warfare agents (BWA), 3–5, 109, 116, 124, 153, 267
BioSense, 293
BioVigilant, 129
Biral, 126, 128
Bird flu, 43
Blastomycosis, 43
Blunt sampler, 70
Bond, 223, 242
  adsorbate–substrate, 231
  chemical, 113, 207, 224
  conjugate, 113, 116
  covalent, 241
  double, 113
  hydrogen, 241
Botulinum toxin, 40, 317
Burkholderia spp., 10

C
CCD camera, 96, 97, 123, 124, 131, 145, 151, 212, 213, 230, 308, 310, 311
C-FLAPS, 122
CHASE, 146
Chemical
  field enhancement, 224
  fingerprint, 144
  matrix effect, 160
Chemical biological radiological and nuclear (CBRN), 4, 191, 192, 198
Chemiluminescence, 15, 17
Chemometrics *see* Data analysis
  artificial intelligence, 103, 104, 109, 285
  artificial neural network (ANN), 106
  discriminant analysis (DA), 233
  discriminant factor analysis (DFA), 233
  Gaussian mixture discriminant analysis (MDA), 233
  hierarchical cluster analysis (HCA), 233
  K-nearest neighbor classifier (kNN), 233
  learning machines, 105, 107, 109
  linear discriminant analysis (LDA), 233
  megavariate data analysis, 161
  multivariate analysis, 234, 235, 281
  multivariate data analysis (MVDA), 161, 163, 171, 318
  partial least squares (PLS), 145, 161–163, 233
  partial least squares-discriminant analysis (PLS-DA), 233
  principal component analysis (PCA), 123, 145, 153, 161–163, 233, 234
  quadratic discriminant analysis (QDA), 233
  spectral angle mapping, 234
  support vector machines (SVM), 107, 233, 285
Chlamydomonas spp., 35
Cladosporium spp., 37
Classification, 15, 56, 64, 65, 67, 79, 106, 111, 281, 296
Clogging, 69, 175
Clostridium spp., 36
  Clostridium botulinum, 40
Cloud condensation nuclei, 45
Coarse particle concentration, 63, 81
Collection efficiency, 68, 80, 131, 197
Colony forming unit (CFU), 231, 282
Computational modeling, 245
Concentration factor, 80, 81, 151
Confirmed identification, 7
Continuous wave (CW), 130, 159, 191
Covariance matrix, 286, 291, 292, 295, 296
Cross section
  absorption, 115, 116, 245
  fluorescence, 114–117, 124, 127, 131, 287, 288, 294
Cyanobacteria, 38
Cytoplasm, 144
Cytoskeleton, 14

D
Data analysis
Dandruffs, 38
Data acquisition, 98, 250, 310
Defence Research and Development Canada (DRDC), 121
Defence Science and Technology Laboratory (Dstl), 125, 128
Defense Advanced Research Projects Agency (DARPA), 123, 129, 130, 317, 319, 320
Defense Threat Reduction Agency (DTRA), 129, 319, 320
Dekati, 129
Deoxyribonucleic acid (DNA), 197, 220, 241–245, 250, 254
Derivatization reagent, 170
DETALAC, 146
Detect to treat, 7
Detect to warn, 7
Detection, 8, 26, 89, 128, 151, 155, 159, 197, 215, 218, 222, 226, 267, 321
  volume, 63, 64, 79, 82, 130, 132
D-Glucose, 195, 196
Differential mobility analyzer (DMA), 103
Differential scattering (DISC), 207, 267
Diffraction,
  effect, 308
  peak, 305
Diffractive target plates, 311
Diffusiophoresis, 72
Dioctyl phthalate (DOP), 13
Dipicolinic acid (DPA), 116, 188–190, 214, 217, 221
Direction Générale de l'Armement (DGA), 150
Dispersion, 3, 51, 52, 54, 59, 66
DNA vaccines, 36, 89, 144, 197, 220, 241–245, 250, 254, 255
Double scattered light, 302
Dugway Proving Grounds (DPG), 20, 278

E
Ebola, 40
Edgewood Chemical Biological Center (ECBC), 123, 317
Effective diameter, 304, 305, 307, 308, 310, 311, 313
Elastic cloud mapping, 270
Electrical effects, 72
Electrical low pressure impactor (ELPI), 103
Electromagnetic field enhancement, 224
Electronic conversion factor, 275, 287
Electrospray ionization (ESI), 170, 194
Endospores, 36, 41, 214, 221, 226, 235
Endotoxin, 36, 43, 193
Enterobacter aerogenes, 172, 217
ENVI BioScout, 127–129
Environics, 129
Epithelium, 197
Ergosterol, 38, 44, 193, 194
Erwinia herbicola, 45, 185, 226, 277, 279
Escherichia coli, 10, 180, 215–218, 225, 226, 231, 254–256
Eukaryotic, 38
European Defence Agency (EDA), 4, 6, 127, 320
European Space Agency (ESA), 157
Excitation-emission matrix (EEM), 279
Experimental, 60, 80, 158, 186, 188, 216, 232, 242, 244–246, 276, 304, 313
  validation, 308, 313
Extinction coefficient, 301–303
  scattering, 303

F
False alarms, 5, 7, 9, 17, 18, 20, 127, 155, 162, 192, 269, 270, 284, 296, 297, 313, 318, 319
Ferrulic acid, 184
FIDO B2, 129
Field-of-view (FOV), 314
Fingerprints, 114, 144, 152, 207, 213, 228, 241, 242, 247
First responder, 319
Flame emission photometry, 318
Flame emission spectroscopy (FES), 143–146, 150–152, 155, 156, 161–163, 319
Flame photometric detector (FPD), 112, 146
Flame photometry, 146, 147
Flavin, 117, 118
Flavin adenine dinucleotide (FAD), 116
Flavivirus, 35
Flawbacterium capsdatum, 217
FLIR, 129
Flow velocity, 64, 67, 70, 73–75
Fluid,
  density, 64
  dynamics, 64, 93
  properties, 64
  viscosity, 64
Fluorescence, 119
  cross section, 287, 288, 293
  decay time, 119, 127
  fluorescent dyes, 21
  intrinsic protein, 14
  quenching, 226

Fluorescence aerodynamic particle sizer (FLAPS), 23, 24, 26
Fluorescence aerosol shape analysis system (FLASAS), 125, 131
Fluorescence applied to biological agents detection (FABIOLA), 127, 131
Fluorescent Aerosol Particle Sensor, 122, 123, 128, 317
Flying insects, 72
Foot-and-mouth disease, 41
Fransisella tularensis, 40, 117
Frequency, 208, 233
  resolved, femtosecond techniques, 221
Fungi, 38
Fur fibers, 38

G
Gas chromatography (GC), 170, 171, 193, 195, 196
Gaussian distribution, 286, 287
GC-MS, 192–196
General guidelines, 78
Glycogen, 221
Gram-negative, 10, 35, 36, 197, 234
Gram-positive, 10, 35, 36, 197, 231, 234

H
H5N1 virus, 43
Health effects, 38, 44
Hemoglobin, 15
High-resolution, 96, 108, 145, 206, 250
Holograph optics, 213
Human immunodeficiency virus (HIV), 226
Hybrid photodiode detector (HPD), 125
Hygiene hypothesis, 44
Hyphae, 37

I
Ice nuclei, 45
ICx Technologies, 129, 317
Identification, 5, 6, 10, 24, 112, 147, 170, 180, 185, 189–198, 217, 220–226, 232–235, 242, 244, 247, 250, 252, 319
Immunoassay, 122, 123, 128, 169
Impactors, 15, 23, 67, 68, 79–81, 103, 128, 129, 150, 151, 192
Impinger sampler, 282
Improved dielectric filters, 213
Inductively coupled plasma (ICP), 143, 147, 161
Inertial deposition, 67, 73, 77, 78
  in a bend, 77
  in contractions, 77
Infectivity, 9, 10, 24
Influenza virus, 41
Infrared spectroscopy (IR), 112, 114, 185, 206, 208, 242, 245, 248, 250, 251, 254, 267, 271
Insect secretions, 38
Instantaneous Biological Analyzer and Collector (IBAC), 128, 129
Intensified charge-coupled device (ICCD), 131, 161, 280, 285, 308, 310, 311, 313, 314
Interferants, 286
Interference filter, 313
Internal reference standard, 147
Ion trap mass spectrometry, 182, 183, 195, 197
Ion trap mass spectrometry, 172
Irradiance, 175, 272, 274, 288
Isoaxial sampling, 70, 71
Isokinetic sampling, 70, 171

J
Jablonski diagram, 112
Jet velocity, 67, 81
Joint biological standoff detection system (JBSDS), 277, 279

L
Laminar
  aerosol flow, 64
  parabolic flow profile, 73
Laser
  Argon laser, 123, 131
  HeAg laser, 131
  HeCd laser, 23, 24, 121, 130
  hollow cathode laser, 131
  KrF laser, 130
  Nd, 24, 122, 130, 155, 157, 159, 185, 212
  YAG laser, 122, 130, 155, 157, 159, 185, 212
  YLF laser, 24, 122, 130
  NeCu laser, 131
  XeCl laser, 130
  XeF laser, 130
Laser Doppler Anemometry, 103
Laser pyrolysis, 171
Laser-induced breakdown spectroscopy (LIBS), 112, 125, 132, 143–147, 155, 157, 161–163, 318
Laser-induced fluorescence (LIF), 111, 122, 123, 147, 157, 163, 191, 298
Latex beads, 13
Legionella pneumophila, 226
Legionnaire's disease, 43

LIBS-LIF, 157
LIBS-Raman, 157
Lichens, 38
Light detection and ranging (LIDAR), 19, 25, 269–278, 280–289, 291–293, 296–298, 308–313
  based biological standoff detection, 279
  transmitter, 273, 274
Light-emitting diodes (LED), 11, 25, 111, 123, 125, 130, 131
Lipopolysaccharides (LPS), 36, 193, 194
Liquid chromatography (LC), 171, 175, 194
Listeria monocytogenes, 10
Long-Wavelength InfraRed, 284
Lorenz-Mie theory, 211, 222
Luminol, 17
Luminescence, 112, 113

M
Mahalanobis distance, 285, 286, 291–293, 295
MALDI matrix, 175, 178, 179, 183, 184, 188, 190, 197
Male-specific bacteriophage 2 (MS2), 276
Mannitol, 38, 195, 196
Mars Science Laboratory (MSL), 157
Mass spectrometry (MS), 146, 161, 198, 318
Matrix effect, 159, 160
Matrix-assisted laser desorption-ionization (MALDI), 170, 185, 188, 190, 197, 198
Maximum energy position, 308
Maxwell's electromagnetic field equations, 91
Megavirus chilensis, 36
Metallic colloidal nanoparticles, 224
Microalgae, 38
Micrococcus spp., 35
Micrococcus lysodeikticus, 172
Microwave induced plasma (MIP), 143
Mid-Wavelength InfraRed, 284
MIT Lincoln Laboratory, 121–123, 125, 132, 318
Mobile aerosol sampling unit (MASU), 19
MODTRAN, 296
Moniteur d'Alarme Biologique (MAB), 145, 150–155, 163
Monochromator, 21, 206, 212, 213, 229
Monodisperse particles, 179
Monte Carlo simulations, 61, 313
Morphological structures, 14
Multichannel detectors, 212, 213
Multidrug-resistant *Staphylococcus aureus* (MRSA), 44
Multiple scattering, 301–303, 307
Multiple-field-of-view (MFOV), 314
Multiple-Field-Of-View lidar, 301, 302, 304, 308, 312, 313
Multi-slit arrangement, 81
Multi-stage construction, 81
Multivariate analysis, 234, 235, 281
Multivariate data analysis (MVDA), 161, 163, 171, 318
Muramic acid (Mur), 193
Mycelium, 37
Mycobacterium tuberculosis, 43
Mycotoxin, 38, 40, 44

N
National Aeronautics and Space Administration (NASA), 157
Naval Research Laboratory (NRL), 121, 125, 131
Newton's second law, 65
Nicotinamide adenine dinucleotide (NADH), 21, 22, 116–119, 270
Nozzle
  circular nozzles, 67, 68, 81
  impactors, 68
  slit nozzle impactor, 67

O
Oligonucleotides, 170
Optical depth, 301, 304, 305, 311–313
Optical parametric oscillator (OPO), 131, 220
Optical particle counters (OPC), 13, 20, 23
Optics transmittance, 296
Orifices, 77, 150, 175, 188
Orthogonal, 318, 319
Ovalbumin (OV), 280

P
Pantoea agglomerans, 234
Particle
  analysis by mass spectrometry (PAMS), 172
  asymmetry, 93, 97
  background, 144
  charge, 102
  counter, 12, 13, 20, 23
  deposition, 64, 70, 73, 82
  refractive index, 13, 90, 91, 243, 246, 250, 253–255, 308
  relaxation time, 67, 75
  shape, 14, 91, 94, 96, 97, 103, 104, 23
  sizing, 11, 12, 58, 67, 69, 71, 72, 96, 103, 112, 122, 124, 129, 132, 159, 175, 178, 179, 198, 231, 243, 301, 302, 305, 308
  transfer, 63

Particles per liter (ppl), 282
Pathogen, 10, 33, 43, 87, 89
Peptidoglycans, 14, 36, 182, 193, 194
Phase Doppler Anemometry, 103
Phase function, 303–305, 308
Phenylalanine, 24, 116, 188, 233, 270
Phospholipids, 170, 221
Phosphorescence, 113
Photomultiplier tube (PMT), 12, 17, 23, 97, 122, 124–126, 131, 280, 318
Physical matrix effect, 160
Picolinic acid (PA), 179, 214
Plank-Einstein energy conservation principle, 204
Plant fragments, 38, 39
Plaque-forming unit (PFU), 282
Plasma, 143–145, 147, 155–163
PM10 inlets designs, 72
Point detector, 7, 8, 63, 69, 280, 319
Poisson distribution, 286, 288–290
Polarization detection module, 310
Pollen, 37
 Populus deltoides (cottonwood), 226
 Sequoia sempervirens (redwood), 226
 apricot pollen, 217
 elm pollen, 312
 grass pollen, 217
 ragweed pollen, 217
 redbud pollen, 35
 timothy pollen, 312
Polydispesed particles, 179
Polymerase chain reaction (PCR), 5, 123, 128
Proengin, 146, 150, 163
Prophylaxis, 10, 269
Provisional identification, 7
Pseudomonas aeruginosa, 226
Pseudomonas fluorescens, 45
Pseudomonas syringae, 45
Pulse repetition frequency (PRF), 122, 130, 131, 144, 284
Pulse shaping, 221
Pyrolysis (Py), 169–171, 190, 197

Q
Quadrupole mass spectrometry, 171
Quantum efficiency (QE), 115, 131
Quantum yield, 24, 115, 116, 275, 287, 293–296

R
Rain, 5, 39, 41, 58, 72
Raman
 Cavity enhanced Raman Spectroscopy (CERS), 228
 Coherent anti-stokes Raman Spectroscopy (CARS), 215, 222
 FAST CARS, 217, 221, 222
 Laser trapping Raman spectroscopy (LTRS), 226
 Microscopy, 213, 214, 220
 Microspectroscopy, 213, 214, 226
 Normal Raman Spectroscopy, 204, 214, 220
Resonances, 127, 206, 241–245, 250, 254, 257, 260
Resonance Raman Spectroscopy (RRS), 206, 215, 218, 229
 Spatially offset Raman Spectroscopy, 229
 Spontaneous Raman Spectroscopy, 203, 210, 213, 214
 Spontaneous Raman imaging, 215
 Surface-enhanced Raman Spectroscopy (SERS), 215, 226
 UV Raman Spectroscopy, 215, 218
 Ultraviolet resonance Raman (UVRR), 217
Rayleigh
 radiation, 204
 scattering, 204, 212, 243
 scattering coefficient, 304
Receiver operating characteristic (ROC), 130, 281, 319
Rectangular slit nozzle, 68, 81
Reflectron TOF-MS, 178
Remote detection, 8, 19, 210, 217, 244
Resonant enhancement, 221
Respiratory tract depositions, 42
Resuspension, 37, 54
Reynolds number, 64, 68, 75, 76, 81
Rhinovirus,
Ribonuclease A, 221
Ribonucleic acid (RNA), 36, 89, 217, 221, 242, 245
Royal Institute of Technology (KTH), 126

S
Salmonella typhimurium, 227
Sampling
 efficiency, 69, 70, 72
 in calm air, 72
 system, 69, 72, 145, 150, 214, 226, 270
Scatterers, 207, 229, 272, 273, 275, 282
Scattering,
 elastic scattering, 112, 122–125, 128, 129, 132, 267, 281, 314
 forward scattering, 94, 302, 303
 inelastic scattering, 203, 204, 214, 273
 light scattering, 33, 45, 108, 204, 223, 224, 231
 light scattering, 94
 Mie scattering, 13, 243

multiple scattering, 301–303, 307
Raman scattering, 114, 127, 203–206, 219, 224, 226, 229
Rayleigh scattering, 204, 205, 212, 243, 304
SCP 1021, 15
Segregation, 10
Semiconducting absorbers, 213
Semiconductor ultraviolet optical source (SUVOS), 123, 317
Settling under gravitational force, 64
Settling velocity, 41, 58, 59, 65, 66, 73, 74
Shape analyser (SA), 14
Sheath
air, 82
flow, 12, 122, 159
Short-wavelength infrared (SWIR), 284
Shot noise, 285, 286, 308, 313
Shot noise factor, 313
Sick building syndrome (SBS), 44
Signal-to-noise ratio (SNR), 130, 131, 175, 212, 247, 277, 280, 281
Simulants, 104, 124, 153, 155, 208, 277, 283, 285, 297
Sinapinic acid (SA), 180, 185
Single particle aerosol mass spectrometry (SPAMS), 192, 198
Single particle fluorescence analyzer (SPFA), 124
Single pulse multiplex, 221
Size classification, 65
Skin fragments, 35, 38, 40
Slip correction, 65
Smallpox, 40
Solution, 117, 119, 131, 146, 147, 159, 175, 207, 208
Sorting, 10, 11, 22, 124, 156
Spark spectroscopy, 143, 147
Spatiotemporal evolution, 277
Specificity, 5–7, 111, 114, 119, 127, 132, 163, 194, 241, 256, 271, 281, 297, 318
Spectra, 24, 117, 119, 153, 172, 179, 180, 182, 185, 188, 189, 276, 277
bacterium flame, 147
CARS, 220, 222
fluorescence
as in fluorescence spectra
LIBS, 161
Raman, 206–208, 215, 217, 218, 226
SERS, 234
Spectroscopy, 24, 133, 163, 260, 235
Standoff
biosensors, 273, 284
detection, 112, 118, 121, 157, 206, 221, 222, 235, 271, 279, 281, 297
monitoring systems, 269
optical trigger, 269
Standoff integrated bioaerosol active hyperspectral detection (SINBAHD), 276, 277, 279, 282, 283
Staphylococcus epidermidis, 218
Stokes drag, 102
Stokes law, 65
Stokes number, 67, 68, 70, 72, 75, 76, 81
Streptococcus spp., 35
Sub-isokinetic sampling, 70
Sub-micrometer cutpoint, 81
Super-isokinetic sampling, 70
Swedish Defence Research Agency (FOI), 126
Swine flu, 43
SWOrRD, 128

T

Tactical Biological Detector (TAC-BIO), 123, 317
Tampere University of Technology (TUT), 127, 129
Targeted ultraviolet chemical, biological, and explosives (TUCBE), 127
Taxonomy, 10
Technological readiness level (TRL), 163
Tee connectors, 77
Teraherz spectroscopy (THz), 112, 114, 260
Terminal settling velocity, 65, 73
Thermophoresis, 41, 72
Thin-walled sample probes, 70
Time-gated approaches, 221
Time-of-flight mass spectrometry (TOF-MS), 185, 186, 189, 191, 197, 198
Time-resolved studies, 203
Transmission, 250, 251, 275
airborne, 40
efficiency, 69–71, 74, 76, 77
optical, 284, 293
Transmittance, 287, 288, 291, 293, 295, 296
Transport efficiency, 70–74, 77
Triangular centroid plot, 99
Tribocharging, 101
Trimethylsilyl derivative, 193–195
Tryptophan (Trp), 24, 116, 117, 183, 188, 270
Turbulent,
deposition efficiency, 75
eddies, 41, 52, 64
Turex (commercial prep. of *Bacillus thuringiensis*), 119
Tyndall
effect, 11
meter, 11
Tyrosine (Tyr), 24, 116, 188, 270

U
UV-APS, 122

V
Valves, 77, 78
Variola major, 40
Vena contracta, 70, 96
VeroTect, 126, 128, 129
Vibrational, 112, 114, 132, 204–208, 214–217, 220–224, 230, 232, 241, 243, 247, 248, 250, 251, 254, 255, 257, 260
  analysis, 203
  assignments, 233
  resonances, 245
  spectroscopy, 114, 208, 213
Vibrio cholerae, 10
Virtual impactor (VI), 15, 23, 79–81, 128, 129, 151
Virulence, 6, 9
Virus, 3, 24, 33, 36, 40–43, 87, 89, 118, 169, 210, 217, 282
Volumetric flow rate, 72, 74, 78, 80

W
Wide issue bioaerosol sensor (WIBS), 126
Wind speed, 37–40, 49–51, 54, 56–59, 72, 89, 102, 105

X
Xenon lamp, 125, 131
XM19, 15, 17, 18
XM2, 15, 17, 18, 23

Y
Yale University, 121, 123, 124, 318
Yersinia pestis, 40, 42
Yersinia pseudotuberculosis, 117